Lecture Notes in Networks and Systems

Volume 1097

The series "Lecture Notes in Networks and Systems" publishes the latest developments in Networks and Systems—quickly, informally and with high quality. Original research reported in proceedings and post-proceedings represents the core of LNNS.

Volumes published in LNNS embrace all aspects and subfields of, as well as new challenges in, Networks and Systems.

The series contains proceedings and edited volumes in systems and networks, spanning the areas of Cyber-Physical Systems, Autonomous Systems, Sensor Networks, Control Systems, Energy Systems, Automotive Systems, Biological Systems, Vehicular Networking and Connected Vehicles, Aerospace Systems, Automation, Manufacturing, Smart Grids, Nonlinear Systems, Power Systems, Robotics, Social Systems, Economic Systems and other. Of particular value to both the contributors and the readership are the short publication timeframe and the world-wide distribution and exposure which enable both a wide and rapid dissemination of research output.

The series covers the theory, applications, and perspectives on the state of the art and future developments relevant to systems and networks, decision making, control, complex processes and related areas, as embedded in the fields of interdisciplinary and applied sciences, engineering, computer science, physics, economics, social, and life sciences, as well as the paradigms and methodologies behind them.

Indexed by SCOPUS, EI Compendex, INSPEC, WTI Frankfurt eG, zbMATH, SCImago.

All books published in the series are submitted for consideration in Web of Science.

For proposals from Asia please contact Aninda Bose (aninda.bose@springer.com).

Rohit Kandakatla · Sushma Kulkarni ·
Michael E. Auer
Editors

Academic Leadership in Engineering Education

Learnings and Case Studies from Educational Leaders Around the Globe

Springer

Editors
Rohit Kandakatla
KG Reddy College of Engineering
and Technology
Hyderabad, India

Sushma Kulkarni
NICMAR University
Pune, India

Michael E. Auer
CTI Global
Frankfurt, Germany

ISSN 2367-3370 ISSN 2367-3389 (electronic)
Lecture Notes in Networks and Systems
ISBN 978-3-031-68281-0 ISBN 978-3-031-68282-7 (eBook)
https://doi.org/10.1007/978-3-031-68282-7

This Springer imprint is published by the registered company Springer Nature Switzerland AG
The registered company address is: Gewerbestrasse 11, 6330 Cham, Switzerland

Foreword

Today's engineering industry demands more from our graduates: an emphasis on critical thinking, a firm understanding (and application) of the principles of sustainability in engineering design, and an increased focus on professional skills such as communication, teamwork, and appreciation for diversity and inclusion. These are all critical components in shaping the skills of tomorrow's engineers.

Drawing on my 40-year career, marked by significant events shaping my worldview, I often think about key "lessons learned" and how I can apply those to future challenges. A common thread emerges—the importance of strong leadership. My journey exposed me to several important aspects of leadership: strategic vision, ability to inspire others, understanding people and their needs, empathy, and most importantly: employing the proper leadership style for each situation.

In this book, we hear from several esteemed professionals leading institutions and programs, focusing on governance, human resource management, institutional identity, branding, and setting quality performance standards. Emphasizing the importance of student outcomes, we explore the assessment of adult learning through evidence-based systems. Additionally, insights from those involved in developing new "knowledge hubs" that empower both faculty and students for the twenty-first century providing them the knowledge, tools, and experiences they need to solve some of the biggest challenges facing our planet today.

ABET is honored to support the development and publication of this important book. I want to thank the many professionals who contributed to its creation—editors, authors, and advisory committee members for their many contributions to this project.

Michael Milligan
CEO
ABET
Baltimore, USA

Preface

Engineering education systems around the world are undergoing a phase of transformation as they work to adapt themselves to the learning needs of students in the modern era, changing trends in the requirements of the industry, and growing concerns about issues related to sustainable development and climate change. Engineering institutions are increasingly focused on transforming themselves to create quality learning experiences for students that will enable them to develop deeper learning skills such as critical thinking, problem-solving, lifelong learning, and the ability to work in teams. It is important that future engineering graduates be equipped to tackle complex problems in society that are aligned with the United Nation's Sustainable Development Goals (SDGs). There are increasing calls for engineering curricula to be made multidisciplinary, innovative, and outcome-driven through the integration of evidence-based pedagogies and learning mechanisms.

For this to happen, it is essential for academic leaders to reimagine their curriculum with significant changes at the administration, governance, and leadership levels. However, it is often observed that individuals holding various academic leadership positions in engineering institutions are unprepared for their role, as most don't have the opportunity to develop leadership skills early in their careers. Most possess a Ph.D. in engineering and are promoted primarily due to their academic achievements and are offered leadership opportunities based on their exemplary competence and contributions as a researcher or teacher. As a result, many are left to learn while on the job as they have little or no training for the leadership responsibilities of a new position. Establishing world-class institutions that meet international accreditation standards requires good academic leadership at multiple levels that can work collaboratively together to achieve the vision and mission of the institution.

This book is an attempt to share key learnings from academic leaders from around the world on important trends emerging in engineering education. The book is divided into four parts, with each part comprising of multiple chapters written by different academic leaders based on their experiences of implementing best practices at their respective institutions. The parts and chapters will be targeted at various stakeholders including Faculty Members, Heads of Departments, Deans, Directors, Registrars, Vice-Chancellors, etc. Each chapter will be presented through case studies from successful programs initiated and pioneered at various engineering institutions across the globe.

We now present an overview for each of the section and list down the list of chapters in the four parts:

Governance and Leadership of Engineering Institutions

Academic leaders are expected to govern educational institutions and lead all stakeholders to collectively achieve their vision and goals. Education in principle is a service-based field as faculty should be viewed as human resources who are being employed to teach and impart knowledge to the students. Effective governance requires systematic human resource management of all stakeholders which include faculty, staff, and students. Faculty are one of the most essential stakeholders and academic leaders therefore need to be strategic with respect to their recruitment, on-boarding, and human resource management. The organizational structure and performance indicators should be aligned to ensure all faculty are effectively and collectively utilized to achieve the goals of the institution. Engineering institutions across the globe are also expected to adhere to quality benchmarks mostly set by national accreditation agencies which are members of the Washington Accord. Academic leaders must hence establish governance models and operational processes supported by sustainable financial management to systemically engage all stakeholders and achieve quality benchmarks.

Beyond the governance of engineering institutions, academic leaders need to build a brand identity and value for their institutions. They are expected to lead all stakeholders to believe in a collective vision and help achieve it through strategic planning and implementation. The academic and learning environment should be nurtured to motivate and encourage all stakeholders to aspire to the vision of the institution. Academic leaders, while leading the transformation of engineering institutions, should learn to adapt their leadership style and change management strategies based on the context. As most transformational efforts will require active engagement with the faculty, the leadership approach should be decided based on the evaluation of the human resources and the goals to be achieved. Transformation of engineering institution would often require change and engineering education leaders be prepared to overcome any of the barriers to change through relevant change management strategies. The purpose of this section is to prepare academic leaders for senior leadership positions that would require extensive governance and leadership of engineering institutions.

List of Chapters

1. Transformation of Institutions with Strategic Vision and Planning to Build Brand Identity
2. Governance Models of Engineering Institutions for Effective Administration and Human Resource Management
3. Establishing and Maintaining Operational Systems and Processes that Meet International Quality Standards, such as those recommended by Washington Accord
4. Finance Management and Revenue Generation for the Sustainability of Engineering Institutions
5. Leadership Approaches and Change Management in Engineering Education: Case Studies from Four Contexts

Creating Quality Learning Experiences

The last few decades have witnessed a tremendous transformation in engineering education with respect to how engineering students have been taught, learnt, and evaluated. The emergence of engineering education as a discipline of its own has led to numerous efforts by practitioners and researchers who aimed to solve critical issues in the education system. Research in engineering education led to the generation and documentation of evidence that justify the need for systematic transformation of the teaching and learning practices in engineering education. Engineering education is in the midst of a paradigm shift from teacher-centric to student-centric education system and engineering institutions must therefore realign their priorities to the learner's interests, needs, and outcomes. The shift of focus to learners has led institutions to transform themselves from an input-based to an outcome-based education system, so that student's success can now be evaluated based on their attainment of the program outcomes before graduation. The entire curriculum and course content should then be designed and structured with a goal to help students achieve the program outcomes.

Engineering faculty must be encouraged to create student-centric learning environments where the assessment and pedagogical practices are aligned with the learning outcomes of the courses. Students must be made active learners in the classrooms particularly to engage them in higher order learning and help build deeper learning skills such as critical thinking, problem-solving, and teamwork. Research in engineering education has plenty of evidence that promotes the value of hands-on and experiential learning and its impact on student's learning outcomes. Institutions could use experiential learning as an approach to engage engineering students in multidisciplinary learning that would be beneficial once they graduate and work in the industry. Strategic collaborations with industries could help engineering institutions to align their curriculum and program outcomes with the expectations of the industry partners. Industries expect students to possess both technical and professional skills, and conscious efforts should be taken to build professional skills such as ethics, empathy, communication, teamwork, and lifelong learning. In the last couple of years, the disruptions caused by the COVID-19 pandemic have pushed all educators to reimagine how engineering could be taught and learnt through the use of technology. All institutions in the future must explore possibilities on how technology can be used as an enabler to improve the access and support provided to students with a goal to enhance learning experiences.

The purpose of this section is to introduce academic leaders to key topics that are essential for creating quality learning experiences for students in engineering institutions.

List of Chapters

6. Transformation to Outcomes Based Education
7. Creating Student-Centric Learning Environments Through Evidence-Based Pedagogies and Assessments
8. Inclusion of Inter- and Multi-disciplinary Experiences Through the Adoption of Experiential Learning in Undergraduate Engineering
9. Reimagining Engineering Education Through Technology
10. Industry Academic Partnerships Leading to Better Preparation of Graduates for Employment
11. Integration of Professional Skills (Communications, Teamwork, Etc.), Ethics, Empathy Lifelong Learning into the Student's Educational Experiences

Preparing Institutions to Become Knowledge Hubs for Research, Innovation, and Entrepreneurship

Engineering institutions which are considered among the best are recognized not only for their ability to teach and impart knowledge to students but also for their capacity to create knowledge that would be beneficial for the betterment of the society. Building world-class engineering institutions requires academic leaders to have the vision to transform their institutions into knowledge hubs, which would enable the development of research and innovations leading to the generation of startups and intellectual property. Faculty in the engineering institutions should be encouraged to engage in quality research through research culture and support systems enabled by strategic collaborations. Each Institution should aim to build its research capacity in specific domains based on the collective technical expertise of its faculty members. Engineering institutions as knowledge hubs could aim to become locally relevant by aligning their research and innovations to solve problems in nearby communities. An increase in local relevance would enable the co-existence of the engineering institution and nearby communities through meaningful partnerships that benefit all stakeholders. While the nearby communities would benefit from engineering solutions that could lead to their socioeconomic development, engineering institutions would be able to provide their students with real-time problem-solving opportunities that could lead to the development of innovations, startups, and intellectual property.

Academic leaders should also build an ecosystem that promotes faculty and students to engage in innovation and entrepreneurship activities. Students with entrepreneurial mindsets and aspirations should be encouraged to develop innovations that are designed to solve critical issues in the society. Students with innovative ideas with potential for entrepreneurship should be further nurtured to start their companies through support from in-house or external incubation facilities. Institutions should provide legal support to faculty and students and help them to protect their innovations through the generation of intellectual property. Intellectual property generated by engineering faculty and students could serve as another source of revenue generation for the institution as the products and services developed to solve critical issues could have the potential to be commercialized for large-scale usage.

The purpose of this part is to provide perspectives to academic leaders on how they can transform their engineering institutions into knowledge hubs. The chapters in the part will provide ideas, suggest opportunities, and help build vision on how to evolve engineering institutions from "imparting knowledge" to "creating knowledge". We invite prospective authors to contribute to the following chapters:

List of Chapters

12. Fostering Research Culture, Collaborations, and Support Systems Leading to the Production of Quality Output
13. Building a Pipeline to Encourage Innovation, Entrepreneurship & Incubation for Faculty and Students
14. Increase Regional and Local Relevance of Engineering Institutions Through Community Engagement Aimed to Support Their Socioeconomic Development
15. Monetization of Intellectual Property Through Commercialization of Products and Services Developed Through Research and Innovation

Empowerment of Faculty and Students for the Twenty-first Century

Engineering education has observed a paradigm shift in the twenty-first century with respect to the role of faculty and students and the relationship between them. Institutions should urge their faculty to realize their change of role from a teacher to that of a facilitator of students learning. Faculty must adapt from the traditional passive modes of teaching to engage students actively in the classroom to tackle the low attention timespan of twenty-first-century learners. Engineering faculty must also realize their role as learner and recognize the importance of lifelong learning. Academic leaders must therefore organize pre-service and in-service capacity building programs to shift the mindset of faculty and help them build necessary knowledge and skills to provide students with quality learning experiences. Pre-service trainings must be conducted for new faculty on teaching and learning in the twenty-first century and introduce them to various evidence-based pedagogies that are needed to be integrated into the curriculum. Faculty should be encouraged to engage in lifelong learning through in-service trainings and programs conducted by the institution and other external organizations.

Engineering students in the twenty-first century are expected to go beyond just gaining knowledge of their discipline and build a variety of skills that include critical thinking, problem-solving, teamwork, project management, and lifelong learning. Twenty-first-century learning needs to be an accumulation of domain-specific knowledge, deep thinking skills, and professional attributes that will help students lead successful careers in the modern workplace. Academic leaders must therefore explore how an ecosystem can be fostered in their institutions that will promote holistic learning and development among students. Apart from providing students with authentic learning experiences inside the classroom, efforts should be taken to engage students in other informal activities and programs outside of their academics that will enable them to self-regulate their learning, engage with peers, manage projects, and build their professional network. The learning environment in engineering institutions must also encourage diversity among faculty and students and ensure their programs are inclusive to everyone.

The COVID-19 pandemic has particularly brought issues related to the physical and mental well-being of all individuals to the forefront. While engineering institutions focus on quality learning experiences, efforts should also be taken to encourage the physical and mental well-being of faculty and students. Academic leaders should foster a culture of well-being and mindfulness so faculty and students can seek out for help when necessary. Institutions must initiate welfare programs and build support systems which can be utilized by all stakeholders to promote physical and mental well-being of faculty and students. Another topic that was at the forefront of conversations during the COVID-19 pandemic is global collaborations. The world has witnessed and recognized the value of global cooperation with the fast development of vaccines as a result of collaboration between many countries. To solve global issues such as climate change and sustainable development goals, engineering students must be equipped with the skills required to collaborate in global, cross-national, and diverse teams. Institutions must therefore nurture their students into global engineers through systematic and strategic collaborations with international universities.

The purpose of this section is to introduce academic leaders with key topics on how engineering institutions could empower faculty and students in the twenty-first century.

List of Chapters

The four parts and the various topics of the book were decided by the editors of the book in consultation with an advisory committee, which included key academic leaders from around the globe. We express our sincere gratitude and appreciation to all the members of the advisory board for their valuable insights during the planning stage of the book.

List of Advisory Committee Members

Maria Joao Viamonte, President, Instituto Superior de Engenharia do Porto (ISEP), Portugal
Michael Milligan, Chief Executive Officer, ABET, USA
Ramiro Jordan, Associate Dean of Engineering, University of New Mexico, USA
Robert Bishop, Vice Chancellor for Engineering, The Texas A&M University System
Stephanie Farell, Founding Head of Experiential Engineering Education Department, Rowan University, USA
Sunil Maharaj, Vice-Principal: Research, Innovation and Postgraduate Education, University of Pretoria, South Africa
Tania Cristina D. Bueno, President of Executive Board, Instituto i3G, Brazil

We also thank our collaborators GEDC and IFEES, especially their secretary general Dr. Hans J. Hoyer for all his support in identifying authors for the various chapters in the book. Their extensive network of academic leaders from around the globe was instrumental in bringing diverse geographical perspectives to the book.

GEDC
Global Engineering Deans Council

IFEES
International Federation of Engineering Education Societies

Finally, we would like to thank our sponsors Accreditation Board for Engineering and Technology (ABET) and CTI Global for their willingness to financially support this project. This book could not have been possible without your trust and commitment to support future academic leaders who will be key to the transformation of engineering education in the future.

Furthermore, we deeply thank Devasmita Dey for the technical editing of this book.

We hope you enjoy reading this book and find it to be a valuable resource in your journey of academic leadership in engineering education.

Hyderabad, India — Dr. Rohit Kandakatla
Pune, India — Dr. Sushma Kulkarni
Frankfurt, Germany — Dr. Michael E. Auer

Contents

Governance and Leadership of Engineering Institutions

Transformation of Institutions with Strategic Vision and Planning to Build Brand Identity

Daniel Bergen, Marshall Sheldon, and Eddy Evans

Abstract Utilizing case studies situated in unique types of higher education settings in three different countries, this chapter emphasizes the "transformation of institutions with strategic vision and planning to build brand identity." The institutions differ in size, location, socio-economic and cultural context; however, they all feature a distinguishing element in their success: shared leadership. According to Morrison et al. (Strategic doing: ten skills for agile leadership. Wiley Publishing Company, New York, 2019), shared leadership is defined as leadership that is carried out by the team [rather] than solely by a single designated individual. There is a growing body of research that points to shared leadership in teams and organizations as being positively associated with team effectiveness and productivity; and, that the relationship between shared leadership and effectiveness is *even stronger* when the team's work is highly complex. Each case study showcases the complexity of the environment as well as the value and effectiveness of utilizing a team approach. Ranging from the establishment of a new "*Renaissance-style*" Canadian engineering school, to the dynamic cultural and strategic shift at a 120-year-old midwestern, United States engineering university, to the government-driven merger of South African universities and Technikons in a post-apartheid era, the authors narrate the opportunities and challenges presented in these transformative moments, as they create, build upon, and/or maintain their brand identities. Academic leadership audiences will benefit from strategic documents, best practices, applicable insights, retrospective lessons learned, and suggested references, as they engage this chapter.

Keywords Strategic planning · Engineering education · Brand identity

D. Bergen (✉)
Milwaukee School of Engineering, Milwaukee, WI, USA
e-mail: bergen@msoe.edu

M. Sheldon
Nelson Mandela University, Gqeberha 6001, South Africa

E. Evans
Lakehead University, Thunder Bay, ON, Canada

R. Kandakatla et al. (eds.), *Academic Leadership in Engineering Education*, Lecture Notes in Networks and Systems 1097, https://doi.org/10.1007/978-3-031-68282-7_1

1 Introduction

This chapter captures the strategic transformations across three unique types of engineering higher education settings in different countries, emphasizing "***vision and planning to build brand identity***." The institutions differ in size, location, socio-economic and cultural context; however, they all feature a distinguishing element in their success: shared leadership. Each case study traces the evolution of the strategic vision, engaging key stakeholders, and ensuring that their brand identity is bolstered.

In the opening case study, Evans and Kozinski detail the step-by-step process of initiating an entirely new School of Engineering at York University, inviting readers to consider what they might do with a blank slate and a tremendous philanthropic gift.

Their answer:

> Solving our greatest challenges takes more than technical talent alone. The world needs engineers with passion and perspective – or as we call them: Renaissance Engineers.

The design of the school speaks directly to the challenges facing the contemporary field of engineering, the curriculum, the brand and its built-environment. It presents a unique, dynamic study in form and function.

The second case study focuses on a 120-year-old university situated in Milwaukee, Wisconsin, a midwestern, US city with a population of just under 600,000. Since its inception in 1903, the Milwaukee School of Engineering (MSOE) has intentionally focused its offerings on industry-responsive programs, including engineering, business, nursing, and more recently, computer science. This laser focus has allowed it to remain nimble in its original mission—delivering top-tier talent to its regional, industry partners.

Bergen, Walz, and Baumgartner, demonstrate how the combination of a new president (only the fifth president in the history of the university), the first-ever campus-wide strategic planning process, and shared leadership and stewardship, set the stage for a cultural and institutional transformation at MSOE. Moreover, the institution's unwavering focus on maintaining its core brand, despite such a rapid transformation, distinguishes it among its peers. Similarly, the study demonstrates how the complexity of a transformative process is effectively mitigated by broad campus collaboration, well-leveraged key stakeholders, and thoughtful communication.

In the final case study, Sheldon places two transformative mergers of a newly formed University of Technology and a Comprehensive University against the backdrop of a post-apartheid, South African democracy. She illustrates how critical the engineering faculty strategy, branding, profile, and positioning is, to attract, retain, and ensure the success of students. Having served in both—a University of Technology and Comprehensive University—Sheldon has been involved in crafting a faculty strategy in two different contexts, and demonstrates that irrespective of the type of institution, the context, or geographical location, it is imperative to co-create a faculty strategy that aligns with that of the institution and includes input from all relevant stakeholders. The study details "educational transformation" both at a structural and curricular level.

2 Background

In the latter part of 2020, KPMG International published a report titled, "*The future of higher education in a disruptive world*" [2]. Stephen Parker states in the opening of the report, "*even before the pandemic, the relentless rise in costs was going to bring matters to a head. The broad support for universities that has been a feature since the 1960s is faltering and real increases in per student funding are unlikely in many jurisdictions*" [2].

Beyond COVID, rising costs, and declining public trust, the authors of this chapter quickly agreed upon a series of additional global challenges occurring within the rapidly evolving landscape of post-secondary engineering education, including changing demographics; online and on-demand education; rising nationalism within the geopolitical landscape; artificial intelligence; and faculty, staff, and student emotional and mental health.

Given the pace of disruption within the sector of higher education, readers can easily contribute additional challenges to this list and, moreover, by the time of publication, new disruptions will have been introduced. Within this environment, there may be a tendency to lean on outdated, "safe," strategic approaches; however, the complexity of the challenges warrants innovative, and in some cases, radical solutions.

The following case studies provide structural and cultural solutions, demonstrating innovative approaches for these rapidly changing times in higher education.

3 Case Studies from Engineering Institutions

3.1 Case Study 1: Creating the Lassonde School of Engineering: Home of the Renaissance Engineer

Evans, Eddy, and Kozinski, Janusz

The Lassonde School of Engineering was created at York University in Toronto, Canada in 2012 with the support of the Government of Ontario and philanthropist Pierre Lassonde.

Before embarking on this $150-million project, the leadership team of the new school engaged with global design thinking agency IDEO to consider the "why" and the "how" of building a brand-new engineering school in the first half of the twenty-first century. With IDEO's user-centric design methodology, they set themselves the ambition of creating a new school that would be based on the evolving needs of students within the context of the economic, environmental, and societal challenges—locally and globally.

The school's leaders began the project with a candid and rigorous assessment of the challenges facing engineering education and the fundamental issues that needed to be addressed with the rare opportunity to start from a blank sheet of paper which forms the basis of Part One.

Their proposed solution, outlined in Part Two, is the outcome of a design-thinking led approach to build an authentic identity that inspired every aspect of the school's foundations.

Finally, Part Three outlines a selection of initiatives the school embarked upon to bring the brand identity to life, and to reach a wider audience, including events, thought leadership and the design of the school's new home.

Part One: The Challenges Facing Engineering Education

Along with collaborators and partners, the school identified five key challenges in engineering education that needed to overcome and addressed in this new project:

1. **Theoretical Curriculum Content**: A typical undergraduate engineering course that a student must complete is focused almost exclusively on abstract concepts and theory without enough context or real-world application.
2. **Outdated Learning Model**: As engineering education has continued to expand in North America, the reflex of universities, seeking to increase student numbers, has been to expand lectures and reduce hands-on practical learning. The lecture model, particularly in the crucial early stages of an engineering degree, reduces the learning experience to a dissemination of knowledge where a student passively ingests information.
3. **Survival of the Fittest Culture**: The experience of engineering students is dominated by an overwhelming battery of testing and measurement. This inevitably leads to a culture of competition and comparison, rather than collaboration and cooperation.
4. **Inadequate Partnerships with Schools and Employers**: Too many students are 'flying blind' when they select a degree in engineering. The first time a student will meet an engineering professor will likely be on their first day of class, and similarly, the first time many students consider their future career trajectory will be in the later stages of their university careers. Much more needs to be done to break down walls and to build new bridges for the benefit of students and their future employers.
5. **Lack of Gender Diversity**: Using almost every measure available, the representation of women in engineering is unacceptably low. Engineering schools have a responsibility to look inwards to confront the systemic prejudices and biases that have remained. They also need to look outwards to organizations that have begun to successfully address this issue.

Part Two: The Solution—Home of the Renaissance Engineer

Having identified the fundamental challenges, the Lassonde School of Engineering community—together with collaborators and partners—sought to design a new, bold, and different approach to engineering education. The outcome was the school's founding identity built upon a compelling proposition:—Solving our greatest challenges takes more than technical talent alone: The world needs engineers with passion and perspective—or as we call them: Renaissance Engineers.

The school centered itself as the academic heart of an emerging Engineering Renaissance where engineers were no longer working in silos or purely focused on technical solutions. Instead, they are leaders, communicators, entrepreneurs, creators, and innovators working across a whole range of fields and industries. They are building not just the technological foundations of the Fourth Industrial Revolution—they are defining the human experience and shaping society more than ever, given the integration of technology into every aspect of life—both individually and collectively.

The newly created School developed a series of 10 Principles. These were designed to both inspire the community—and guiding decision-making and determining priorities to be relevant to the everyday functions of the school. The brand identity—the home of the Renaissance Engineer, could not simply be a slogan or a superficial gloss.

For a brand to build value and to be authentic over time, these 10 Principles needed to guide everything the school did and every experience it created. The 10 Principles were the bridge between the aspirational brand and the everyday experience of students and the whole community.

1. **An enlightened curriculum**: Renaissance Engineering has its roots in the highly valued technical acumen instilled by an accredited engineering education. But its essence comes from a broader perspective: an awareness of oneself and the world around, and above all, the passion and ability to contribute positively to society. The curriculum will evolve to reflect the complexity of the individual instead of a one-size-fits-all approach, which is typically imposed on engineering schools by the regulatory requirements of a professional program.
2. **50:50 Gender Balance**: The Lassonde School of Engineering has set a challenge to become the first engineering school in Canada to reach a 50:50 gender balance.

   This will be a School that looks 50:50—student enrolment and new faculty hires must have an equal gender balance.

   The school will feel 50:50—all genders feel confident, safe, and valued in an inclusive culture.

   The school will act 50:50—decision-making on compensation, promotion, and leadership appointments is consistent with the commitment to achieving a gender balance.
3. **Hands-on Learning**: Classroom technology is opening up the possibility of more varied learning experiences where students can learn by creating. Students will watch many more of their lectures anywhere, anytime and then come to class to discover answers for themselves under the guidance of their instructors.
4. **Putting Community First**: The Lassonde School of Engineering will be a community of people who love to apply the principles of engineering and science to challenge the world around them. It will be a formative experience where

people discover the passions that will drive the rest of their lives and where collectively we accomplish more together than we could alone.

5. **Trusting Our People**: The school believes that if our people are trusted to take ownership of projects, they will have the confidence to be creative and take risks. It will offer a safety net to give people the freedom to take risks, rather than hardwiring their every move. They will hire people who fit the emerging culture and understand from the outset that roles and teams will change as priorities change.
6. **Researching With Purpose**: The school aspires to be a place where different passions and perspectives intertwine to develop adaptive solutions to complex problems. The driving force behind this is that big global challenges—the likes of climate change, water scarcity and cyber-security—don't respect national borders or cultural divides.
7. **Partnering Without Borders**: The greatest asset an engineer can have in our global economy is the ability to work with others. The greatest achievements are accomplished by multidisciplinary teams, not lone wolves. Engineering educators, with the help of students, must do more to eliminate these inward-looking tendencies and create a learning experience that's relentlessly outward-looking, team-focused, and open to outside influences.
8. **Educating for Ingenuity**: Knowledge flows from the university to the populace around it in the form of embodied knowledge: well-educated graduates who make up a talented workforce. The most valuable output of the Lassonde School's research enterprise will be the minds of our graduates. The school doesn't just want to create world-changing research; it wants to nurture researchers who will change the world many times over.
9. **Traveling Together**: As a professional school, Lassonde has a responsibility to prepare graduates to be practitioners. Their professional designation grants them the license to operate as individual entities, and we must equip them with the confidence and the competencies to do so.
10. **Investing Wisely**: The formation of a new academic unit creates a rare opportunity to redesign the financial foundation of a professional school within a large, multidimensional university. An activity-based budgeting model eliminates ambiguity around resource allocation and helps local decision-makers understand where efforts need to be made.

Part Three: Evolving and Projecting the Brand Identity

The school's identity needed to leap off the page into the real world. To help take the principles from abstract to reality, the school needed to develop a series of events and initiatives to build engagement from the community and from external audiences—to project the school's identity.

- *Documentary Film—"Let Me Do It"*: A feature-length documentary "Let Me Do It" was commissioned by the school and produced by an independent filmmaker who spoke to the innovators, students, and institutional leaders in engineering education across the world about their visions for the future.

- *The Lassonde Debate Series*: The Lassonde Debate was designed to bring together futurists, academics, thought leaders and entrepreneurs from a variety of backgrounds to discuss the major challenges facing technology.
- *Bergeron Centre for Engineering Excellence*: The School's new building was designed from the outset to reflect and project the new institution's brand identity. The building was designed around the student journey—not the convenience of academics or administrators. That included zero lecture halls—replaced by flexible classrooms, learning studios and student-led makerspaces.

All these initiatives were inspired and informed by the 10 principles. They brought the brand identity to life as a meaningful and authentic vision around which to build a new academic community.

A decade after the Lassonde School of Engineering was created, the "how" and the "what" continue to evolve responsively to reflect the needs of students, employers, and the broader society, while remaining true to the "why"—the founding vision and principles that took this project from a blank sheet of paper to a thriving educational institution at the heart of the Engineering Renaissance.

3.2 Case Study 2: "Revealing the Gem": Cultural Transformation Through Strategic Planning

Bergen, Daniel, Walz, John, and Baumgartner, Eric

> All signs tend to confirm the flattering prediction often made that Milwaukee is to become the manufacturing metropolis of the West. [3]

It was against a backdrop of city-wide optimism that 23-year-old Oscar Werwath leveraged his German technical and engineering education to establish the Milwaukee School of Engineering (MSOE) in 1903. Designed to meet the growing need for workers in the city, Werwath, as quoted by Langill [4], emphasized "practical work supplemented by the study of mathematics and the theory of electricity." Within the initial decade, enrollments would grow consistently, initially drawing from Wisconsin, before expanding its reach regionally, nationally, and internationally. With its reputation well-established, Werwath created the first Industrial Advisory Committee in 1911. Beyond supporting curricular design, the group was well-versed in the latest trends in innovation and technology occurring around the country, giving the school an edge in the education landscape, and setting a strong foundation for industry and employer relations that has remained a cornerstone of the institution's brand identity for 120-years.

Currently, as Milwaukee realizes the late nineteenth century "flattering prediction" cited above, boasting the "number one concentration of manufacturing employment in the U.S.", it positions itself at a new frontier, being named the "number five 'up-and-coming' tech city" in the country, according to *Business Facilities Magazine* [5] and *Forbes Magazine* [6].

Once again, elevated optimism around the city's central role in technology, along with a rapidly changing industry landscape, provides the institution with the opportunity to embrace its core identity, while transforming itself to respond to its contemporary environment. In 2017, under the leadership of the fifth president, Dr. John Walz, MSOE undertook the first ever campus-wide, collaborative strategic visioning and planning process. Out of this experience, *Extraordinary Together* was created, a strategic plan touting three commitments, nineteen strategies, and 81 actions.

While the development of a strategic plan is not unique, the process, coupled with an exceptional discipline on execution, and an openness to evolving, reflects a strategic cultural transformation for the university.

The Process of Creating Extraordinary Together

> One of the things that I heard a lot when I came here was 'MSOE. You guys are a hidden gem; And I always say, 'Well, I like half of that. I like being a gem, but it doesn't help us to be hidden.'—John Y. Walz, [7]

In 2017, MSOE embarked on a year-long process to develop the first comprehensive, campus wide strategic plan for the university. President Walz deliberately waited until his leadership team was formed to move into the execution of the process. In developing it, the team leveraged Academic Leadership Associates, LLC, an external consultant, along with a collaborative cross section of MSOE faculty, staff, students, alumni, and Regents.

Previous leadership had successfully utilized a centralized style to set the direction and vision for the institution, but the rapidly changing higher education landscape demanded a new approach. The group convened on August 28 and 29, 2017, to begin the process of establishing a vision, mission, and strategic plan for the next five years of MSOE. The experience seeded a cultural transformation that would not only generate a significant amount of buy-in from all stakeholders but also strengthen trust and community among colleagues.

Throughout the next two months, members of the planning team utilize the strategic framework to establish a draft mission, set of values, and distinctive capabilities and measures. Simultaneously, a series of task forces are developed to address key implementation elements, including strategies and actions, around the following strategic areas: Education Programs, Scholarship, People, External Relations, and Internal Operations.

Eric Baumgartner, executive vice president of Academics, recalls:

> I was surprised by the number of participants leading – 44 members in total – I had never seen a plan developed like that. It set a tone of inclusivity from the beginning. Further, it became a cathartic opportunity for community members to address and voice their challenges. As a leader, Dr. Walz stated 'your voice is as important as my voice' and faculty and staff appreciated that approach.

On November 3, 2017, members of the broader MSOE community responded to recommendations and provided feedback. The energy was palpable as the campus community and other key stakeholders increasingly coalesced around the mission, vision, and strategic plan for the future of MSOE. Throughout the following six months, countless meetings involving faculty, staff, students, alumni, and other key stakeholders occurred. The agenda was to continue the process of information gathering, brainstorming, feedback, analysis, careful thought, and revision. While the mission, values, and plan to take a more formal shape, the impact of the cultural shift occurring across the institution cannot be understated. In fact, according to Russ Meier, professor of Electrical Engineering and Computer Science,

> Across our campus, change is evident. Driven by our institutional strategic plan, Extraordinary Together, and our Diversity, Equity, Inclusion, and Belonging Strategic Plan, our campus is vibrant with men and women pursuing their dreams in varied majors, collaborating with each other, and respectful of the paths each has chosen as a career. I sense a new community-wide understanding of how MSOE reaches its most impactful moments when we are working to be Extraordinary Together rather than living in the organizational silos that once isolated programs from each other.

At the June 2018 Board of Regents meeting, the board reviewed and discussed *Extraordinary Together*, along with a set of strategic priorities for the upcoming academic year. Following the adoption and launch, a strategic implementation team composed of key leaders and stakeholders across campus was assembled, and execution of the plan began immediately in the Fall of 2018. With the plan in place and the team assembled, MSOE began the process of revealing the long-hidden gem of its institution, setting it on the path towards greater recognition for its historical contributions to the city of Milwaukee and beyond.

The Discipline on Execution of Extraordinary Together

To execute the plan, leadership knew that it would require an engaged team of members, invested in the prioritization of the strategies and actions. Dubbed the "Implementation Team" this group, comprised of nearly 20 faculty, staff and senior leaders. They were tasked with the following objectives:

- Prioritize and lead the implementation of the actions.
- Identify new opportunities.
- Track our performance.
- Keep our constituents updated.
- Engage with other strategic planning groups across campus.

Utilizing this structure allows for two key outcomes: alignment on communicable, annual priorities across campus; and a commitment to the success of these priorities, including the removal of any barriers that may pro habit them from moving forward.

Priorities are given to champions who lead the development of action plans, reporting out on progress monthly to the broader team. The action plans are frequently composed of "micro-commitments," or small, short-term tasks that can be achieved quickly with a relatively low time investment. Report-outs become trust-building, dynamic conversations. As Morrison et al., states:

> In meeting adaptive challenges, we learn by doing. So, we need conversations that lead us to action, to experimentation, so we can generate new insights into these complex situations (2019).

Bi-annual campus updates in November and May are milestones for the completion of strategies and actions, ensuring that lines of campus communication remain consistent, and the plan does not grow stale. Within the first 18 months of the launch of the plan, a campus climate survey has been completed by 81% of faculty/staff, a new institute has been formed to support the integration of experiential learning across campus, and new positions have been created to steward corporate and K-12 relations, respectively. As Donald Sull and Charles Spinosa suggest, "***promises kept, even small promises, result in increased trust. One of the reasons organizations face difficulty in strategy execution is the inability to develop clear commitments in a systematic way. This weakness is a function of the inability of leaders to build trust across an organization or a community***" [1].

Despite the global pandemic, the fruits of the campus efforts in development and early-stage execution are not lost. By early 2021, Walz assembled members of the Board of Regents, as well as the implementation team, to discuss revisions. *Extraordinary Together*, was relaunched stating that—"***modifications and enhancements should be made to account for the changes in higher education, technology and the way the world interacts post-pandemic***."

In January and April 2021, campus wide community forums were held, and by Summer 2021, *Extraordinary Together (2021–2026)* was adopted.

The updated plan considers the impact of online education, as well as the role technologies, including augmented and virtual reality, will play in the future of engineering. Following the adoption of the revised plan, the leadership team hires the senior advisor to the President, a new position designed to steward the strategic plan and by April of 2022, a new implementation team assembled. 2022–23 priorities were established, and once again, execution began in earnest.

The Dynamic, Industry-Responsive Evolution of Extraordinary Together

At the time of publication of this case study, over 75% of the 81 action items in *Extraordinary Together* were in progress or have been completed.

The plan has been an excellent tool, guiding the community, but it has never been rigid, or stagnant, nor has it wandered from the institution's core vision to "***set the standard for preparing leaders to solve the diverse technical challenges of the twenty-first century***."

Arguably, the diversity and complexity of those technical challenges are evolving more quickly than ever, demanding that MSOE be increasingly nimble to maintain its identity as an industry-responsive institution. Here are a few examples of how MSOE has responded to these demands over the past six years:

- The donation of a $35 million-dollar, state of the art computational science building in 2017 by MSOE alum Dwight Diercks.
- The launch of a computer science program in 2018, which now boasts the second highest number of majors at MSOE.
- The development and launch of a fully online master's program in Machine Learning in 2022.
- The opening of a Center for Professional Education in the Fall of 2023.

- A transition from a quarter-based system to a semester academic calendar in Fall of 2023, allowing for greater ease of transfer, as well as anticipated increase in retention and persistence.
- The full integration of the MSOE Mindset, which seeks to develop graduates as leaders of character; responsible professionals; passionate learners; and value creators.

Not all the examples are named actions in *Extraordinary Together*; however, cultural transformation stewarded by leadership and cultivated in the creation of the plan has allowed for MSOE to respond nimbly, thrive, and adjust despite the rising challenges facing higher education. To maintain its competitive edge, the university has continued to do what it has done so well for 120-years—focusing on the industries it serves. However, evolving has required a strategic cultural transformation in the delivery of outcomes and, according to Michael A. Diamond, senior partner, Academic Leadership Associates, a consulting firm focused on supporting higher education strategic planning and execution, "*[Dr. Walz is] a model on how to do effective implementation and how to make the most of the strategic planning process.*"

Whether supporting industry partners in making Milwaukee the western "metropolis of manufacturing," or delivering the talent that will support it as "an up-and-coming tech city," MSOE is realizing its full potential in new and dynamic ways, and the gem is only just beginning to be revealed.

3.3 Case Study 3: A South African Perspective: Unifying Diverse Paths and Forging Unity

Transformation and Restructuring of the South African (SA) Higher Education Sector

In August 1997, the SA Government gazetteed a White Paper—A programme for Higher Education Transformation [8], to strengthen transformation in higher education (HE) focusing on equity and redress; quality; effectiveness; efficiency; academic freedom; institutional autonomy; and accountability. The white paper, coupled with the National Plan for Higher Education released in March 2001 [9], served as the precipitating events for the transformation and restructuring of a novel HE landscapes. In 2002, the minister of Education gazettes the merger of institutions and the rationalization of programmes. The redefined institutional landscape provided opportunities for contributing to the national goals and priorities in a democratic society that continues to grapple with unemployment, inequality, and poverty.

In the process three types of institutions were established:

- Comprehensive Universities:—merger of a University and Technikon that offers both career-focused and general academic programs.
- Technikons:—It was later renamed to Universities of Technology; and
- Universities.

The author of this case study served as Dean of Engineering at a University of Technology (UoT) and a Comprehensive University. As will be shown, the transformation of these institutions was complex, and the success of these transitions was due to the engagement with various stakeholders in the strategic visioning and planning processes.

Thirty (30) years into our democracy, SA HE is still facing a multitude of challenges. One of the most profound was the #FeesMustFall, a student-led protest movement for free, quality, and decolonized education in mid-October 2015. The protest demonstrated the collective frustration of students with the rising cost of tuition fees, which led to financial and access barriers to HE for many students in SA. The movement led to (1) no fee increases in 2016 and 2017; and (2) a change in the National Student Financial Aid Scheme (NSFAS) government allocations for student fees, accommodation, study material, and living expenses.

Many would argue that despite this progress, some students are still excluded from HE for several reasons, and that much still needs to be done [10]. In fact, the 2023 Human Sciences Research Council (HSRC) report on the state of transformation in SA's public universities stated that although there have been great advances and achievements toward transformation, many SA higher education institutions continue to face deeply entrenched inequalities and challenges in many domains [11]. The focus for engineering leaders shifted to providing access and opportunities to previously marginalized groups with diverse socio-economic backgrounds, including women.

In today's rapidly changing HE environment, ethical leaders focusing on both transformative and transformational leadership, are required in the creation of a strategic vision. The following two narratives capture the critical moments of transition and transformation within two SA institutions over the past 20 years. The case studies demonstrate the importance of an engineering faculty's leadership and shared governance in advancing "educational transformation."

Part 1: Cape Peninsula University of Technology (CPUT)

Unifying Diverse Paths: The Transition to a University of Technology

In January 2005, Cape Technikon and Peninsula Technikon merged to become Cape Peninsula University of Technology (CPUT) and transitioned from the Technikon environment to a University of Technology (UoT) (for which no institutional memory existed). CPUT faced many challenges through the merger process given the cultural differences between the two historic institutions. A study was conducted by Garraway and Winberg—two researchers within CPUT—focusing on reimagining futures of the UoT that had a lack of clarity and identity. Against this backdrop, CPUT went through multiple phases [12]:

1. *CPUT 1.0—Vision 2010*—under the leadership of the first Vice Chancellor, the institution focused on formulating common practices and forming a baseline for a new UoT.

2. *CPUT 2.0—Vision 2020*—focused on four strategic thrusts (i) teaching and learning; (ii) research, technology, and innovation; (iii) sustainability and efficiency; and (iv) the student experience.
3. *CPUT 3.0—Vision 2030* (*One Smart CPUT*)—informed by two dimensions: *Oneness* and *Smartness*; including three foci: i.e., Learning and Teaching Smartness; Research, Technology, and Innovation Smartness; and Smartness in operations and support.

The process of crafting the "*One Smart CPUT*" vision was an extensive consultative process involving various engagements with different stakeholders. It considered the impacts of the #FeesmustFall student movement on various aspects of the institution including all the key infrastructure destroyed during this period; as well as the extensive Higher Education Qualification Sub-Framework (HEQSF) realignment and re-curriculum designing process, Gazetted in October 2014 [13]. The crafting of the new vision took place through multiple workshops; and provided the building blocks for crafting the new Vision 2030. This was important not only to ensure that the various voices were heard but to ensure buy-in from the start and building trust; hence the focus on building "*Oneness*"—creating one institutional culture and a sense of belonging; and "*Smartness*"—working collectively, taking shared responsibility; breaking silos; and working across departments, faculties, and disciplines; and staying responsive to a rapidly changing environment.

Engineering One Smart Faculty

On 5 November 2018, the Senate approved a faculty name change to the "*Faculty of Engineering and the Built Environment (FEBE)*" to reflect all the qualifications offered. In November 2020, in line with CPUT 3.0—*One Smart CPUT*, the process of crafting a new faculty strategy commenced under the leadership of the new Dean, who is the author of this case study; a team consisting of the faculty management, that comprised of two Assistant Deans; eight Heads of Departments; faculty portfolio coordinators; and faculty research entity leaders. They functioned through a range of workshops. After each workshop, the draft strategy was discussed within departments and at the Faculty board for input from the broader faculty staff and circulated through all faculty structures. The strategy considered the institutional strategy of being *One Smart CPUT*; the impacts of the new Higher Education Qualification Sub-Framework (HEQSF) alignment process and future programs; the new Engineering Council of South Africa (ECSA) qualification standards, the impacts of the #feesmustfall movement; and a full faculty situational analysis. This was a collective and consultative process to ensure contributions and commitment from everyone for the implementation of the plan. On 23 May 2022, the Senate approved the Engineering and the Built Environment Faculty's strategic plan for becoming "*One Smart Faculty*".

Part 2: Nelson Mandela University

Unifying Diverse Paths: The Transition into a Comprehensive University

The transition of Nelson Mandela University is captured here through a partial experience having joined the institution only in October 2022. Nelson Mandela Metropolitan University, a Comprehensive University, established during the nationwide restructuring and merger of Port Elizabeth (PE) Technikon, the University of Port Elizabeth (UPE), and the Port Elizabeth campus of Vista University (Vista PE), opened its doors on 1st January 2005. In July 2017, with approval from the Nelson Mandela Foundation, the institution was officially renamed the Nelson Mandela University, the only Higher Education Institution, named after the first democratically elected president of SA, the late Nelson Rolihlahla Mandela. This name change was significant in that it defined a new trajectory for the institution inviting it to reposition and rebrand itself to pay tribute to the values and ethos of its namesake and honor his name.

The late Nelson Mandela was dedicated to changing the world and one of his most famous quotes is:

> Education is the most powerful weapon which you can use to change the world.

Nelson Mandela University's process of crafting Vision 2030 linked to the purpose of being "*In service of society*", under the leadership of the Vice Chancellor, included a listening campaign and focus group discussions with employees, organized labor, middle management, student leaders, alumni, and stakeholder engagement. The situational analysis considered the global impact of the COVID-19 pandemic; online learning and the digital divide; decolonization of higher education in South Africa which came to the fore as part of the #Feesmustfall movement; the state of unemployment and access to higher education and healthcare; the future world of work and the future graduate attributes; and the state of the University. The reflections included "*what Nelson Mandela as a University is good for and what it is good at*" (Nelson Mandela University's Vision 2030 Strategy) [14].

Its core values are respect for diversity; excellence; social justice and equality; ubuntu; integrity; and sustainable stewardship.

Nelson Mandela University's Vision 2030 is an articulation of its strategic intentions and aspirations with four strategic focus areas: (i) to liberate human potential through humanizing, innovative learning and teaching experiences; (ii) impactful research, innovation and internationalization; (iii) transformation and engagement through equalizing partnerships to co-create responsible solutions; and (iv) student access for success through student-centric approaches and practices that provide life-changing experiences within and beyond the classroom [14].

Engineering a Faculty "In Service of Society"

In 2020, as part of the institutional organizational redesign, the School of Architecture was incorporated into the Faculty. Consequently, the Faculty name was changed from "*Faculty of Engineering, the Built Environment, and Information Technology—EBEIT*" to "*Faculty of Engineering, the Built Environment, and Technology—EBET*" consisting of four schools—Architecture, Built Environment and Civil Engineering, Engineering and Information Technology.

At the time of writing this case study, the process of developing the faculty's new strategic plan was underway. This process, under the leadership of the new Dean (author of this case study) with the support of a facilitator, evolved through a faculty-wide, inclusive, collaborative process consisting of workshops including the Faculty Management (Dean, Assistant Dean, four Directors of Schools; and 15 Heads of departments). Each workshop had a specific theme and focus area:

1. One workshop focused on crafting a collective new Vision and Mission with invited representatives of the Institutional Strategy unit; and a full faculty situational analysis,
2. Subsequent workshops each on a Strategic Focus Area related to the institutional vision 2030 of:
    - humanizing and innovative learning and teaching with additional invited members from the faculty learning and teaching committee and relevant institutional support units.
    - impactful research, innovation, and internationalization with invited professors, heads of faculty research entities, and members of relevant institutional support units.
    - student access for success, a specific session was held with students representing all the schools and departments; as well as the faculty representative from the central student representative council (SRC).

During the process, interesting and thought-provoking discussions emerged, such as the interpretation of "humanizing pedagogies" within the EBET faculty context. As a new Dean I had assumed that since the concept was already integrated into the institutional strategy, all staff members were well-versed in, and understood this pedagogical approach. However, it became evident that this was not the case.

The faculty's research and engagement themes and focus areas were aligned with the Institutional Research themes; as well as the local, national, continental, and global imperatives (i.e., Sustainable development goals). A very important factor was the location of the faculty within the Eastern Cape Province and responding to local needs of industry, government, communities, and being "*in service of society.*" Specifically, having a program qualification mix that is responsive, relevant, and sustainable was critical. To this end, the faculty's latest annual research and engagement publication, 2023 *iDEATE volume 3*, is themed "***Pursuing Sustainable Futures for a Better World***" [15].

4 Key Takeaways for Academic Leaders

- Transformative strategies and brands invite campus-informed, collaborative and community invested leadership. Leaders should utilize their campus communities to thoughtfully advance their plans to ensure successful execution.
- Demographic shifts, and specifically increases in historically underrepresented populations, including women and racially-minoritized students, in the STEM fields, demand thoughtful and strategic consideration. Leaders should be intentional in their engagement of these populations to ensure the sustainability of their brands.
- "Discipline on plan and brand" is critical to navigating the current disrupted sector of higher education. Leaders may be distracted by flashy trends, and tempted to shift strategy, but must remain focused on the community-informed, institutional strategy and brand.

5 Conclusion

The case studies demonstrate the unique intersection of responsibility that higher education holds during the contemporary dynamic global environment. The sector is expected to educate and deliver quality engineers who embody what the United States KEEN Network defines as the Entrepreneurial Mindset, or "technical skills [accompanied by] curiosity, connections, and creating value" [16]. At the same time, it is being asked to undergo perhaps the most rapid organizational evolution in its history.

Whether building and launching a new renaissance engineering school, transforming a culture using a shared strategic plan, or merging institutions under the direction of the national government, the "speed to market" of each of these initiatives is reflective of the new environment within which higher education is operating. Interruptions to business continuity require an urgent response leading to changes in educational modes of delivery combined with the use of technology. Subsequently, engineering leaders must adapt their approach, faculty/staff must adapt their learning and teaching pedagogies. It is of utmost importance to create a conducive environment for learning and teaching, and impactful research and innovation. However, it cannot be done alone. The global landscape of higher education is rapidly transforming, and our future success and brand identity is dependent upon the collaboration of forward thinking, impactful strategic academic leader.

References

1. Morrison E, Hutcheson S, Nilsen E, Fadden J, Franklin N (2019) Strategic doing: ten skills for agile leadership. Wiley Publishing Company, New York. ISBN: 978-1-119-57840-6

2. KPMG (2020) The future of a higher education in a disruptive world. https://assets.kpmg.com/content/dam/kpmg/xx/pdf/2020/10/future-of-higher-education. Last accessed 4 Apr 2024
3. Still B (1948) Milwaukee, the history of a city. State Historical Society of Wisconsin, Madison
4. Langill E (2003) MSOE: Milwaukee School of Engineering: the first 100 years centennial 1903–2003. MSOE Press, Milwaukee, Wisconsin
5. Business Facilities Magazine. https://www.choosemilwaukee.com/
6. Forbes Magazine. https://www.choosemilwaukee.com/. Last accessed 15 Sep 2023
7. Walz JY (2023) President, MSOE, Milwaukee Journal Sentinel. https://www.jsonline.com/story/news/education/2023/05/17/msoe-president-wants-to-raise-the-profile-of-milwaukees-hidden-gem/70047205007/. Last accessed 4 Apr 2024
8. Republic of South Africa (RSA). White paper on A Programme for Higher Education Transformation. Government Gazette No 18207 of 15 August 1997
9. Republic of South Africa (RSA) (2001) National plan for higher education. https://www.justice.gov.za/commissions/feeshet/docs/2001-NationalPlanForHigherEducation.pdf. Last accessed 4 Apr 2024
10. Mutekwe E (2017) Unmasking the ramifications of the fees-must-fall-conundrum in higher education institutions in South Africa: a critical perspective. Perspect Educ 35(2):142–154
11. Luescher TM, Mncwango B, Fongwa SN, Oppelt T, Mthombeni Z, Paterson M (2023) The state of transformation in South Africa's public universities. Research report of the Ministerial Oversight Committee on Transformation in the South African Public Universities (TOC). TOC/Department of Higher Education and Training, Pretoria
12. Cape Peninsula University of Technology. https://www.cput.ac.za/about. Last accessed 14 Sep 2023
13. Republic of South Africa (RSA) (2014) Higher education qualifications sub-framework. Government Gazette No 38116 (17 October 2014)
14. Nelson Mandela University. https://www.mandela.ac.za/About-us. Last accessed 14 Sep 2023
15. Faculty of Engineering, the Built Environment and Technology. https://ebet.mandela.ac.za/. Last accessed 24 Oct 2023
16. Kern Entrepreneurial Engineering Network (KEEN). What is KEEN? | Engineering unleashed. Last accessed 24 Oct 2023

Governance Models of Engineering Institutions for Effective Administration and Human Resource Management

Archana Mantri, Sheldon Marshall, and Rohit Kandakatla

Abstract In an ever-changing landscape of Higher Education across the globe, governance of HEIs remains most stressed and has always been a challenging preposition. On one hand, the administrators need to keep pace with the demands of dynamic change, and on the other, there are huge capacity building requirements of the human resources. In sectors where machines are doing the job, it is a matter of time that new ones are built or reprogrammed to cater to the changing market requirements. But, in the Education sector, it is the huge mass of human resources which needs to adapt and adopt the change. This chapter dwells upon the current scenario of higher education, challenges, opportunities and goes on to discuss the case studies from an autonomous college and a state-private university—both from India, in addition to that of a University in South Africa. In the conclusion, principles of effective governance are proposed, opening doors for further deliberations and discussions.

Keywords Governance model · Effective administration · Human resource management

1 Introduction

A central issue in any organization—higher education institutions included—is governance.

It is challenging because it determines how the institution would function or dysfunction. Governance in Higher Education Systems lies at the cross-section of

A. Mantri
Current: Anurag University, Earlier: Chitkara University, Punjab, Hyderabad 500088, India

S. Marshall
Nelson Mandela University, Gqeberha 6031, South Africa

R. Kandakatla (✉)
KG Reddy College of Engineering and Technology, Telangana 500075, India
e-mail: rohit.kandakatla@kgr.ac.in

R. Kandakatla et al. (eds.), *Academic Leadership in Engineering Education*, Lecture Notes in Networks and Systems 1097, https://doi.org/10.1007/978-3-031-68282-7_2

internal and external environments and is complex in nature as it must juggle through the ever-changing state-level gamut and the more subtle nature of organisation's culture and ethos. The internal environment is the cultural, social, organisational foundation and processes of the institution which takes its own shape depending upon the socio-cultural-economic profile of the agents (students, faculty, administrators) and their interconnecting equations. On the other hand, the external agents are its geo-political milieu, socio-economic environment, national and international policies, regulatory requirements, and others depending on the institution.

No matter how much the institutional governing body may want to manage matters at its optimum, the external environment keeps casting its shadow on the internal environments—sometimes for advancement and at other times as confusion. The prime intention of governance in institutions, thus, should be to continuously mitigate the risks emerging from too frequent changes in the external environments and preserve the ethos of the institution in line with the vision and mission of promoting quality education.

It can be conveniently noticed that those institutions, which make long-term policies and strategic plans, that are flexible enough to absorb the vibrations caused by external agents, can survive and perform better, while those making too rigid plans end up suffocating their stakeholders to non-performance and hampering the potential and innovation of individuals and teams. The process of governance should be participative for the formulation of policies and procedures and decentralized for implementation. There should be more accountability and flexibility instead of more hierarchical and tight regulations. For Institutions to survive in the disruptive world of the twenty-first century and beyond, it is advisable to make societal and sustainable priorities related to environmental, social, economic, and governance strategies; bring research, innovation, engagement, and academic freedom and autonomy into the prime focus of all actions to build a green and digital culture.

As Faculties/Schools of Engineering continue to operate in a disruptive state of flux leaders must manage in a state of VUCA (volatile, uncertain, complex, ambiguous). This, combined with some country's slow economic growth; and continuous decrease in government funding; results in pressures to diversify funding sources (i.e., such as increase third-stream income). Deans are required to do more with less. The landscape of engineering education has clearly shifted, and our ability to adapt, and leverage new opportunities are set to define our success.

These changes call into question the role and purpose of governance and leadership models and structures while at the same time maintaining accountability and responsibility, and delivering on the strategic vision, mission, and mandate. In today's rapidly evolving world, where technology, innovation, and globalisation drive change at an unprecedented pace, the role of ethical and responsible governance has become ever more crucial. Leaders and managers must embrace new models of governance to transform these institutions. Recognition of human resources as the most essential stakeholder in academic institutions will play a pivotal role.

2 Background

2.1 Governance Models

By 'Governance model' we mean the organizational structures and strategies through which policies and regulations are formulated, processes are established, and decisions are made for effective and efficient management [1]. These models depict how institutions ensure effective interaction and collaboration among human resources to achieve the vision, mission, and goals of the institution. Dobbins et al. [1] summarized the European governance models into (i) state centered; (ii) academic self-governance; and (iii) market-orientated models.

There were further attempts to develop three ideal types which takes into consideration the role of the state, external stakeholders, and the university governance.

A few commonly adopted governance models in engineering institutions are:

- *Traditional University Model*—a model where the engineering institution functions as part of the larger university with a centralized governance structure. All decisions with respect to the academic programs, budgets, human resources, and policies are taken centrally at the university level. The senate and/or council composed of elected representatives make key decisions related to curriculum, academic policies, and faculty matters. A Vice-Chancellor/Rector/Principal is appointed to manage the institution.
- *Board of Trustees or Board of Governors Model*: The governing board, comprising of nominated external and internal members of eminence provides key insights and strategic direction. The board may appoint a president or chancellor to manage day-to-day operations, including academic and research activities.
- *Semi-Autonomous Model*: The institution operates as a semi-autonomous entity within a university. The engineering institution has its own dean or director who oversees academic programs, research, human resources, and administrative matters. While some decisions are made locally, there would be many decisions that need to be taken in consultation with the university's central administration.
- *Decentralized Model*: Some engineering institutions adopt a decentralized governance approach, where, except for very few strategic issues involving finance and outreach, all other strategies are made by individual departments or centers/entities. In this model, the departments/centres/entities have a high degree of autonomy and accountability over curriculum, research, engagement, human resource, and administrative matters.

Effective governance models ensure transparency, accountability, and the alignment of institutional decisions with the institution's long-term goals and values.

2.2 Organizational Structure

Worldwide, Engineering institutions adopt varied organizational structures to ensure effective and efficient functioning of operations and management of human resources such as faculty, students, and others. The choice of organisational structure depends on the choice of the relevant founding members, or the relevant Institutional Statute, Council, Executive Management, aligned with the goal for which the Institution was established, and compliance with regulatory requirements. Changes in the organisational structure are very rare but sometimes become necessary to address the changing requirements of government, industry, and society. Some of the different organizational structures are (but not limited to):

- *College/Faculty/School structure*—The institution functions as a College/Faculty and is further organized into schools, each representing a discipline. The College or Faculty is governed by a Director, Executive Dean, Dean, or Head of School. In this structure, each school is responsible for its own programs and functions under the umbrella of the College/Faculty of Engineering.
- *Department structure*—The Institution is organized into multiple departments each representing a particular discipline. In this structure, the departments are governed by the Head of the Department and function as per the regulations framed by a central body of the engineering institution.
- *Matrix Structure*: Engineering disciplines are organized into departments while inter-disciplinary research and projects are facilitated through cross-functional teams. This structure encourages collaboration across disciplines while maintaining specialized departments.
- *Research-Center or Research Entity Structure*: The Institution establishes specialized research centers/entities focused on specific areas of engineering. These entities may be inter-disciplinary and bring together faculty, researchers, and students from different departments to work on common research goals.

The choice of governance and organizational structures depends on factors such as the institution's vision, mission, size, culture, and legal structure. Effective governance structures ensure transparency, accountability, and the alignment of institutional decisions with the institution's long-term goals and values.

2.3 Human Resource Practices in Academic Institutions

By human resource practices, we mean recruiting, training, and retaining a talented, skilled, knowledgeable, specialized, and motivated workforce to support the institution's vision, mission, and goals. These practices should foster a supportive, diverse, and inclusive work environment that enables faculty and staff to excel and contribute to the institution's success. Some important human resource practices that engineering institutions often implement are:

- Recruitment and Selection
  - Recruitment Strategies: This includes deciding and strategizing on the frequency of recruitment, equal opportunity and access, transparency on job roles and responsibilities, and compensation packages to attract top talent. The adverts are professional and unambiguous. Both digital and print media can be used to advertise job openings.
  - Rigorous Selection Process: Implement thorough interview and evaluation processes to assess candidates' qualifications, skills, and cultural fit.
- Orientation and onboarding
  - New Employee Orientation: Provide comprehensive orientations to introduce new hires to the institution's culture, policies, and resources.
  - Mentorship Programs: Pair new faculty and staff with experienced colleagues to facilitate a smooth transition and foster professional development.
- Faculty development, training, and support
  - Professional Development Opportunities: Offer workshops, conferences, and training programs to enhance teaching, research, and leadership skills; to let the individuals set their personal and professional goals and devise strategies to achieve them.
  - Research Support: Provide resources, funding, and facilities to enable faculty to conduct cutting-edge research and secure external grants.
- Performance Management
  - Performance Metrics: Define key performance indicators (KPIs) for faculty and staff, based on their roles and responsibilities.
  - Performance Appraisals: Conduct regular performance assessments to provide feedback on individual and professional goals and reward the achievements.
- Compensation and benefits
  - Competitive Salaries: Ensure that compensation packages for faculty and staff are competitive within the industry and commensurate with their contributions.
  - Indirect Benefits: Offer comprehensive benefits, including health insurance, retirement plans, and leave policies.
- Employee growth, engagement, and well-being
  - Promotion and Tenure: Establish clear criteria and processes for faculty promotion, tenure, and performance evaluation.
  - Employee Recognition: Recognize and celebrate faculty and staff achievements through awards and recognition programs.
  - Wellness Programs: Offer wellness initiatives, mental health resources, and stress management programs.

These six pointers discussed above have an umbrella cover that considers the quality of the faculty, their progress, productivity, and future. The leadership team's responsibility here is not just decision making and good governance as and when called for, but to synchronize the whole process so that it runs seamlessly and progressively.

It's a continuous learning process for leaders in academia and every day is as much a challenge as an opportunity. The endeavor should be to share best practices from across the world through communication and collaboration, learn new leadership skills from all verticals and keep students' career and future as the central focus.

In the next section, we present three case studies of engineering institutions from India and South Africa. In each case study, we discuss the governance model adopted by the institution, the organizational structure being followed for effective administration, and the human resource practices being implemented to recruit, empower, and support all the stakeholders involved.

3 Case Studies from Engineering Institutions

3.1 Case Study 1—KG Reddy College of Engineering and Technology, India

3.1.1 Governance Models

KG Reddy College of Engineering and Technology (KGRCET) is an autonomous institution in the south of India that offers undergraduate and postgraduate programs in engineering and management.

In India, higher education institutions can be divided by their governance models into three categories—university system, autonomous system, and affiliated system.

In a university system, the entire institution's governance is taken care of within the university, and degrees are offered by the same university. Universities can be established by public or private entities through recognition from the state government (State Private Universities) or from the central government (Deemed to be Universities) [2]. In autonomous and affiliation systems, the institutions are associated with a parent public university which is responsible for offering the degrees to all the graduation students [3]. Autonomous and affiliated systems differ in their governance models with respect to the extent of autonomy given to them to function. Most institutions in India are established as affiliated entities which over time evolve to Autonomous and University status based on when they meet respective eligibility criteria recommended by state and national regulatory authorities.

KGRCET was established in the year 2008 as an affiliated institution of Jawaharlal Nehru Technological University Hyderabad (JNTUH) and has evolved into an autonomous institution in 2021.

In an autonomous system, the institution is allowed to internally establish governance mechanisms as prescribed by the University Grants Commission (UGC), one of the national regulatory bodies for higher education in India. As per the guidelines of the UGC, KGRCET has formed four statutory bodies that monitor and ensure effective governance of the institution at various levels—Governing Body, Academic Council, Board of Studies, and the Finance Committee [4]. The members in each of these statutory bodies are nominated from a list of diverse stakeholders such as the governing trust of the institution, eminent academicians, researchers, industry representatives, principal, deans, heads of departments, senior faculty members, students, and alumni of the institution. JNTUH nominates its faculty to all the statutory bodies to ensure the functioning of the KGRCET does not deviate too much from the parent university.

The governing body is considered to be the highest governing authority of the institution and is responsible for overseeing the progress of the institution in line with its mission and vision statements, approvals of new programs, review of strategy plans, recruitment of faculty and staff, and approving the annual budget and expenditure. The Academic Council of the institution is responsible for the finalization of the academic regulations based on the recommendations received from the Board of Studies for various programs being offered. The academic regulations include the decisions on the total number of credits required for graduation, models of assessment, and pedagogical methods to be implemented. The Academic Council will also prescribe regulations for admissions and recommend strategies to be adopted to encourage research, innovation, co-curricular, and extra-curricular activities in the institution. A Board of Studies is formed for each of the departments available in the institution and they're responsible for finalization and recommendation of the curriculum of the programs being offered to the academic council. The finance committee as the name indicates is responsible for planning and monitoring all the finance-related activities in the institution. By hierarchy, the Board of Studies reports to the Academic Council, who along with the Finance Committee reports to the Governing Body.

3.1.2 Organizational Structure

KGRCET follows a department where the different disciplines of engineering and management are divided into multiple departments. Heads of Departments are appointed for each of the departments who oversee its day-to-day functioning, serve as Chairperson of their Board of Studies, and members represent their respective departments in the Academic Council. All decisions taken by the Head of Department should adhere to academic regulations approved by the Academic Council of the institution. All the Head of Departments report to the principal who acts as the Head of the entire institution. KGRCET also appointed Deans for Academics, Research, Internal Quality Assurance, Industry Relations, and Human Resources at the institution level who work in cross-section with all Head of Departments to monitor their progress and support the principal. The day-to-day functioning of the Deans and Heads of Departments intersect at various levels with an aim to improve accountability and monitoring of their respective roles and responsibilities. KGRCET

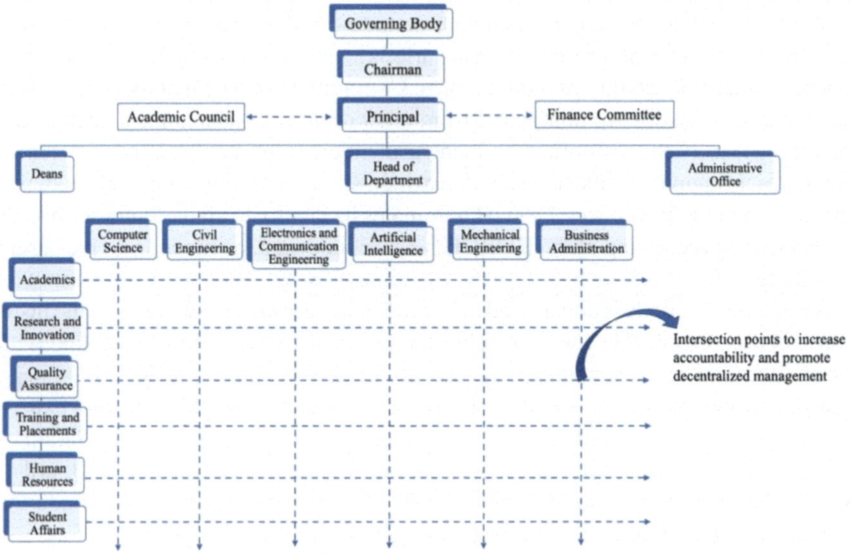

Fig. 1 Central organizational structure of KG Reddy College of Engineering and Technology

has also adopted a matrix structure where specific Centers of Excellence are established strategically to help achieve the vision of the institution. Independent heads are appointed for each of the centers to plan and oversee the day-to-day activities The Principal is supported by a large administrative office whose head oversees all day-to-day administrative tasks in the institution (Fig. 1).

An internal organizational structure is also implemented at the department level to streamline all the operations and ensure the effective utilization of human resources. Multiple committees such as the Department Development Committee (DDC), Program Assessment Committee (PAC), Project Review Committee (PRC), and Finance Committee are established to decentralize the administration in the department and make the functioning more process oriented. The Head of Department by default serves as the head of all the committees and is supported by their faculty who serve as coordinators and members of different committees (Fig. 2).

3.1.3 Human Resource Practices

KGRCET has an HR policy that has been thoroughly drafted and approved by the governing body. The HR policy includes specific recommendations on the recruitment process, expected qualifications of faculty and staff at multiple levels, onboarding process, leave policies, welfare measures, performance appraisal process, promotion policy, and resignation process. We elaborate below on some of the best practices from the HR policy.

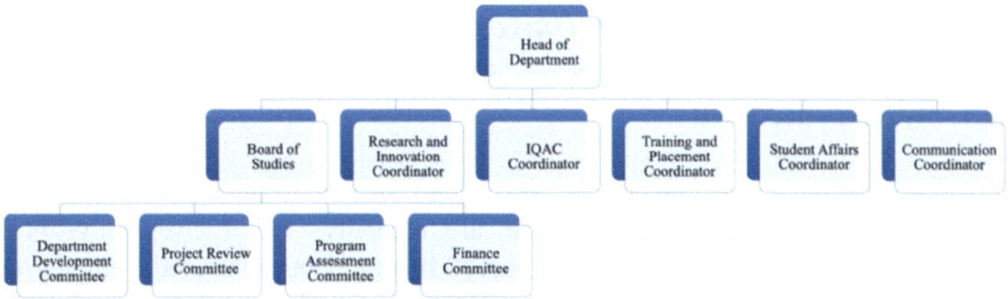

Fig. 2 Department level organizational structure of KG Reddy College of Engineering and Technology

- Performance Appraisal System

The institution follows a systematic performance appraisal system as we believe it enables faculty and staff to be motivated and self-accountable towards their roles and responsibilities. Faculty at KGRCET are expected to carry out three types of responsibilities—teaching, research, and administration at the department and institute levels. The expectations from faculty for each of the three categories depend on their cadre such as assistant, associate, and full professors. Further expectations from each of the categories i.e. teaching, research, and administration are specified in the performance appraisal process through appropriate weightage to highlight the importance of each sub-parameter.

At the start of the year, faculty are expected to set their goals for the various parameters and sub-parameters and meet with the performance review committee (consisting of Principal, Deans, and Head of Departments) to get their yearly goals approved. Mid-year reviews are conducted to monitor the progress of faculty's performance and suggestions are provided for improvement. Faculty at the end of the year must submit their analysis on the progress of goals set at the start of the year with relevant proof to validate their claims. A cumulative score is provided to each faculty based on their performance, which indicates their annual increment and promotions.

- Welfare Measures

KGRCET has initiated "Monthly Spotlight Awards" to recognize faculty for their outstanding service highlighting different attributes such as innovations in teaching, innovation and research, community engagement, etc. which are essential to the growth and vision of the institution. The institution organizes KGR Socials, a monthly meeting to provide the Spotlight Awards to deserving faculty, celebrate birthdays in the month, and organize recreational activities focused on team building. Individual departments and centers of excellence also organize semester-level retreats where the respective teams meet outside of the institution with the aim of building personal bonding and connections. The retention of faculty and staff is a critical factor that influences the growth and success of any institution.

We have witnessed that the acknowledgment and recognition of human resources in the institution significantly impacts their self-morale and retention.

3.2 Case Study 2: Chitkara University, Punjab, India

3.2.1 Governance Model

Chitkara University was founded in 2002 as an affiliated engineering college to State Technical University (Punjab Technical University—PTU). Over the years it added one faculty each to become a group of Institutes but remained affiliated to PTU. In 2010 it was granted the status of State Private University by the University Grants Commission (UGC) of India. The University is governed by Chitkara University Act which was passed by the Legislative Assembly of Punjab (the state in which the University is situated). The act defines some statutory bodies such as the Governing Body, Board of Management, Academic Council, Research Advisory Board, Finance Committee, Planning Board, and Board of Studies. The constitution of these bodies and their respective roles are well defined in the University Act. Some of these statutory bodies have representatives of Local government and others have representatives from Alumni, industry and other academic institutes, social bodies, and Non-Government Organisations (NGOs). Some of the key positions such as Vice Chancellor, Registrar, Dean Examinations, Chief Finance Officer are defined in the governance structure; the rest of the positions can be created as per the functioning requirements.

Chitkara University has a clearly stated Vision, Mission, Charter, and well-defined Values, reflected in its academic and administrative governance. Ethics and cultural values are duly considered while framing policies, systems, Standard Operating Procedures (SoPs) and feedback for work. The University is a value-based, well-connected, process driven and people-oriented organization wherein all the decisions are collectively taken. The idea of decentralization and participative decision-making ensures that each improvement in process is quickly translated to the bottom of the pyramid and to all stakeholders. External eminent personalities are appointed as members in the apex bodies to ensure transparency, diversity, and excellence in governance. The leadership philosophy at the University is to 'lead by example' and 'lead from the front'. The University is governed by its ordinances. Each policy also has a feedback mechanism in place for continuous improvement. The Vice-Chancellor is the functional head of the University; however, appropriate delegation of authority is ensured to various office bearers. The administrative structure gives flexibility, and autonomy with accountability in its functioning to key leadership positions. A 7-year strategic plan has been put in place to establish the principles of participation and transparency. The University adopts a well laid-out strategy for mobilization of funds and its optimal utilization. Annual financial budget prepared before the start of every fiscal year ensures the inflow and outflow of funds in a timely manner. Internal and external audits are done periodically to ensure transparency and compliance. The University has a fair staff welfare policy, appraisal system and promotion policy, befitting the guidelines set by UGC. 360-degree feedback mechanism is successfully operational yielding continuous improvement in functioning.

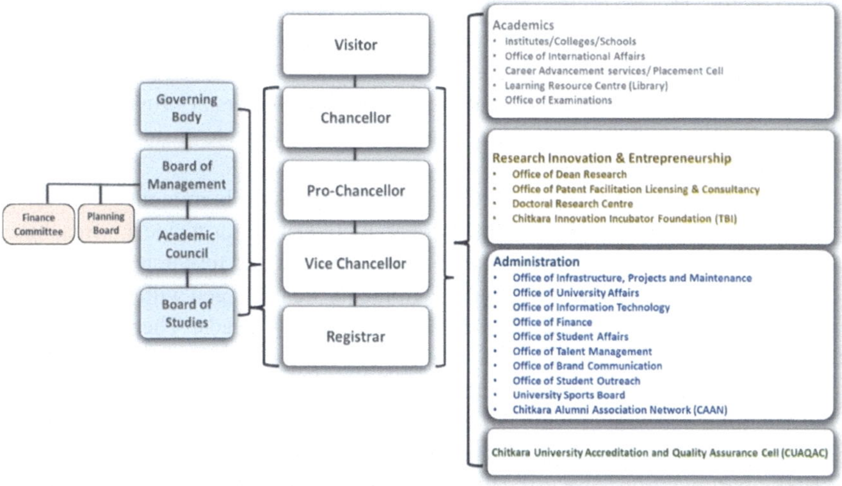

Fig. 3 Organizational structure of Chitkara University

3.2.2 Organizational Structure

Chitkara University has built a flat organization structure so that any 'issue-to-decision' would require at most 3 levels of escalation. In this process the University has taken conscious steps to recruit, train, and empower the people for taking decisions. Once those decisions are taken, the people are made accountable for their actions. The energy in follow up, thus, is minimized.

In operations, the University follows a bottoms-up approach—the actions are strategized at the grassroot levels, so that the people at the bottom of the pyramid feel responsible for their actions. This is in contrast to the top-down approach where the vision of the top leaders is usually not transformed to the last person in the human chain. The people at the bottom of the pyramid thus feel either thrusted upon with the task or do not feel connected to it. For smooth functioning the organization has lots of central departments which take care of the administrative issues so that the academic units can focus on the academic issues and can innovate with free mind on achieving learning outcomes. A pictorial representation of the organization chart is shown here (Fig. 3).

3.2.3 Human Resource Practices

The University accords its Human Resource as the most crucial and important resource of the University and allocates adequate avenues for letting them explore their potential. It has rewards and recognition policies in place for their extraordinary contribution(s). In addition to a structured system for professional development of the staff, the University has implemented well-defined service rules, promotion policies, leave regulations, welfare schemes and grievance redressal mechanisms. Achievements of faculty and staff are timely assessed and accorded due recognition through financial and/or non-financial incentives. The University provides lucrative welfare schemes to all the employees to keep them motivated to ensure that they are working with zeal and efficiency.

Annual Appraisal System and Corresponding Salary Increments: Appraisal system of teaching staff is based on the four parameters of performance, i.e., teaching, self-upgradation, students' feedback, their achievements and research output.

For the non-teaching staff, the annual increments are packaged with special emoluments based on recommendations of their department heads. There are several Medical and Superannuation benefits such as Contributory Provident Fund ESI Group Accidental Insurance and Gratuity. Beside several types of paid leaves, subsidized accommodation is provided to those joining from other locations and to those who are required to stay in the Campus. In addition, the administration has acquired 2/3 BHK apartments on lease in the vicinity of the University to offer accommodation to the employees on subsidized and sharing basis. The facility of creche is available to toddlers of the University staff. Financial aid is provided to support staff to improve the working conditions, provide social security, and raise their standard of living. Specific measures are employed for increasing engagement of employees through Team bonding activities. Free Transportation is provided to all the employees of the University.

Deans' Excellence Awards (for excellence in leadership), Teacher's Excellence Awards (excellence in teaching), Research Excellence Awards (excellence in research), Samman Samaroh (for loyalty and long-term association with the University), Joy of Giving (for contributing to social cause) are organised annually to recognise, reward, and motivate employees to excel more and more in their respective fields.

The leadership rigor here is 'achieving excellence through acknowledgement of the human resource'. Recognising that a motivated faculty and staff can prove to be the backbone for an educational institution, is the key component towards its growth and development.

3.3 Case Study 3: A South African Perspective

3.3.1 Governance Models

This case study is limited to two types of public Higher Education Institutions (HEI) in South Africa (SA); one a University of Technology (Cape Peninsula University of Technology—CPUT) and the other a Comprehensive University (Nelson Mandela University), as explained earlier in the book in Chap. 1.

Irrespective of the type, all public HEI in SA is governed through the Department of Higher Education and Training (DHET).

The DHET provides for and oversees higher education policy development, supports research, and regulates the private higher education system.

Governance in SA public higher education is therefore managed through a combination of government bodies, university councils, and academic senates [4]. These governance structures aim to ensure representation of various stakeholders while maintaining academic quality and financial sustainability. The Higher Education Act 101 of 1997 (as amended from time to time) provides for the establishment, governance, and funding of all public higher education institutions in SA. Chapter 4 of the Act specifies the HEI's governance structures including the appointment of the chancellor; council; senate; principal; vice-principal; student representative council; institutional forum; and such other structures and offices as may be determined by the Institution's statute. The Council and Senate may establish sub-committees to perform some of their functions but are not divested from responsibility for their performance of these functions as delegated or assigned to the relevant committee.

As set out in the Act, individual HEI have their own governing councils that are responsible for the governance of the institution subject to this Act, any other law, and the Institution's statute gazette by the Council with the approval of the Minister of Higher Education and Training.

The statute promotes effective governance and management of the institution in respect of matters not governed by law such as the appointment, functions, term of office, and vacation of office of the Vice-Chancellor, Deputy Vice-Chancellors, Registrar, Executive Deans, Executive Directors, and other employees. It also sets out the Statutory Structures, composition, functions, election or nomination of members, chair, and deputy chairperson appointments of the Council, Senate, Institutional Forum, Student representative council (SRC), Faculty Board, Student services council, and Convocation. Institutional Councils govern the university subject to the Higher Education Act and are responsible for strategic decisions, policies, and oversight over all functions of the university. Each HEI also has an approved delegation of authority matrix, to establish a structured decision-making process, delegation of powers, and assignment of duties and functions to certain offices or structures within the institution. The Senate is accountable to the Council for matters related to learning and teaching; research functions; and any such functions as delegated by the Council.

This governance model and structure aims to balance government oversight, institutional autonomy, and to align national educational goals. This model further seeks to balance academic freedom and accountability (Fig. 4).

<u>Faculty committee structures</u>

The faculty board is the highest academic decision-making structure within the faculty, the composition, and functions, of which are stated in the institutional statute or relevant institutional governance framework. The governance, committee frameworks, and policies also specify other committee structures, composition, and functions such as for example a Faculty Management Committee (FMC), Faculty Committees for Faculty Learning and Teaching (FLTC); Research and Innovation (FRIC); Faculty Postgraduate studies committees (FPGSC) or Faculty Higher

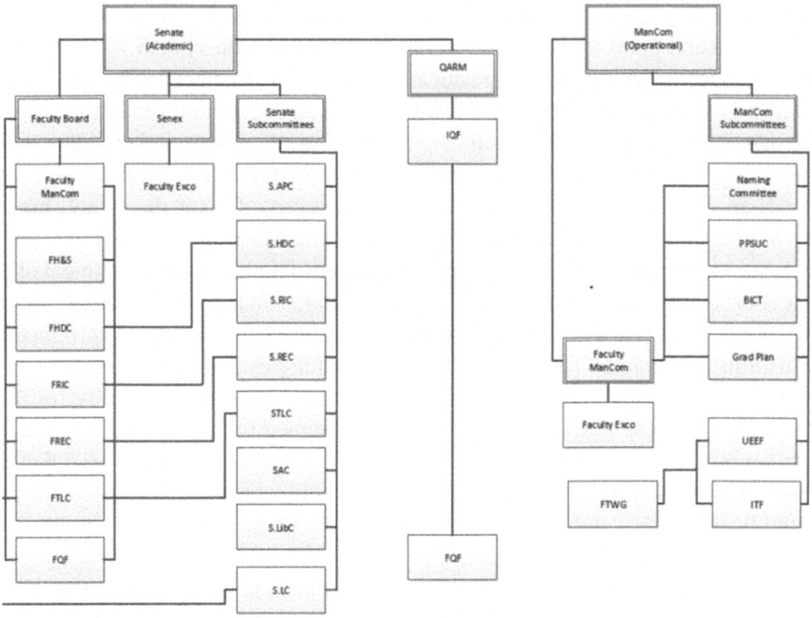

Fig. 4 Example of Cape Peninsula University of Technology governance structures and committees

Degrees Committee (FHDC); Faculty Health and Safety Committees (FHSC); etc. and the appropriate approval processes. This therefore leaves some but very limited flexibility for Deans of Faculties of Engineering or Directors of Schools to decide on in terms of governance models, and structures as it relates to academic and operational matters and the appointment of staff. There is some flexibility in appointing portfolio coordinators that align with strategic goals and objectives, but also through the approval process. The Faculty Board and committees normally recommend academic matters for approval.

3.3.2 Organizational Structures

Within SA Public Higher Education Institutions, the structure for Engineering is either through what is called a "Faculty of Engineering"; or "School of Engineering." In some instances it is combined with other disciplines such as at the Cape Peninsula University of Technology which is named the "Faculty of Engineering and the Built Environment" headed by a Dean. At Nelson Mandela University, it is named the "Faculty of Engineering, the Built Environment and Technology" headed by an Executive Dean. The structure at Nelson Mandela University includes a "School of Engineering" headed by a "Director of School" which is further organised into departments each with a "Head of department". The Dean/Executive Dean is supported by a Deputy/Assistant Dean of Learning and Teaching; Research and Innovation; or coordinators and champions for different portfolios. As explained, these Faculties and Schools of Engineering are not autonomous and are subject to the institutional governance structures, frameworks, and policies that provide for faculty structures, their purpose, composition, and functions. The research, innovation, and engagement are managed through the establishment of entities with a Director/Manager depending on the type of entity. They have their own advisory boards, or oversight committees, and the leaders reporting to the relevant Faculty committee.

For the management of the Human resources in the Faculty/School, the line managers are guided by the institutional policies related to talent acquisition (the process and requirements for the appointment of academics and non-academic staff); promotion; retention; succession planning; employee relations; etc. The institutional policies determine the composition of the recruitment panels of staff appointments. The panel would make a recommendation for approval through either an institutional committee or through the line function such as the Deputy Vice-Chancellor or Vice-Chancellor for approval.

In addition, the Engineering Profession Act (46 of 2000), mandates the Engineering Council of South Africa (ECSA) to provide for the registration of staff as professionals, candidates, and specified categories in the engineering profession; the regulation of the relationship between the ECSA and the Council for the Built Environment; and matters connected therewith [5]. ECSA became a signatory to the internationally recognised Washington Accord in 1999; a founding co-signatory to the Sydney Accord in 2001; and the Dublin Accord in 2002; to recognise substantive equivalence in engineering qualifications and its accreditation. ECSA provides for policy development and oversight of accreditation standards, qualification standards, curriculum development, and graduate attributes; in conjunction with academic institutions to ensure quality engineering education. All engineering programmes must be accredited every five years. Section 21 of the Engineering Act also empowers the ECSA to develop a policy on the Identification of Engineering Work (IDoEW). Most recently on 26 March 2021, a board notice on the Identification of Engineering work regulations was recorded in a gazette. While the objective of the IDoEW was to protect the public and the environment, it has a major impact on academic staffing implications for implementation (Identification of Engineering work regulations, Gazette no 44333).

3.3.3 Effectiveness of the Governance Model and Organisational Structures for Human Resource Management and a Changing Staff Demographic

The human resource management in the Faculty/School is subject to central institutional governance oversight related to staff functions, recruitment, retention, succession planning, training, development, promotion, performance, workload, and labour relations matters.

Given the changing nature of our landscape, there is an increasing need for human resource practices to be adaptable to various cultural, social, and economic factors, while complying with institutional governance structures, policies, and rules.

Within the SA historical context transformation plays a huge role in shaping strategies to promote diversity, equity, and inclusivity; as well as creating a conducive working environment for staff and a learning environment for engineering students.

The profile of engineering staff has evolved in response to various factors, including the National Development Plan (2023), the Department of Higher Education, and government policies such as the Employment Equity Act (No. 55 of 1998) [6] and continued efforts to increase diversity and inclusivity. This includes efforts to recruit and retain individuals from historically disadvantaged groups, especially women. But in many of the engineering disciplines, this still proves to be a challenge. It is due to the scarcity of SA candidates with the appropriate combination of qualifications, experience, and research and innovation track record. In some disciplines, this therefore necessitated the recruitment of staff from other countries in scares skills areas, which then, at the same time, encouraged international collaborations.

The most recent 2023 SA Science, Technology and Innovation Indicators Report provides an overview of the STI human resources in SA higher education institutions which showed that less than half of academic staff had doctoral qualifications in 2021.

Higher education institutions currently have 10.4% of staff over the age of 60 and will lose about 10% of staff with doctoral qualifications due to retirement in the next five years. There is a huge need to identify and intensify the support and nurturing of young academic staff to become the next generation of academics using the SA Government's instruments and programmes such as the NESP (nurturing emerging scholar programme) and nGap (new generation academic programme). While the proportion of female academics has increased from 27.8% in 2011 to 37.5% in 2020, much work is still needed to close the gender gap, especially in the engineering disciplines [7].

The requirements of the IDoEW imply that all engineering academics involved in the offering of the programmes to be registered with the ECSA. However, the impact of this regulation and the serious implications for the management of academic staff will now require much more careful considerations, and management, and further complicate the planning of human resources; especially at UoTs and Comprehensive Universities where fewer numbers of staff are currently registered with the ECSA.

Careful implementation plans with financial and other support will have to be put in place that consider further development, and training requirements for current staff not yet registered to meet the required registration standards. In some instances, this will require more rigid ECSA registration requirements as part of the recruitment process, which might reduce the potential pool of candidates, or in other instances might require eligible candidates to register within a certain period after joining the institution.

While the central oversight, in terms of human resource management, provides for consistency, transparency, and fairness, the centralised governance structures and organisational models can become cumbersome and time-consuming to retain and attract talent timeously.

3.3.4 Exploring the Changing Student Demographic

The profile of engineering students varies from one university/institution to another depending on the location and context and will continue to evolve over time. Many attempts have been made over the years to ensure the engineering student demographic within the faculty is diverse and inclusive. Even though progress has been made to include students from historically disadvantaged groups, there are still disparities, especially as far as the gender balance is concerned, since engineering fields have historically been male dominated. Engineering undergraduate programme students are typically 18–25 years old, while some might be older, especially those doing postgraduate studies and students studying part-time due to financial constraints and other reasons. Depending on the engineering discipline, females form as low as 30–35% of some programmes. The faculty has a percentage of international students, further contributing to the diversity of the student profile.

A recent study by the ECSA on the Engineering skills pipeline showed that of the thousands of learners that enter the basic education schooling system, only a limited number meet the required threshold to continue with engineering programmes at higher education institutions and even fewer register as engineering professionals, with a significant loss of talent along the way. The study (focusing specifically on a cross-sectional analysis of the 2020 numbers, i.e., not a longitudinal cohort study) showed that of the total number of 725 034 learners that wrote the National Senior Certificate (NSC) in 2020 only 41 103 learners with Mathematics and Science achieved 50–100% eligible to enter engineering qualifications or any other qualifications requiring Mathematics and Science. This is further broken down into 15 501 learners achieving between 50–59% for entrance into an engineering technician qualification; 10 129 learners achieving between 60–69% for entrance into an engineering technologist qualification; and 15 573 learners achieving between 70–100% for entrance into a professional engineering degree qualification; with only 1.6% achieving professional registration status in 2020 (Engineering Council of South Africa, 2022).

While significant efforts have been made to increase the number of and support students who study Mathematics and Physical Science at the Basic Education level (such as the "STEM-in-Action" programme in the faculty at Nelson Mandela University), to encourage more women to pursue STEM (Science, technology, engineering, and mathematics) and specifically engineering programmes; and to support and mentor students to be successful (such as the Women in Engineering Leadership—"WELA" programme in the faculty at the Nelson Mandela University), much more effort is needed to further transform.

4 Key Takeaways for Academic Leaders

- When Institutions move from being governed or regularised by a central body to being autonomous, the responsibility suddenly grows many fold (reflected in Chitkara University case study). Till the time, the institutions are regulated or governed by somebody else, every failure could be attributed to the affiliating university or regulator, however, as soon the institution becomes autonomous, they become responsible for their actions.
- As the Universities grow in size, shape, and complexity, the governance, management and leadership have to transform. The SA case study highlights the complexity of the governance structures in the higher education sector and the challenges and opportunities. The case study also shows how to strike a balance between government regulations and institutional autonomy. It also ensures adherence to the academic and operational policies and guidelines while allowing some limited flexibility for the Dean of Faculties, Directors of Schools, and Heads of Departments.
- It is very important to brainstorm periodically what kind of organisational structure is to be put in place, what policies are to be framed/revamped/introduced, and what strategies are to be decided so that people take ownership of their actions. The faculty and students, through their representatives, should be a part of the decision-making process. A decentralized approach to governance could be one approach in such a situation.
- It is also important to hire the right talent and keep them motivated to perform and excel. The connection of the students with the faculty members happens only when both are motivated and engaged. Reward and recognition policies should be framed and continuously revamped to meet the needs of prevailing times. To recruit and support young academics (early-career researchers) should continue to be a priority and government support programs (viz., NESP, nGap and future professors' programs in SA) should be explored, specifically in the scarce skills and disciplines.
- In the strategic plan priorities should be defined and more should be added based on prevailing societal requirements. Next rung of leaders should be prepared so that the governance remains process and performance oriented rather than person and people oriented.
- Continued efforts and interventions will be required to address the challenges related to the engineering skills pipeline including efforts to attract, access, and support students for success (graduation); and finally, to become professionally registered. Ongoing efforts will be required to bridge the gender gap in engineering.

5 Conclusion

Effective governance of engineering institutions requires a balance between strategic vision, collaboration, adaptability, and continued commitment to academic excellence. Leadership of Engineering is a dynamic and evolving role. Future leaders of Engineering faculties/schools/departments should therefore be encouraged to embrace diversity and inclusion; stay informed and adaptive; prepare for changing staff and student demographics; intensify engagement, collaboration, and partnerships with external stakeholders including industry, government, and communities to make an impact. In conclusion, thus, six clear principles emerge for effective governance:

- Maintain diversity and redundancy.
- Manage connectivity.
- Encourage learning and experimentation.
- Broaden participation.
- Promote polycentric governance.
- Foster understanding of complex adaptive systems.
- Manage slow variables.

References

1. Dobbins M, Knill C, Vögtle EM (2011) An analytical framework for the cross-country comparison of higher education governance. High Educ 62(5):665–683. https://doi.org/10.1007/s10734-011-9412-4
2. "2998278_University-Grants-Commission-(Institutions-deemed-to-be-Universities)-Regulations,-2023.pdf." [Online]. https://www.ugc.gov.in/pdfnews/2998278_University-Grants-Commission-(Institutions-deemed-to-be-Universities)-Regulations,-2023.pdf. Accessed 19 Nov 2023
3. "0367475_UGC-(Conferment-of-Autonomous-Status-upon-Colleges-and-Measures-for-Maintenance-of-Standards-in-Autonomous-Colleges)-Regulations,-2023.pdf." [Online]. https://www.ugc.gov.in/pdfnews/0367475_UGC-(Conferment-of-Autonomous-Status-upon-Colleges-and-Measures-for-Maintenance-of-Standards-in-Autonomous-Colleges)-Regulations,-2023.pdf. Accessed: Nov. 19, 2023
4. "Higher Education Act 101 of 1997 | South African Government." [Online]. https://www.gov.za/documents/higher-education-act. Accessed 19 Nov 2023
5. "Engineering Profession Act 46 of 2000 | South African Government." [Online]. https://www.gov.za/documents/engineering-profession-act. Accessed 19 Nov 2023
6. "Act—Employment Equity 1998.pdf." [Online]. https://www.labour.gov.za/DocumentCenter/Acts/Employment%20Equity/Act%20-%20Employment%20Equity%201998.pdf. Accessed 19 Nov 2023
7. "STI Indicators Reports—National Advisory Council on Innovation." [Online]. https://www.naci.org.za/index.php/studies/. Accessed 19 Nov 2023

Establishing and Maintaining Operational Systems and Processes that Meet International Quality Standards

S. A. Rajala, D. Iacona, M. I. Ruiz-Cantisani, V. Lara-Prieto, and B. Kanmani

Abstract This chapter describes the Washington Accord, a mutual recognition agreement among accrediting bodies of tertiary-level engineering educational programs and its global relevance and context. Case studies are presented from India, Mexico, and the United States. The studies describe the engineering educational programs' experiences in establishing and maintaining operations systems and processes that meet the international quality standards prescribed by the Washington Accord. The case studies presents not only their successful strategies and recommendations but also highlight the challenges they encountered. Creating and maintaining operational systems and processes that meet international quality standards requires full commitment and a culture of accreditation among all stakeholders, ultimately yielding benefits for both students and the institution.

Keywords Quality assurance systems · International standards · Mutual recognition agreements · Higher education · Engineering education

S. A. Rajala (✉)
ABET, Baltimore, MD, USA
e-mail: sarah.rajala@gmail.com

D. Iacona
International Engagement and Governance, ABET, Baltimore, MD, USA

M. I. Ruiz-Cantisani · V. Lara-Prieto
School of Engineering and Sciences, Tecnologico de Monterrey, Monterrey, Mexico

B. Kanmani
Department of Electronics and Telecommunication Engineering, BMS College of Engineering, Bengaluru, India

R. Kandakatla et al. (eds.), *Academic Leadership in Engineering Education*, Lecture Notes in Networks and Systems 1097, https://doi.org/10.1007/978-3-031-68282-7_3

1 Introduction

This chapter will explore case studies of engineering educational programs' experiences in establishing and maintaining operations systems and processes that meet international quality standards, such as those recommended by the Washington Accord. The programs at the following institutions have been accredited by different signatories of the Washington Accord, who are accrediting bodies of tertiary engineering education in their respective jurisdictions:

1. ABET—United States—Iowa State University
2. Consejo de Acreditación de la Enseñanza de la Ingeniería, A. C., (CACEI)—Mexico—Tecnologico de Monterrey
3. National Board of Accreditation (NBA)—India—BMS College of Engineering (BMSCE)

1.1 The Washington Accord

The Washington Accord is a mutual recognition agreement (MRA) among accrediting bodies of tertiary-level engineering education representing 23 jurisdictions. The Washington Accord recognizes the accreditation systems of its full signatories as being substantially equivalent to one another. Therefore, it also recognizes the programs accredited by these bodies as being substantially equivalent to one another. The Washington Accord is one of three education-focused MRAs that come under the umbrella of the International Engineering Alliance (IEA). The other two are the Sydney Accord (engineering technologist programs) and the Dublin Accord (engineering technician programs). The IEA also includes four professional agreements, which focus on the recognition of professional licensure/registration: International Professional Engineers Agreement (IPEA), APEC Agreement, International Engineering Technologists Agreement (IETA), and Agreement for International Engineering Technicians (AIET) [1].

The Washington Accord was established as a mechanism to facilitate the global mobility of engineering graduates by providing recognition of their education for licensure, employment, and higher education purposes. It was established in 1989 by six founding signatories (accrediting bodies of engineering programs):

1. ABET (United States)
2. Engineers Australia
3. Engineers Canada
4. Engineering Council United Kingdom
5. Engineers Ireland
6. Engineering New Zealand

The Accord steadily grew over the years to include the following organizations [2]:

7. The Hong Kong Institution of Engineers (1995)
8. Engineering Council South Africa (1999)
9. Japan Accreditation Board for Engineering Education (2005)
10. Institution of Engineers Singapore (2006)
11. Accreditation Board for Engineering Education of Korea (2007)
12. Institute of Engineering Education Taiwan (2007)
13. Board of Engineers Malaysia (2009)
14. Association for Evaluation and Accreditation of Engineering Programs (Turkey) (2011)
15. Association for Engineering Education of Russia (2012)
16. National Board of Accreditation (India) (2014)
17. Institution of Engineers Sri Lanka (2014)
18. China Association for Science and Technology (2016)
19. Pakistan Engineering Council (2017)
20. Instituto de Calidad y Acreditacion de Programas de Computacion, Ingenieria y Tecnologia (Peru) (2018)
21. Colegio Federado de Ingenieros y de Arquitectos de Costa Rica (2020)
22. Indonesian Accreditation Board for Engineering Education (2022)
23. Consejo de Acreditación de la Enseñanza de la Ingeniería (Mexico) (2022)

1.2 Relevance and Context

The relevance and importance of the Washington Accord have increased over time with the continued increase of globalization. With the increased global mobility of students and professionals, and the expansion of multinational corporations, the Washington Accord facilitates the mobility of graduates of recognized programs. It helps employers, higher education institutions, and licensing bodies assess the educational base of applicants. Recognition by the Washington Accord indicates that the graduate attained the attributes required for entry into the engineering profession.

The IEA Washington Accord, through the collective work of its members, has developed a set of exemplar Graduate Attributes and Professional Competencies for each of its Accords and Agreements. Recently, these Graduate Attributes and Professional Competencies have been updated in collaboration with the World Federation of Engineering Organizations (WFEO) [3]. The Washington Accord Graduate Attributes specify the recommended skills and knowledge that graduates of accredited engineering programs should possess.

Attainment of these exemplar Graduate Attributes is a key component that ensures graduates are prepared to contribute to the solving of the many challenges we are facing as a planet, as have been identified by, for example, the United Nations Sustainable Development Goals.

2 Case Studies

2.1 ABET Case Study at Iowa State University—USA

Iowa State University is a public land-grant research university located in Ames, Iowa. The university is a community of students, faculty, and staff who work together in their teaching, learning, and research to solve the most pressing challenges facing our world.

The university's strategic plan is forward-thinking, thus allowing the university to innovate, take advantage of new opportunities, and share stories of its progress. The new vision, mission and values support these efforts [4]. For example, in 2023 the university was awarded a new five-year NSF project focused on creating new technology and growing the advanced biomanufacturing capacity in Iowa.

Vision: Iowa State University will create, share, and apply knowledge to make our students, Iowa, and the world a better place.

Mission: Iowa State University will advance the land-grant ideals of putting science, technology, and human creativity to work.

Iowa State University is governed by the Board of Regents, State of Iowa [5]. The university has four divisions—president, academic affairs, student affairs, and operations and finance. The academic affairs division includes the eight academic colleges—agriculture and life sciences, business, design, engineering, human sciences, liberal arts and sciences, veterinary medicine, and graduate college. As of Fall 2023, Iowa State University had a total student enrollment of 30,177.

Iowa State University has a long-standing commitment to continuous improvement and outcomes-based assessment that ensures excellence in its academic programs. It achieves this academic excellence through accreditation and the program review processes. This begins with university-level accreditation under the auspices of the Higher Learning Commission (HLC), which is one of the seven regional accreditors of degree-granting post-secondary educational institutions in the United States [6]. The university participates in the HLC Open Pathway accreditation process and successfully completed its 'Year 4 Assurance Review' in February 2020. In Spring 2021 the university began working on an 18-month Quality Improvement Initiative, which was approved by the HLC in April 2021. The core of this initiative was selected by the American Council on Education as one of its Learner Success Laboratories (LSL) [7]. The goal of the LSL is to increase student success by systematically integrating evidence-based practices throughout the university [8]. The recommendations of the initiative were used in the development of the new strategic plan and will be reported to the HLC during the next review cycle.

All academic programs are reviewed every seven years according to the Board of Regents, State of Iowa policy for the purpose of providing guidance on their development on a continuous basis. The goals of the review are to evaluate the program status, effectiveness, and progress; to help identify the future direction, needs, and priorities; and to articulate a set of future action plans for the program. The review process [9] includes the preparation of a self-study, a site visit by an external review team, and submission of the response and action plans to the administrative leadership and the Board of Regents, State of Iowa. To ensure the success of the action plans, the reviews must be aligned with the university's strategic plan and allocation of resources.

Iowa State University College of Engineering. The College of Engineering (COE) is one of the eight academic colleges at the university. The college leadership, faculty, and support staff are committed to providing the best educational experience to its students and to making Iowa and the world a better place. The college pursues excellence by strategically investing in their students, faculty, and staff, so that they are inspired to think creatively and passionately on this incredible journey. Its vision and mission support these goals [10].

Vision: We will be a premier engineering college among the land-grant institutions, globally recognized as a destination for excellence and value in education, research and engagement.

Mission: We will achieve broad-based impact by:

- Fostering an inclusive and creative environment that promotes scholarship and learning.
- Providing a high-quality, experiential education that prepares students to meet the global challenges of the 21st century.
- Conducting discipline-bridging, innovative research.
- Engaging broadly with partners in Iowa and around the world.

To achieve its goal of providing high-quality, experiential education that prepares graduates to meet global challenges, COE is committed to continuously improving the learning opportunities for its students. Accreditation of its academic programs is an important part of the process. ABET accreditation provides assurance that the engineering programs meet the quality standards of the engineering profession. In addition, the first step towards professional licensure is to take and pass the Fundamentals of Engineering examination, which in most U.S. states requires a student to be enrolled in or have graduated from an ABET-accredited program. The College of Engineering offers Bachelor of Science (B.S.) degree programs in aerospace, agricultural, biological systems, chemical, civil, construction, computer, cybersecurity, electrical, industrial, materials, mechanical, and software engineering, which are all accredited by ABET [11]. Recently, the college introduced two new B.S. degree programs in biomedical engineering and environmental engineering. Once the programs have awarded degrees, they will be eligible to request an accreditation review by ABET.

ABET. ABET was founded in 1932 as the Engineers' Council for Professional Development (ECPD) as a nonprofit, non-governmental organization whose purpose was to serve as the engineering professional body dedicated to the education, accreditation, regulation, and professional development of engineering professionals and students in the United States. After more than 80 years, ABET's standards continue to play this fundamental role and have become the basis for quality for STEM disciplines all over the world. The standards ABET sets, and the quality they guarantee inspire confidence in those who aim to build a world that is safer, more efficient, more comfortable, and more sustainable.

Today ABET accredits college and university programs in the disciplines of applied and natural sciences, computing, engineering, and engineering technology at the associate, bachelor's, and master's degree levels. ABET's accreditation is voluntary and to date, academic programs in 40 countries have received ABET accreditation and over 200,000 students graduate from ABET-accredited programs each year. The ABET criteria are developed by technical professionals from its 35 member societies [12] and focus on what students experience and learn.

Strategies for success. The success Iowa State University has achieved in receiving and maintaining ABET accreditation of its engineering programs is due in great part to a university-wide commitment to continuous improvement and belief in the importance of accreditation to the university and its academic programs. To ensure success requires support from the leadership at all levels across the university. The president, vice presidents, academic deans, department heads were all engaged in continuous improvement and accreditation efforts. However, success is not guaranteed unless the faculty and support staff are also actively engaged in the process; there is coordination among accreditation and program review processes and expectations; and adequate resources (e.g., space, funding, technology) are provided at all levels. Above all, open communication among all stakeholders is of utmost importance.

Iowa State University must meet the expectations of the Board of Regents, State of Iowa for academic programs, the regional accreditation by the Higher Learning Commission at the university level, and various academic program accreditation bodies across the university. In the College of Engineering ABET accredits the B.S. engineering programs, as well as the B.S. computer science program in the College of Liberal Arts and Sciences (LAS). The software engineering program is administered as an interdisciplinary program between COE and LAS. The Department of Agricultural and Biosystems Engineering (ABE) is jointly administered between COE and the College of Agricultural and Life Sciences. ABE offers both ABET-accredited agricultural engineering and biological systems engineering programs, and ATMAE [13] accredited agricultural systems technology and industrial technology programs.

If not carefully coordinated, this commitment to outcomes assessment and accreditation could overwhelm the academic programs and their people. For example, the Board of Regents review cycle for academic programs is seven years and requires programs to prepare a self-study as part of the review process [9]. In addition to providing an update on the status of the program(s), the self-study needs to be forward looking, addressing strategic directions and strategies for improving the academic program and student success. Additionally, the review process should result in well-defined action plans for future development of the academic program. The ABET accreditation review cycle is six years and also requires the preparation of a self-study, but the purpose is quite different. The ABET review process focuses more on the student achievement of outcomes, the outcomes-based assessment process, the curriculum, and the faculty. The university-wide accreditation done by the HLC has a 10-year accreditation cycle, with a comprehensive review at both years 4 and 10 in the Standard Pathway and an assurance review at year 4 in the Open Pathway. Although the HLC accreditation process focuses on the institution, their accreditation criteria include evaluation and improvement of teaching and learning [14].

Over the years, the College of Engineering has established new B.S. engineering academic programs, each with the desire of future ABET accreditation. The four most recent programs to be added are: software engineering, cyber security engineering, biomedical engineering, and environmental engineering. Software Engineering was established in 2007 as a joint program between the Department of Computer Science and Department of Electrical and Computer Engineering. The program has been accredited by ABET since 2011. There were issues that had to be addressed in its initial attempt to obtain accreditation related to the administrative structure of the program. The faculty supporting the program came from two departments of two different colleges with two different histories and academic cultures. Coming to consensus on a joint program takes careful planning, open communication, and good leadership. The Board of Regents approved the establishment of the B.S. Cyber Security Engineering in November 2018 with an official start date of Fall 2019 and the program has been ABET accredited since 2019. The program resides in the Department of Electrical and Computer Engineering, which has a long history of individuals serving as ABET program evaluators and/or commissioners. The faculty leader for the program was a part of the ABET team that developed the criteria for cyber security engineering. The faculty involved in developing the program had a clear understanding of the ABET criteria and expectations for successful accreditation. Because the department had been teaching cyber security courses for several years, they had considerable expertise in the area. The program was well-prepared for its ABET review and achieved accreditation on its first attempt.

What works. One of the most important factors for success is the university-wide commitment to continuous improvement. Another important factor is a commitment to open communication and planning. This is evident in the university strategic planning process, as well as the strategic planning done within the academic colleges. Processes, priorities, strategies, and ultimately resource allocation are aligned. One of the key strategies is the work done between the Division of Academic Affairs and the academic colleges to identify ways in which self-study materials can be shared among the various program reviews. Innovation is a way of life at Iowa State University. This encourages faculty and students to be creative in the development of new teaching and learning strategies, as well as in research and entrepreneurship.

Engineering faculty have a long-standing commitment to providing students with a high-quality education and understand the importance of continuous improvement, outcomes assessment, and ABET accreditation. Leadership within the college supports and encourages faculty to serve as ABET program evaluators and commissioners. This is a key factor to successful accreditation of its academic programs.

Possible Roadblocks. Success does not happen without a commitment to open communication and a desire to continuously improve. Even with the best intentions, there can be challenges. Agreeing on what continuous improvement means, and how it impacts academic programs, administrative processes, resource allocation, technical support, and people is not easy. To be successful, it takes good leadership, ongoing commitment to success, good planning, and open communication. This can be quite challenging at a highly diverse and distributed university environment. If planning is done at the university level without the engagement of faculty, staff, and students who are directly impacted by planned changes, success is not likely. Even when there is engagement in planning at all levels, if plans are not broadly communicated or opportunities for discussion and feedback are not provided, success may be hampered. Faculty, staff, and students each have different backgrounds, perspectives, and priorities. Sharing information among stakeholders, allowing adequate time to review and discuss, and providing the resources and training to support change are all critical to ensuring success.

2.2 CACEI Study Case at Tecnologico de Monterrey, Mexico

Tecnologico de Monterrey. Tecnologico de Monterrey was founded in 1943 in Monterrey, Mexico. It is a private, non-profit university committed to the quality of higher education in Mexico. Today, our university has 26 campuses in Mexico and 18 offices around the world. It currently has 94,424 students (high school, undergraduate, and graduate), of which 60,169 are undergraduate students. 54% of our students have a scholarship or some sort of financial aid. We have 10,896 faculty members. Internationalization is a key element for our university and students. Tecnologico de Monterrey has academic agreements with 706 universities in 46 countries around the world.

The 2030 vision at Tecnologico de Monterrey is: leadership, innovation, and entrepreneurship for human flourishing.

Aware of the speed with which the world is changing, our university presents a new stage in the transformation of the institution, maintaining its essence in terms of leadership and entrepreneurship, and focusing on a more humane and innovative future, centered on the well-being of our community.

At Tecnologico de Monterrey we are convinced that universities should prepare professionals to work in companies that don't yet exist, to use technologies that haven't been invented, and to solve problems that we don't even know are problems yet.

The vision is highly ambitious. Work in that direction started with a recently launched Tec21 Educational Model, which was fully implemented in 2019.

Tec21 Educational Model has four main pillars: Challenge-Based Learning, flexibility in when and where the learning experience takes place, inspiring faculty, and a memorable student experience.

Our model focuses on transversal and disciplinary competencies development through working on the solution of real challenges, in real contexts, with a strong liaison with our educational partners.

Regarding the accreditation culture, Tecnológico de Monterrey has been accredited by the Southern Association of Colleges and Schools Commission on Colleges (SACSCOC) since 1950. SACSCOC is the body for the accreditation of degree-granting higher education institutions in the United States for the Southern States and Latin America. The Tecnológico de Monterrey's accreditation includes its educational programs and services at the undergraduate and graduate levels, on campus as well as via distance learning. This accreditation is reaffirmed every 10 years. The most recent reaffirmation was obtained on December 11, 2018, for the period 2018–2028. International rankings are also an indicator that reflects the institution's academic quality. Tecnologico de Monterrey is ranked #184 in the QS World University Rankings, #26 in the world for employability in the QS Graduate Employability Rankings 2022, and #4 in the QS World University Rankings: Latin America and The Caribbean 2024.

School of Engineering and Sciences. The School of Engineering and Sciences is the biggest School at our university, with 43% of the student population, that is 26,000 students at the undergraduate level and 3,200 postgraduate students, distributed in the 25 campuses located throughout the country. The School of Engineering and Sciences offers 18 undergraduate engineering programs, organized into four main areas of expertise: Computing and Information Technologies, Innovation and Transformation, Bioengineering and Chemical Processes, and Applied Sciences.

Our institution is deeply committed to fostering a robust culture of internationalization. 60% of our alumni gain invaluable international experience before they graduate. These experiences can take the form of intensive summer/winter courses, a semester or even a full year spent at foreign universities, or hands-on internships within companies and research centers located in different countries.

In addition to institutional accreditations, the School of Engineering and Sciences has various engineering programs accredited by national and international accreditation boards. These include renowned entities such as ABET (Accreditation Board for Engineering and Technology), IFT (Institute of Food Technologists), CACEI (Council for the Accreditation of Engineering Education), and CIFHRS (Interinstitutional Commission for the Formation of Health Human Resources).

Our School is committed to the accreditation of its programs as a cornerstone of maintaining academic excellence. Accreditation processes help to maintain the academic quality of the engineering programs, meet international standards, have international recognition, and facilitate the international acceptance of our alumni's degrees. These accreditations are also relevant in forging international partnerships with other universities worldwide. Having a culture of accreditation within the school contributes to the continuous improvement of our programs.

CACEI. The Council for the Accreditation of Engineering Education is a non-profit organization that develops accreditation processes with engineering education programs to promote the quality of education to students enrolled in universities. CACEI is currently recognized internationally by the Washington Accord (WA), which is part of the International Engineering Alliance (IEA). CACEI's mission is:

> To contribute to the improvement of quality in engineering education through the preparation of academic cadres that carry out the evaluation and accreditation with international recognition of undergraduate higher education programs in engineering taught by public and private institutions, both in the country and abroad, with quality, transparency, confidentiality, professionalism and honesty [15].

As part of the accreditation ecosystem, CACEI has a team of evaluators from different disciplines, who are constantly trained in the current framework as well as in the evaluation processes and the use of rubrics, to align criteria. These evaluators are professionals who volunteer for the work of evaluators and are committed to education. They work with the conviction that through quality trainings and graduates committed to the environment, society can be progressively impacted. Faculty members of our institution are part of this group of volunteers and participate in evaluation processes at other universities.

CACEI Background. In 1994, the Council for the Accreditation of Engineering Education was created as a Civil Association, with the participation of associations representing higher education institutions and professional associations such as National Association of Engineering Schools and College (ANFEI) and National Evaluation Center for Higher Education (CENEVAL), the federal government and the productive sector.

CACEI became the first accrediting institution in Mexico and seeks to support Mexican society in promoting the growth of the country through quality training of engineers graduating from certified programs, as well as to foster a culture of continuous improvement by complying with minimum international standards.

Integration to the Washington Accord. Based on the culture of continuous improvement and the importance of international linkage, CACEI seeks to become a signatory of the Washington Agreement, to offer the institutions an international recognition that allows the graduate profiles of its engineering programs to be "substantially equivalent among the programs of the member countries". And with this, promote the mobility of engineering professionals because they are accredited and recognized in the countries that are part of the Accord.

As part of the integration with the Washington Accord came the design of the new 2025 framework for CACEI, which changes its structure to prioritize the curriculum, educational objectives, and Egress attributes to ensure program quality (see Table 1).

An arduous process of training in the new 2025 framework aligned with the WA, two simultaneous processes have been developed as part of the CACEI strategy:

- Pilot program of application of the new framework, where staff of the institutions are preparing to document aligned to the new requirements and forms. This program covers academic institutions and programs in both public and private institutions.
- Training programs for evaluators and the institutions themselves that will participate in the pilot.

Table 1 Comparison between the 2018 and 2025 frameworks

Reference framework 2018	Reference framework 2025
Academic staff	Students
Students	Study plan
Study plan	Educational objectives
Continuous improvement	Egress attributes
Infrastructure and equipment	Academic staff
Institutional support	Institutional support
	Continuous improvement
	Program areas of expertise

Study Case: Actions to Achieve CACEI (Washington Accord) Accreditation

Strategies for Success. It is particularly important to be conscious of the diverse accreditation requirements for our programs, both for our faculty and leadership teams. One way of doing this is having our teams trained on the benefits of accreditation processes, the main accreditation criteria, and the internal processes we follow to maintain academic excellence. Having an accreditation culture helps everyone to get on board and take the accreditation criteria in mind on the strategies and decision-making processes. Having a documentation culture is equally important to have a record of our processes and evidence of their implementation.

Our institution is deeply committed to engaging faculty members in a range of institutional efforts, with the primary aim of driving continuous improvement in our academic programs. Some of these efforts are the National Faculty Reunion (RNP) that gathers national full-time and part-time faculty to get training, updating, and collaborative discussions aimed at enhancing our academic offerings. There are also National Faculty Meetings to exchange best practices and suggest course improvements as part of our continuous improvement system to ensure quality education. Although these gatherings and collaborative sessions were not initially planned as part of an accreditation culture, they are now aligned with the accreditation processes.

Our university defines a strategic plan every 10 years to ensure the progress of our university, considering the needs of our stakeholders and society in general. The vision is updated and aligned with the strategic plan. The academic programs include these updated elements during the study plans design process and update to ensure the consistency of the Program's Educational Objectives with the mission of the institution. Our institution has a strong innovation, leadership, and entrepreneurial culture, committed with the environment and the society, as well as with the personal integral well-being for human flourishing. It has a strong liaison with companies, non-governmental organizations, and society. It is considered as an educational partner through challenge-based learnings applied in the curricula.

There is a solid institutional structure for student advising and career guidance, starting from the admission process and career election, throughout their studies and after graduation. This structure includes the Entrance Director, the

Academic Program Head, the Mentor, the Academic Mentors of Excellence, Tutoring, Academic Improvement, Wellbeing and Counseling, and the Career Center.

What Works. Our university and School leadership team are convinced of the importance of being accredited by national and international accrediting bodies. With their support and commitment, the accreditation process becomes a priority for everyone involved and that helps with the continuous improvement strategies and the gathering of information when preparing for an accreditation visit. Tecnologico de Monterrey has 25 campuses and most of them have engineering programs. Even though the institutional information and study plans for a particular engineering program are the same nationwide, the accreditation process is per program per campus. There is a national structure to support the accreditation processes and to maintain institutional communication to all the corresponding campuses. There is a National Program Director for each engineering program who prepares all the information related to the study plan. Each campus has a Program Head and a Regional Department Director that are responsible for preparing all the information that corresponds to a particular engineering program at their campus.

Our university is divided into five regions: Monterrey, North, West, Mexico City, and Central-South.

Each region has a faculty member in charge of the accreditation processes of the region, coordinated by the National Director of Liaison and Internationalization. The regional representative also acts as the project manager of the accreditation process, ensuring the corresponding information and the visit are prepared promptly.

A key element is the university's commitment to continuous improvement, including competencies and outcomes assessment, educational innovation research to improve the teaching–learning experience, course analysis, and improvement with new activities, resources, and redesign, among others.

Every 4–5 years, a direct revision/evaluation of program's educational objectives and student outcomes are performed and appropriate changes to the basic curriculum is conducted. Normally this revision is the ground for updating and proposing new curricula for all academic programs. For this revision, feedback from alumni, employers, and faculty is used to define any required adjustments to the program's educational objectives according to the university mission and vision. The accreditations criteria are considered during this redesign process of the new curricula.

Challenges for Accreditation. Even though our institution has a solid structure and strategic planning that supports accreditation processes, there are several challenges when preparing the documentation for the accreditation or re-accreditation process. It definitely involves a great amount of work for the main roles involved (i.e. Academic Program Head and Academic Department Head). The workload becomes intense during the report preparation and the accreditation visit, in addition to the normal operational responsibilities, which can be overwhelming. Training on accreditation criteria is essential, particularly when there is a rotation of key roles within the process. It should also be clear whose responsibility each part is as well as the expected deadlines.

The accreditation and documentation culture are not permeated across the university administrative departments and our information systems are not prepared to generate the reports as required by the different accreditation agencies. Therefore, human resources are required to gather this information and analyze it to generate the required results. In general, the institution must allocate financial, time, and human resources for the accreditation processes.

2.3 NBA Case Study at BMS College of Engineering (BMSCE)—India

Introduction. The National Board of Accreditation (NBA) represents India, a recognized signatory to the Washington Accord (WA) from 2014 [16].

In India, institutions are classified in two major categories: Tier-I Institutions with Academic Autonomy, that have the responsibility of designing the curriculum, its effective delivery and conducting examinations, leading to award the engineering degree.

Tier-II Institutions, affiliated to a university and deliver the curriculum designed by the university and prepare students for the examinations to be conducted by the university.

In view of this difference, presently, only accredited programs of Tier-I Institutions are covered by the WA.

Initially during 2013, the NBA had defined 12 Graduate Attributes (GAs) and allowed programs to define Program Outcomes (POs), that are aligned to the GAs. However, from 2016, NBA has defined 12 POs for all engineering programs which match the 12 GAs of WA; and are applicable to both Tier-I and Tier-II Institutions. In addition to the POs defined by NBA, programs can define 2 to 4 Program Specific Outcomes (PSOs).

Outcomes Based Education (OBE) is developing every PO and PSO in all engineering students. Effective implementation of OBE happens through a well-designed curriculum, together with suitable pedagogy and relevant assessments. While the POs and PSOs are developed through the curriculum, additional development of skills is through student association with co-curricular and extra-curricular activities on campus [17]. Programs effectively implementing OBE are accredited by NBA.

The Accreditation Process. There exists a well-established accreditation process by NBA which can be described in four phases: the initial phase; the pre-assessment phase; the assessment phase; and the post-assessment phase [16].

A minimum of two cohorts of students needs to have graduated from the program for it to become eligible for accreditation.

Eligible programs apply for accreditation by NBA, in the Format of First Cycle in Tier-I Format or the Tier-II Format, as applicable [16]. Programs accredited for a period of six years, can apply for re-accreditation of their program six months before the expiry of the accreditation in the Second-Cycle format [16]. Programs that are not accredited can apply for accreditation when ready.

An Example Institution. BMS College of Engineering (BMSCE) was established in 1946. It was the first private sector initiative for engineering education in India [18].

The academic processes and initiatives of the college are guided by the Vision:

> Promoting Prosperity of mankind by augmenting Human Resource Capital through Quality Technical Education and Training.

More than 40,000 engineers have graduated from the college and are pursuing their career within and outside of India. Today over 5,000 students are pursuing undergraduate/postgraduate courses in conventional as well as emerging areas. In addition, more than 350 research scholars are pursuing Ph.D. degrees in science, engineering, and management. A single parameter that summarizes BMSCE management policies is demonstrated through faculty retention of more than 95%.

Association with IUCEE: BMSCE has been actively associated with the Indo Universal Collaboration for Engineering Education (IUCEE) since 2008 and is one of the consortium members [19]. This association enabled faculty of the college attend Faculty Leadership Initiatives (FLIs) by well-known international experts in core engineering domain as well as engineering pedagogy, which led to improvements in the teaching learning process. As a specific example, the outcome of having attended the one-week FLI on 'Advanced Signal Processing' helped update knowledge in the domain, resulted in faculty uploading content modules on a platform with global access, and introducing the 'Signals and Systems Concepts Inventory' as one of the internal assessments in the relevant course [20]. Similarly, faculty attended FLIs focused on changes in pedagogy, active learning, flipped classroom, problem-based learning, project based learning, and collaborative learning to enhance student learning [21, 22].

Several faculty members of the college are IIEECP certified, and hence are able to contribute effectively to course design, delivery and assessment [23].

Association with WOSA: Since 2014, NBA has organized the World Summit on Accreditation (WOSA) and includes keynote addresses by academic leaders from member countries of the WA. The summit includes presentation of accepted papers on topics aligned to the theme of the conference, and hence is an opportunity for faculty from engineering institutions to present and share their best practices on accreditation parameters. BMSCE has always encouraged faculty to participate in the summit and a few faculty have published their work on OBE in WOSA. Most importantly, the WOSA conference proceedings are uploaded on the conference website and are a repository available for ready reference [24].

ICTIEE: In 2015, BMSCE and IUCEE jointly hosted the second International Conference on Transformations in Engineering Education (ICTIEE), in collaboration with: AICTE, ASEE, GEDC, IFEES, NBA, IEEE and SPEED [25–30].

The conference brought engineering educators from across the world to share best practices for preparing engineering graduates to address global challenges [31].

The Academic Process: The first component to ensuring academic quality is the design of a curriculum, together with the Course Outcomes (COs) and relevant assessments to help evaluate the POs defined by NBA. Every program has a well-established academic audit process, at different levels. For example, in the Department of Electronics and Telecommunication Engineering, at the beginning of the academic year, they survey the students, parents, and faculty, to capture the expectations of stakeholders, and set targets for PEOs.

The professional career of the graduates is represented by Fig. 1, which summarizes the career path immediately after graduation and three years later [32, 33].

In the OBE framework, curriculum design is the first major task, and this is followed by defining the COs that map to POs, accompanied by relevant assessments.

Most courses are accompanied by laboratory experience, either through exclusive credits of the course or through the pedagogy adopted. The availability of programming tools makes it possible to adopt experiential learning as the pedagogy for engineering concepts [34]. Of course, this requires classrooms equipped with students having individual computing facilities to enable course instructors to engage students in hands-on learning during regular class/tutorial sessions. This pedagogy has collapsed the existing rigid boundary between theory sessions and laboratory sessions.

The organization structure ensures academic processes towards effective course delivery and assessments [35]. At the beginning of the semester, the Department Academic Committee (DAC), ensures the assessments of all courses are well defined, made known to the students, and evaluations are spread across the semester. During

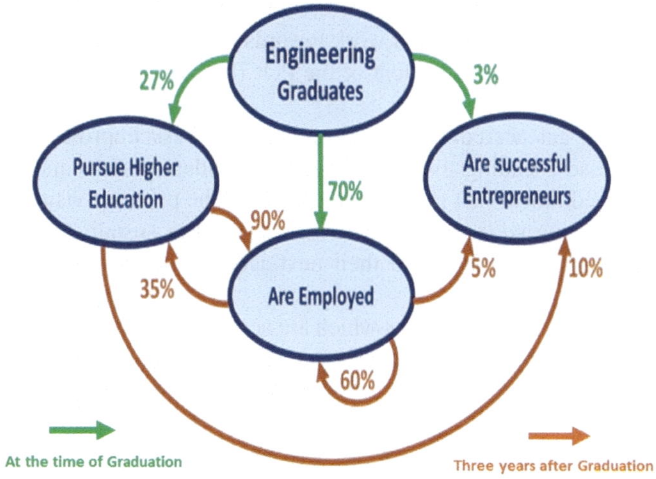

Fig. 1 Graphical representation of the professional career of graduates

the semester, the department Academic Audit Committee (AAC), reviews all internal assessments before the conduction of every in-semester assessment. Their job is to ensure that the quality of questions, and the mapping of questions to the COs are justified. At the end of the semester, the department conducts the Course End Survey. The DAC summarizes the contribution from every course of the curriculum, together with various surveys, and computes the attainments of POs/PSOs for the graduating cohort. The attainments, the action taken based on attainments and the students' performance for the graduating cohort are presented during the annual DAB (Department Advisory Board) meeting. The DAB includes external members (academic, industry and alumni experts) and internal members (faculty along with a few students). The DAB meeting provides an opportunity to present the PO attainments and include any reforms in academic processes towards ensuring improvement in quality of the graduates. In addition to these processes at the department level, the IQAC (Internal Quality Assurance Cell) conducts the academic audit at the institutional level.

Useful Practice: The Department of Electronics and Telecommunication Engineering has established the practice of organizing a felicitation function and releasing a Magazine by the graduating Batch.

The magazine includes significant activities organized by the college/department beyond curriculum. It also showcases individual student achievement in co-curricular and extra-curricular events. The felicitation function involves recognizing every student for his/her achievements (publications/awards/performance in competitive examinations/placements/higher education details). In the OBE framework, this practice is useful, as most of the student's parameters in accreditation criteria are consolidated through the Magazine and Felicitation function; both of which are uploaded on the department website [33].

Reflections: In India, accreditation decisions by NBA for a program are of three types: (i) accredited for six years; (ii) accredited for three years; (iii) not accredited.

Programs that are not accredited work towards establishing academic processes that demonstrate student achievement of the POs (the GAs of WA), and eventually secure accreditation.

Programs that get accredited for three years, address improvements in their academic processes, leading to improvements in student performance, and apply for the next accreditation visit (within three years of the previous visit).

Programs securing six years of accreditation need to sustain their established academic processes and prepare for their next accreditation six years later; a very long-time gap.

It may be useful to have programs which are accredited for six years to upload on the eNBA portal a few selected quantitative student and faculty parameters, including infrastructure details every year. This would reinforce the need to sustain established academic processes.

On retrospection, the OBE process focused on demonstrating the student attainment of GAs by WA, could be considered as broad guidelines for defining academic processes that ensure the quality of engineers. The OBE process eventually translates to student performance metrics like publications, awards, placements, and performance in competitive examinations.

Since institutions, students, and faculty each have unique identification numbers, one could explore ways to connect student and faculty accomplishments with the unique identification numbers to document the necessary performance metrics. This should then reduce the huge burden institutions have in consolidating quantitative data for accreditation.

3 Key Takeaways

The main key points in terms of accreditation processes common to the cases analyzed lead us to consider that a culture of learning and continuous change is encouraged. There is always a pursuit for the quality of academic programs and institutions to have a positive impact on the development of engineering professionals.

The main common learnings are:

(i) The leaders' commitment regarding accreditation plays an important role in the development of this culture of continuous improvement. The leadership team must endeavor to have compliance with the requirements of the accrediting agencies and be convinced of their impact on the quality of the academic programs offered to students.
(ii) The continuous improvement of the academic programs is conducted by the faculty, which is why the commitment and involvement of the faculty is vital for the achievement of the quality objectives and compliance with the accreditation standards. Huge leadership skills of collaboration and progressive thinking is required by all factions of the team to bring home the accreditation.
(iii) Documentation processes, use of technology, and communication with stakeholders are the elements that contribute to the success of the continuous improvement and accreditation processes.
(iv) A culture of continuous improvement supported by accreditation is important to develop through training, faculty meetings and documentation, evaluation, and action processes.

4 Conclusions

Engineering Education has gone through transformation with innovations in pedagogy and assessments. The focus was on addressing the graduate's attributes defined by the Washington Accord.

This transformation in engineering education has provided the ability for graduating engineers from accredited programs to compete, perform, and pursue their career in countries who are members of the WA.

As seen in the case studies, it takes an institutional commitment, and leadership skills to pursue a continuous improvement and educational excellence; to negotiate the practice of transparency and foster the culture of effective communication at all levels of the institution. It also calls upon a dedicated faculty and staff to ensure ongoing success.

Although there are differences in the processes, implementation, and evaluation formats, since the accreditation body of every country ensures graduating engineers from accredited programs have the GAs defined by WA, global acceptance and recognition of engineers is ensured.

References

1. International Engineering Alliance Homepage. https://www.ieagreements.org/. Accessed 15 Oct 2023
2. International Engineering Alliance Homepage. https://www.ieagreements.org/accords/washington/signatories/. Accessed 15 Oct 2023
3. International Engineering Alliance Homepage. https://www.ieagreements.org/about-us/gapc/. Accessed 15 Oct 2023
4. Iowa State University Strategic Plan 2022–2031 Homepage. https://strategicplan.iastate.edu/wp-content/themes/strategic-plan/pdfs/StrategicPlan.pdf. Accessed 31 Oct 2023
5. Board of Regents, State of Iowa Homepage. https://www.iowaregents.edu/. Accessed 31 Oct 2023
6. Higher Learning Commission Homepage. https://www.hlcommission.org/. Accessed 31 Oct 2023
7. American Council on Higher Education Learner Success Laboratory Homepage. https://www.acenet.edu/Programs-Services/Pages/Professional-Learning/Learner-Success-Laboratory.aspx. Accessed 31 Oct 2023
8. Iowa State University, Student Success Recommendation from ACE Learner Success Lab Project Homepage. https://www.provost.iastate.edu/academic-programs/high-impact-practices-and-student-success-initiatives/student-success-recommendations-from-ace-learner-success-lab-project. Accessed 31 Oct 2023
9. Iowa State University, Academic Program Review: Guidelines for Self-Study and Review Foci Homepage. https://www.provost.iastate.edu/academic-programs/accreditation/academic-program-review--purpose-and-process/self-study. Accessed 31 Oct 2023
10. Iowa State University, College of Engineering Strategic Plan 2018–2023 Homepage. https://www.engineering.iastate.edu/dean/files/2017/08/StrategicPlan-2018-2024.pdf?pdf=CoE-Strategic-Plan. Accessed 31 Oct 2023
11. Accreditation Board for Engineering and Technology (ABET) Homepage. http://www.abet.org. Accessed 31 Oct 2023
12. Accreditation Board for Engineering and Technology (ABET) Member Societies Homepage. https://www.abet.org/member-societies/. Accessed 31 Oct 2023
13. Engineering (ATMAE) Homepage. http://www.atmae.org. Accessed 31 Oct 2023
14. Higher Learning Commission Standard and Open Pathways for Reaffirmation Accreditation Homepage. https://www.hlcommission.org/Accreditation/pathways.html. Accessed 31 Oct 2023
15. Consejo de Acreditacion de la Ensenanza de la Ingeneria, A.C. (CACEI). http://cacei.org.mx/index.php. Accessed 31 Oct 2023
16. National Board of Accreditation (NBA). https://www.nbaind.org/. Accessed 25 Oct 2023

17. Kanmani B (2020) Effective implementation of OBE leading to accreditation. Indian Society for Technical Education (ISTE), under the Working Professionals Learning Project, an AICTE Project. ISBN: 978-81-89731-23-6
18. BMS College of Engineering, Bengaluru, India. www.bmsce.ac.in. Accessed 25 Oct 2023
19. Indo Universal Collaboration for Engineering Education. https://iucee.org/. Accessed 25 Oct 2023
20. Wage KE, Buck JR, Wright CHG, Welch TB (2005) The signals and systems concept inventory. IEEE Trans Educ 48(3):448–461. https://doi.org/10.1109/TE.2005.849746
21. Felder RM (1988) Learning and teaching styles in engineering education. Eng Educ 78(7):674–681
22. Felder RM, Brent R (2003) Designing and teaching courses to satisfy the ABET engineering criteria. J Eng Educ 92(1). https://doi.org/10.1002/j.2168-9830.2003.tb00734.x
23. IUCEE International Engineering Educator Certification Program for faculty recognized by (International Society for Engineering Pedagogy (IGIP, Austria). https://iucee.org/engineering-educator-certification-iieecp/. Accessed 25 Oct 2023
24. World Summit on Accreditation (WOSA) by NBA. https://www.nbaind.org/Conferencetrainingprogrammes. Accessed 25 Oct 2023
25. All India Council for Technical Education (AICTE). www.aicte-india.org. Accessed 25 Oct 2023
26. American Society for Engineering Education (ASEE). www.asee.org. Accessed 25 Oct 2023
27. Global Engineering Deans Council (GEDC). www.gedcouncil.org. Accessed 25 Oct 2023
28. Institute of Electrical and Electronics Engineers (IEEE). www.ieee.org. Accessed 25 Oct 2023
29. International Federation of Engineering Education Societies (IFEES). www.ifees.net. Accessed 25 Oct 2023
30. Student Platform for Engineering Education Development (SPEED). www.worldspeed.org/. Accessed 25 Oct 2023
31. Schedule of Pre-Conference Workshops during ICTIEE 2015. https://ictiee.org/ictiee2015/workshop_2015_new.php. Accessed 25 Oct 2023
32. Kanmani B (2014) Program educational objectives: A graphical representation. In: 2014 International conference on interactive collaborative learning (ICL), Dubai, United Arab Emirates, pp 623–626. https://doi.org/10.1109/ICL.2014.7017844
33. Magazine of Batch graduating 2021. https://bmsce.ac.in/reports/TE/Konnect_2021a.pdf. Accessed 26 Oct 2023
34. Buddhi K (2023) A pedagogy for engineering concepts focusing on experiential learning. In: Guralnick D, Auer ME, Poce A (eds) Creative approaches to technology-enhanced learning for the workplace and higher education. TLIC 2023. Lecture notes in networks and systems, vol 767. Springer, Cham. https://doi.org/10.1007/978-3-031-41637-8_8
35. Organization Structure of the Department. https://bmsce.ac.in/home/Electronics-and-Telecommunication-Engineering-Organizational-Structure. Accessed 26 Oct 2023

Finance Management and Revenue Generation for the Sustainability of Engineering Institutions

Sushma Kulkarni, Thanikachalam Vedhathiri, and Sandeep Sancheti

Abstract Financial management and revenue generation for any academic leader in a university is a critical responsibility and hence developing financial acumen is a major task in front of them. Financial planning, financial delegation and discipline is an important function of the Head of HEI. The Faculty in the role of academic leader with an aspiration to hold higher positions in HEI should be able to comprehend Finance management, so that in the future they are prepared to lead their school, institute or University. Higher education institutes have an academic ecosystem where faculty members' primary focus is academic, research, and administrative roles. As they progress, their exposure to financial planning and management is limited, and they are expected to possess financial acumen when assuming the responsibility of academic leadership. There is a big gap dependency of the academic leader on the finance and accounts Staff. This may bring limitations because the growth and development of HEI is dependent on how revenue is generated and expenditure is planned through various heads to bring the Institutional development as per the strategic plan of HEI. This chapter elaborates the key aspects of financial management and revenue generation for the sustainability of the HEI. It advocates for a proactive and adaptive approach to financial management, emphasizing the importance of leveraging technology, fostering innovation and revenue generation, and continuously reassessing strategies to thrive in the ever-evolving financial landscape.

Keywords Financial management · Revenue generation · Financial literacy · Resource allocation · Fund raising · Fee strategies

S. Kulkarni (✉)
NICMAR University, Pune, India
e-mail: sushma.kulkarni@pune.nicmar.ac.in

T. Vedhathiri
NITTTR, Chennai, India

S. Sancheti
Marwadi University, Rajkot, India

R. Kandakatla et al. (eds.), *Academic Leadership in Engineering Education*, Lecture Notes in Networks and Systems 1097, https://doi.org/10.1007/978-3-031-68282-7_4

1 Introduction

In recent years, higher education institutions have undergone significant transformations in their operating environments, encompassing various dimensions such as political, demographic, socio-economic, and technological spheres. The literature widely acknowledges several trends resulting from these changes. These include financial constraints, especially within the public sector and higher education; a continual rise in student enrolment due to escalating societal demand; heightened expectations from governments for accountability and transparency in the higher education market, involving students, parents, and employers; a shift towards greater institutional autonomy and the delegation of responsibilities from the State to individual institutions and from central bodies to basic units within institutions. It gives rise to the need of a quality revenue management system and financial literacy of all internal stakeholders of the higher educational institutions.

This chapter is an attempt to create awareness amongst the Vice Chancellor, Director, Principal, Deans, Head of Department, center head, and faculty in general, who aspire to hold higher positions in HEI. They should understand and comprehend Finance, Accounting, income-expenditure statements, Revenue Generation, various financial terms like debit, credit, Balance sheet, budgeting, planning, etc. so that in the future they are prepared in this important function to lead their school, department, institute or University.

It is also essential to learn the methods of accounting, fee collection, vendor payments and other day-to-day operations using the ERP and advanced banking services like payment gateways to improve the efficiency and effectiveness of people and the tasks to be completed in the given time frame. But, the most important point that needs to be highlighted is that financial skills and expertise cannot be developed overnight after getting into the responsibility of the said role. Indeed, developing proficiency in financial management skills is a gradual process that involves a combination of education, experience, and ongoing learning. It involves a combination of academic training, on-the-job experience, exposure to diverse financial scenarios, and a commitment to staying informed about evolving financial trends and regulations.

This is an attempt to emphasize the need for financial literacy among all internal stakeholders and the vision of revenue generation along with administrative policies and process to promote it as academic leader in higher education institutions. It is well known that investment of time, money, energy, and planning are the four major pillars on which a new organization/institution can be built. Amongst these parameters, financial planning is the most critical and without which everything remains only on paper or pipeline. For new projects, an upfront investment by the promoter(s) in the form of seed money is essential. The requirement of such an amount depends upon various factors like overall size, diversity of operations and expectations from the initiative. The norms for these cannot be defined uniquely as the minimum requirements as prescribed by the regulatory agencies keep changing from time to time. These are also impacted by the nature of the institution being established, meaning thereby, whether it is a degree/diploma/training institution or it is an affiliated institution or a university etc. On the same lines, geographical location (city, town and village) can matter hugely in terms of the amount of expenditure involved. There are several features which can decide the success or failure of an educational institution of which financial management is integral to every such parameter. Accordingly, a good investment should initiate certain good and important financial practices right from the beginning.

- Financial transparency (like purchase processes, well defined salary structure).
- Delegation of financial powers (Financial autonomy in a structured fashion).
- Annual, short term, long term budgetary processes including the features of cash flow, credit limits and loan repayment etc.
- Internal/External/Statutory audit processes to be started from early stages.
- Creation of Corpus funds, wherever possible from the beginning itself. Rules and regulations for utilization of these funds should also be pre-defined such that they add to stability by countering financial fluctuations.

1.1 Fund Generation

Fund Generation is always an important element for a functional institution. However, it is also among the most challenging aspects, influenced by numerous parameters, many of which are beyond the control of the institutions themselves. Some of these are:

- The tuition fees fixation.
- The admission cycles/timings/intakes.
- Number of admissions offered/taken.
- External competition.

In light of the above challenges, an institution should carefully plan the availability of funds before starting its operations. To begin with, funds can be made available by the seeding agency or promoters. However, these are mainly meant for some initial costs and infrastructure creation. These may be considered as one-time investment and subsequently in a 5- and 10-year period, institutions are expected to earn funds out of it and sustain its operations, expansion, maintenance, modernization, diversification etc.

A slight positive part of education institution revenue generation is that unlike other industries/businesses where the payments are credited much later after services/ supplies are rendered, the educational sector typically operates on a model where tuition fees for an entire academic year/semester are paid in advance. This prepayment system ensures a steady and predictable cash flow, allowing institutions to plan and allocate financial resources more efficiently. It also permits extra income by investing a certain portion of initial funds to yield additional interest income out of it.

2 Finance Management

Revenue generation and financial management are crucial for the sustainability, and growth of all higher educational institutes and universities. The responsibility for financial management in higher educational institutes is typically shared among several key stakeholders like governing body members, department heads and deans, administrative offices, innovation and research centers managing grants, and internal and external auditors. It is an essential part of Academic Leadership to be able to apply the concept of Finance and accounts for the governance of the HEI.

2.1 Importance

As a leader, in the academic and administrative role while in the position of Vice Chancellor/Director/Principal/Dean, it is essential to be able to correlate the strategic development of the HEI with knowledge of funds management. Budgeting, allocation of resources to various expenditure heads along with the sources of income. The specific roles and responsibilities may vary depending on the organizational structure and policies of the institution. A leader in higher education institute will be able to transform and bring change and drive new initiatives, it is important to pay attention to the details of improving the sources of income and also plan and budget for requirements according to the new initiatives. Later, resource allocation and accounting will be entrusted to finance department.

Our Higher education Institutes have an academic ecosystem designed in such a way that when a faculty joins the Institute as an Assistant professor and rises through various positions to Associate professor, Professor, and then Principal, Director and Vice chancellor, the prime focus remains as academic, research and the exposure to administration as dean/head is limited in terms of financial planning and management. Then suddenly the Professor is expected to do the Finance management function when entrusted with the responsibility of academic leadership. There is a big gap and hence this function remains neglected, and the Head of the institute depends on the finance and accounts department staff for it. This may bring limitations because the growth and development of HEI is dependent on how revenue is generated and expenditure is planned through various heads to bring the Institutional development as per the strategic plan of HEI. At the same time, it is important that the leaders practice financial probity.

Financial management for any academic leader in a university is a critical responsibility and hence developing financial acumen is a major task in front of them. Financial acumen pertains to the capacity to comprehend and analyse financial information, utilizing it to make well-informed decisions to judiciously allocate resources, adeptly navigate risks, and capitalize on opportunities that contribute to overall profitability for further growth and development of HEIs. The leaders play a key role in overseeing the university's financial health and ensuring that resources are allocated efficiently to support the institution's mission and goals by carrying out proper financial planning, fee fixation per student, financial discipline, allocation of budget to every department for various expenditure heads as defined at HEI level, half yearly reappropriation of budget, follow-up meeting with department heads for the budgeted expenditure in time. In short financial planning, financial delegation and discipline is an important function of the Head of HEI. Let us take an example of preparing the budget for the next financial year for the HEI. This exercise involves planning the revenue collected through student fees, consultancy, revenue generated through internal and external sources and allocating the proposed expenditure for various recurring and non-recurring heads like Salary for teaching and non-teaching staff including guest, visiting faculty, new recruitments (which involves detailed workout on number of teaching staff requirements and non teaching staff for next academic year). Similarly, the requirements have to be worked out on different heads like Research and development, Industry Interactions, Student training and development, Faculty and staff Training, Equipment and machinery, building maintenance and taxes etc. which requires future planning for HEI. Generally, academics get the prime importance and focus on a higher percentage of work time allotment but if the Head of the institute is empowered for financial management along with delegation of authority for financial approval at deans and heads level (as also mentioned by the NAAC and NBA) it will in true sense help in the development of HEI.

Financial Planning, financial budgeting, fee fixation, etc. requires holistic planning for the HEI in terms of prioritizing the activities, expenditure heads like salary, recurring expenditure, capital expenditure, depreciation, loans Impact of Inflation and other subheads for detailing to plan for annual expenditure. The constraints have to be considered while financial planning in terms of revenue generation and expenditures thus prioritizing the allocation of resources is essential while budgeting. It is equally important to study the percentage utilization of allocated resources for each head and also that proper correction measures that can be taken for holistic development of HEI e.g. If the budget allocated for R & D is X amount but only 10% of X amount is utilized, it indicates that R & D activities as planned and budgeted haven't happened and the concerned deans should check why the 90% of the planned expenditure for R & D activities have not happened. Based on this data corrective measures can be taken.

Purchase procedures should be clearly defined with the purchase policy. This helps faculty to plan their purchases for equipment, furniture, stationary recurring items, Machinery, or any items required for the functioning of the Department.

2.2 *Financial Literacy*

Financial Literacy is of utmost importance to lead any HEI through the organization structure for a Vice chancellor/Director/Dean/Department Head/Centre Head (any academic leadership role as a faculty). If an individual has minimal personal knowledge of finances and budgeting, it is sure to impact their private money matters.

But when an educational leader does not have professional-level expertise in fiscal matters, it will have a negative impact on the institution, faculty, staff, and the students.

Despite all that financial leadership and stewardship requires, many educational leaders are not prepared for this part of the job. As stated by Sorenson and Goldsmith (2018) "***in far too many instances, school leaders possess limited background, experience, or expertise with the budgeting process.***"

The good news is that these skills can be developed, even by adolescents and young adults, so don't be discouraged. Despite how you may feel about math, spreadsheets, or money management, you can develop the skill. It is necessary to be a successful financial leader and steward… plus the alternatives are too risky!

3 Key Aspects of Financial Management

Here are some key aspects of financial management for an Academic Leader in HEI.

3.1 Budgeting

The Leader should be actively involved in the budgeting process. This involves working together to co-create the budget with the finance team and overseeing the university's annual budget, that aligns with the institution's strategic plan. The LEADER should work closely with the finance team and academic departments to ensure that budget priorities are set and met.

3.2 Resource Allocation

The Leader must make decisions about how to allocate the university's resources effectively. This includes determining funding levels for various academic programs, research initiatives, and administrative functions.

3.3 Financial Planning

Long-term financial planning is crucial. The Leader should be involved in developing financial projections and strategies for the university's future. This might involve forecasting enrolment, faculty requirements, and fundraising goals, and identifying potential risks or challenges. This may be achieved through developing a strategic plan for a period of five years and then can be broken down to each year.

3.4 *Analysis and Interpretation of Financial Results, Balance Sheet, Cash Flow Statements, Auditing, Budgeting etc.*

It helps to analyze the percentage utilization of allocated budget versus the actual expenditure. The process guides the allocation for the next year and the leader will know under which heads the planned expenditure is not done and the initiatives for future development is lacking. This will draw attention to focus on the weak areas for development.

3.5 *External Research Grants, Internal Revenue Generation, Consultancy*

Leaders play an important role in creating the research ecosystem by motivating the faculty to do research by providing seed funding, encourage innovation, start-ups, patents, makers space etc. by budgeting for it. Also promoting and encouraging IRG and consultancy which will help to increase the revenue.

3.6 *Fundraising*

Leaders often play a key role in fundraising efforts from external sources. This includes engaging with donors, alumni, CSR (corporate social responsibility) funds of Industries and other stakeholders to secure philanthropic contributions and grants to support the university's programs and initiatives. This is all the more important for self- financed institutions, as they have to manage all activities with the student fees which will not be sufficient to develop a quality institution. Keeping the New education policy in view where we are talking of cluster universities, research universities, and interdisciplinary courses, a lot of funds will be required, indicating a dire need for fundraising.

3.7 *Cost Management*

Controlling costs is essential for maintaining financial sustainability. The Leader should work with various departments to identify cost-saving opportunities while ensuring that academic quality is not compromised. Institutional processes and workflows should be regularly reviewed to identify the areas for efficiency improvements. Encourage staff at all levels to contribute ideas for streamlining operations and reducing unnecessary expenses, as the initiative should reach the last level in the institute. This may include, for example, energy-saving and sustainability initiatives to reduce operational costs. The leader may propose energy-efficient infrastructure, waste reduction programs, and environmentally friendly practices. Technology solutions can also be implemented that can streamline administrative processes and improve efficiency.

3.8 Compliance and Reporting

Ensuring compliance with financial regulations and reporting requirements is vital. The Leader should oversee the preparation of financial reports and audits, providing transparency to stakeholders and demonstrating responsible stewardship of university funds.

3.9 Investment Management

Managing the university's investments, endowment, and other financial assets is a critical responsibility. The Leader should work with investment advisors and committees to make sound investment decisions that align with the university's financial goals. An important aspect is creating a corpus fund. The leaders should make provisions for annual salary and recurring expenditure through Fixed deposits (FD) in case the fees are paid by students in instalments. These FDs can be planned to mature per month as per the budgeted expenditure. The interest earned can be used for building up the corpus fund.

3.10 Risk Management

Identifying and managing financial risks is essential. This includes addressing potential economic downturns, changes in government funding, and other external factors that could impact the university's finances.

3.11 Fee Strategies

Setting tuition and fee levels is a delicate balancing act. The Leader should consider affordability for students while ensuring that the university generates enough revenue to cover operating costs and surplus enough for future development of the HEI.

Setting tuition fees for students in higher education requires a strategic approach that balances the financial needs of the institution with considerations for accessibility and affordability. The vice chancellor/Director along with the key people involved should consider several factors when determining tuition fees. Firstly, leaders should conduct a thorough financial analysis to understand the institution's budgetary requirements. This analysis forms the basis for establishing a tuition fee structure that ensures the institution's financial sustainability. Simultaneously, it is crucial to consider the economic backgrounds of the student body. A fair and equitable tuition policy should take into account the financial constraints that students may face. Implementing need-based financial aid programs and scholarships can help mitigate the burden on students from lower-income families, thereby promoting inclusivity and diversity within the student population. Additionally, an assessment of the competitive landscape is essential to ensure that the institution's tuition fees remain competitive within the broader higher education market in the vicinity. Benchmarking against peer institutions and considering regional economic factors can help strike a balance between competitiveness and financial stability. Moreover, involving stakeholders, such as faculty and administrative staff, in the decision-making process can enhance transparency and garner support for the proposed tuition structure. This collaborative approach allows leaders to consider various perspectives and address concerns, fostering a sense of shared responsibility.

Regular reviews and adjustments to the tuition fee structure are essential to adapt to changing economic conditions, enrolment trends, the evolving needs of the institution, and provision of scholarships for economically weaker students. Leaders should be prepared to communicate the rationale behind tuition decisions clearly and consistently, emphasizing the institution's commitment to providing quality education and maintaining fiscal responsibility.

3.12 Technology and Financial Systems

Utilizing modern financial systems and technology can improve efficiency in financial management. The Leader should support the implementation of robust financial systems that provide real-time financial data and analytics. Every institution should use ERP and relevant software for improving the efficiency and effectiveness of financial systems. Partnering with the banks for payment gateways, collection of student fees, vendor payments, international transactions etc. will reduce the manpower requirements and improve efficiency.

3.13 Collaboration

Collaborating with the leaders, including academic deans, department heads, and administrative staff within HEI, is crucial for effective financial management.

The Leader should foster a culture of financial responsibility, delegation, and accountability throughout the institution.

3.14 Communication

Keeping stakeholders, including the university's governing board, faculty, staff, and students, informed about financial matters is essential. Transparent communication helps build trust and support for financial decisions.

3.15 Regular Meetings of Statutory and Non-statutory Committees

Financial Planning and Tuition Fee Fixation will be under the Finance Committee of the Board of Governors. But Purchase Committee, Building and Construction Committee constituted by the head of the institute will follow the decisions of the Board. Purchase of equipment, tools, consumables, maintenance, and taking modern equipment on long term lease will be based on the budget allotted by the Board of Governors. Further, obsolete equipment will have to be auctioned to save space. Sometimes, needed equipment can be added to the research projects and they may fetch funds from the research grants. Purchase committee will strictly monitor that those approvals are taken, quotations are called for and comparative statements are prepared before vendors are summoned for negotiations to do the purchase.

4 Financial Training and Development

Faculty members aspiring to advance into higher leadership roles within higher educational institutes can take deliberate steps to enhance their financial skills. First and foremost, participation in professional development programs focused on financial management in higher education is crucial. These programs offer insights into budgeting, resource allocation, and financial decision-making. Pursuing advanced degrees in areas such as business administration or educational leadership provides a solid academic foundation in financial principles. Actively seeking collaborative opportunities with financial officers and serving on budget committees allow faculty to gain practical experience and a deeper understanding of institutional financial dynamics. Engaging in grant-writing and fundraising activities not only secures external funding but also sharpens financial acumen. Taking on administrative roles within academic departments provides first-hand experience in budget oversight.

Networking with financial leaders, seeking mentorship, and participating in interdisciplinary collaborations are additional avenues for skill development. Continuous learning through conferences, literature, and webinars ensures faculty stay abreast of evolving financial trends.

Training entry-level faculty in higher education for financial skills to develop involves a comprehensive approach that addresses both theoretical knowledge and practical application. Begin by offering workshops or seminars that cover the basics of budgeting, financial planning, and resource allocation within the context of higher education institutions. These sessions should be tailored to address unique financial challenges. Additionally, provide hands-on exercises and case studies to enhance their ability to apply financial concepts to real-world scenarios.

To further develop leadership skills, incorporate elements of strategic financial management. Encourage faculty to think critically about the financial implications of academic decisions and strategic planning. This may involve analyzing enrolment trends, assessing program costs, and understanding the financial impact of different curricular and operational choices. Facilitate discussions on financial leadership, emphasizing the importance of collaboration, effective communication, and data-driven decision-making. Consider mentorship programs where experienced financial leaders guide entry-level faculty in understanding the nuances of financial management. Moreover, continuous learning through ongoing professional development opportunities, such as webinars, conferences, and networking events focused on financial leadership in higher education. Encourage faculty to stay informed about evolving financial trends, emerging technologies, and best practices in financial management. By fostering a culture of financial literacy and leadership development, institutions can equip entry-level faculty with the necessary skills to navigate the complexities of higher education finance and contribute to the institution's long-term success.

5 Case Study of a Leading Autonomous Technical Institute at Maharashtra, India

This case study provides a summary of the key fiscal responsibilities and processes carried out at an Institute in Maharashtra, India.

Each employee is responsible for financial controls relevant to his/her role at the HEI. There are set basic guiding principles and required departmental controls. This includes the responsibilities at different organizational levels, and are well communicated to all those concerned including the delegation of financial powers to sanction bills up to a certain amount well defined.

The budgeting procedures encompass the schedule, methods, and actions involved in establishing an annual budget for an institution. This discussion specifically delves into the context of the institute, exploring the primary sources of income and expenses shaping institutional budgets.

5.1 *Revenue and Expenditure Statement*

A significant income stream is tuition fees. As no state/government funding is available, the institute must depend on the fees collected from students. The fees are fixed by the Fee Regulatory Authority by state government. It is based on salary, Recurring and Capital Expenses, Depreciation, Inflation impact etc. However, even with the revenue generated from tuition fees and development fees, it is just sufficient to cover the cost of education. It necessitates other sources of income known as through auxiliary services.

Within institutional spending, the largest share of 60–65% of tuition fees is allocated to salary head and remaining 35–40% is for other recurring expenses and capital expenditure. Those are for routine maintenance, taxes, and utilities, depreciation and obsolescence. Capital infrastructure pertains to the upkeep of buildings and other capital assets.

5.2 *The Process of Budgeting at the Institute Is as Below*

When the departments are framing their requirements, the head of the department works out the budget based on priorities and initiatives in line with the vision and mission of the department. It also has reference to recurring and capital expenses for past 3 years. This indicates that every faculty, head of department, Dean, and Director along with the administrative office has to participate in the budget process, requirements, and allocations under various heads, which necessitates the development of financial acumen till the lowest level. Utilization percentage of actual expenditure versus the budgeted expenditure is a measure of the correctness of financial budgeting under each expenditure head (Fig. 1; Table 1).

6 Revenue Generation for the Sustainability of Engineering Institutions

6.1 *Introduction: Vision and Mission for Revenue Generation*

Throughout the world, almost all well-performing universities generate additional funds through industrial services, research grants, royalty on publications, and renting out many resources when not used. This additional income is used to pay scholarships, purchase new equipment, and software, and share the project gains with the faculty members who generated the resources. The university gets recognition for

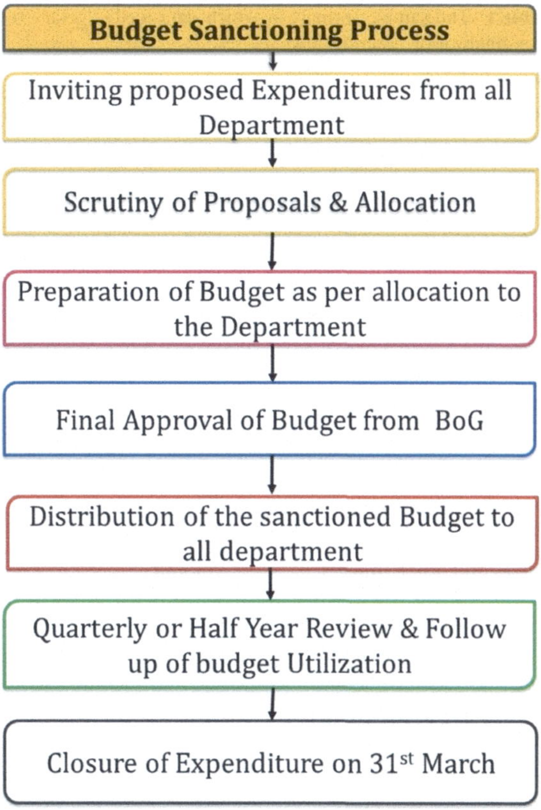

Fig. 1 Process of budget sanctioning

solving many complex problems. Of late the Ministry of Education advises the institutions to generate additional funds. Revenue generation is an important activity for engineering institutions [1–6]. Even though the government provides grants-in-aid for meeting a certain percentage of expenditure, this grant has to be supplemented through planned methods of additional revenue generation. During deficiency of budget provision, the Ministry of Education has recommended the institutes generate needed funds.

It is essential to develop a strategic plan to generate additional revenue by effectively using the resources and the high-performing faculty members. Many institutes have generated funds through consultancy projects and training the executives of various companies. High-performing faculty members need to be constantly updated to save them from burnout. Technology is growing at a rapid speed and employers need industry-ready graduates to take up assigned tasks. They have to be trained using advanced tools and equipment. Institutes need additional funds to modernize the laboratories and workshops. Additional workshops and laboratories will also be constructed as and when needed to accommodate the additional equipment. For undertaking all developmental activities, engineering institutions have to develop an

Table 1 The sample formats in which data for budgetary requirements is sought and disbursed after finalization are as follows

Account heads					
Salary		Amount	Other recurring expenditure		Amount
1	Teaching staff		1	Stationery and printing	
2	Instructor staff		2	Travelling and conveyance	
3	Guest staff		3	Advertisement	
4	Ministerial staff		4	Maintenance and repair	
5	Menial staff		5	Bank commission	
6	Management contribution to PF		6	Function	
7	Incentive		7	Teaching aid	
	Total		8	Refreshment/ hospitality	
Taxes and other charges		**Amount**	9	Admission expenditure	
1	Audit fees		10	Mgt. journal, magazine, newspaper	
2	Affiliation and registration fee		11	Consumable	
3	Administrative expenditure		12	General M/R	
	Total		13	R/C department	
Services		**Amount**	14	Accreditation	
1	Postage and telegram		15	ARA/FRA processing charges	
2	Telephone bill			**Total**	
3	Internal Exp		**Staff training and welfare**		**Amount**
4	Electricity charges		1	Staff training	
5	House keeping Exp		2	Faculty development training	
6	Security Exp		3	BoS meeting Exp	
	Total			**Total**	
Gymkhana		**Amount**	**Research and development**		**Amount**
1	Gymkhana		1	Research and development	
2	Student placement			**Total**	

(continued)

Table 1 (continued)

Account heads					
Salary		Amount	Other recurring expenditure		Amount
3	Student training		**Other expenditure**		**Amount**
4	Cultural activities		1	IRG consultancy	
5	Vision		2	Miscellaneous expenditure	
6	Medical aid			**Total**	
	Total			**Grand total**	

Revenue budget for the year			
Sr. No.	Particular	Budget	Receipt
A	**Revenue receipt**		
1.	Tuition Fees		
2.	Other Revenue Receipt		
	Total		
B	**Revenue expenses**	**Budget**	**Receipt**
1.	Salary		
2.	Taxes and other charges		
3.	Services		
4.	Other recurring expenditure		
5.	Student welfare		
6.	Staff training and welfare		
7.	Research and development		
8.	Other expenditure		
	Total		
C	**Surplus/deficit**		
Accounts officer		**Registrar**	
Director			

appropriate vision and mission to generate revenue using the expertise of the faculty members through appropriate consultancy works. This will create linkages with the fast-growing companies in the region which will assist the departments in modernizing the curriculum and train the students in needed advancement of industrial planning, designing, developing prototypes, testing, choosing efficient manufacturing methods, marketing, and maintenance.

6.2 Administrative Policies and Norms for Faculty Members' Commitment towards Revenue Generation

If the institutes are governed by a board of governors, appropriate policies are required for framing norms for generating revenue using the resources of the institute. The faculty members need a guidance manual for preparing technical and financial proposals to bid for the development projects. The National Education Policy 2020 provides suggestions for institutional development, introducing multidisciplinary programs, and creating effective linkages with the companies in the state, region, and nation. The All-India Council for Technical Education (AICTE) also provides needed norms for sharing the project gains from the consultancy projects. Hence, engineering institutions must prepare appropriate institution-specific norms, standards, and guidelines for preparing technical and financial bids so that they can undertake consultancy projects under various private companies, government engineering departments, and International Development Agencies like the United Nations Development Program, World Bank, Asian Development Bank, etc.

Creating an Ecosystem for Planning and Developing Faculty Members towards Offering consultancy programs: Most engineering institutes have to develop and transform into institutes of innovation. The leaders have to create a sustainable and outstanding institutional culture in engineering education to facilitate the growth of high-performing faculty teams. This requires leaders with equity, integrity, ethics, humility, and an outstanding culture to transform engineering institutions. When high performing faculty members are recruited, they need sufficient delegation to innovate. They need to be scaffolded to undertake complex consultancy projects under various national and international development agencies. The administrators must establish a Center for Offering Consultancy Services to various engineering departments, International Development Agencies (IDAs), and micro, medium, and small enterprises.

6.3 Goals for Revenue Generation

Goals can be outlined for fixing the revenue target per faculty per year, or department based on the number of qualified faculty members, budding faculty members, resources available in the department, institute, previous achievement records, and current projects.

The goals should be achievable and the needed supports is to be provided. In some cases, one may employ external experts for the project.

6.4 *Integrating a Strategic Plan with the Goals of Revenue Generation*

It is desirable to integrate the strategic plan with the goals of revenue generation. In a fast growing country, there will be many opportunities to bid for global projects under various International Development Agencies like the Asian Development Bank (ADB), the Danish International Development Agency (DANIDA), the United Nations Development Project (UNDP), United States Development Agency (USAID), etc. The institute can also invite emeritus faculty members, adjunct faculty members, or scientists of national research laboratories to join the project team.

6.5 *Faculty Development in the Art of Revenue Generation*

All faculty members must be trained in preparing technical proposals, financial proposals, curriculum vitae to suit the project, negotiation skills, winning the project, preparing bills as per the agreement, and completing the project.

6.6 *Process of Revenue Generation*

There are many avenues for generating revenue from the companies and the public. Engineering institutes can choose suitable methods based on their resources, high-performing faculty members, and their linkages with various organizations. In the World Bank assisted projects in Technical Education, the participating institutes are encouraged to generate additional funds to prove the capacity of the trained faculty teams. Some of the significant methods are presented in the following section:

Space Utilization: Many institutes have substantial space resources like hostels, classrooms, auditoriums, multipurpose halls, seminar halls, conference rooms, etc. During holidays, the space can be utilized for generating income. The companies and the public can be allowed to use the auditorium, hostel facilities, classrooms, and sports field by paying rent to the institute.

Royalty from Books: Whenever new curricula are developed, they need suitable textbooks, drawing manuals, laboratory manuals, etc. Institutes can establish a publication center for the production and sale of needed instructional materials. Senior faculty members can be commissioned to write textbooks based on the revised curricula. Institutes can establish a marketing center through which books can be sold. Copyrights can be with the authors. The books can be sold to students, libraries, and the public. To meet the printing, transport, and marketing costs, the selling prices must be fixed. The royalty can be around 20% of the sale price. Every year the books can be revised based on the change and development in the external world. Teachers who wrote the books can be offered 50% of the royalty earned.

Testing Services: Most of the engineering institutes are equipped with modern equipment for testing various materials for teaching and research purposes. Many small and medium enterprises (SMEs) may not invest in testing facilities and they expect professional service from well-established engineering institutes. Further, qualified faculty members can test the materials as per standards and prepare test reports and also provide additional guidance to improve the performance of SMEs. Most of the laboratories are modernized and the departments can prepare a list of testing services for micro, small, and medium enterprises. Testing charges can be determined based on the norms and 50% of the income can be distributed to the faculty and the technical support staff and the balance can be credited to the corpus fund of the institute. By conducting an open house display on the website and sharing brochures, the clients can know the testing services offered.

Contract Manufacturing: Most of the institutes have modern machines and when not in use, they can be outsource the ancillary components. Hence, the institutes can utilize contract manufacturing of subcomponents. Detailed browsers on contract manufacturing can be prepared and distributed to companies. Micro, small, and medium enterprises can utilize such services. 50% of the income generated after deducting the cost of raw materials, machine cost per hour, and the direct cost can be worked out. The students can be included in this venture which will enhance the skills and attitudes of the students and they can be trained in the Six Sigma principle, and Kaizen in quality manufacturing.

Planning and Implementing Company-Sponsored Short-Term Graduate and Postgraduate Programs: Many companies and employees are interested in acquiring further skills through either part-time or full-time programs. Some companies are interested in jointly planning an industry-specific program. All these can be jointly planned and conducted. The companies involved has to meet the expenditure which includes the cost of utilization of resources and faculty expertise. Such sponsored courses will bring good reputation to the institutes. Institutes can plan on-the-job training for regular students. Engineering Construction Corporation (ECC) has jointly planned a postgraduate program in construction management with the Indian Institute of Technology (IIT), Madras. IIT Madras, National Institute of Technology, Trichy, Tamil Nadu, and Surathkal, Karnataka are offering this program. The students get the assistantship from ECC. Further ECC is paying the tuition fees to these institutes. Institutes have to get feedback from the sponsored participants and analyze the strengths and weaknesses in the planned programs with an eye to improvement.

Sponsored Research and Development Projects: Most of the institutes have well accomplished faculty members, advanced laboratories, and up-to-date software. After conducting postgraduate and interdisciplinary doctoral programs, there will be adequate spare time for using these resources for income generation. They can establish linkages with industries and bid for sponsored research projects. The whole cost can be borne by the companies besides paying research assistants. The institute can charge the cost of human resources, energy utilized, and maintenance cost of tools and equipment. This is a win-win solution. The institute can also license the companies to use the intellectual properties for industrial use and collect fees or royalties from the users.

Consultancy Services to Government Engineering Departments: Many government engineering departments need well-performing and well-accomplished faculty members to offer consultancy services. These faculty members have to be permitted to undertake consultancy projects and they may be advised to share 30% of the fees with the institute.

Consultancy Services to Multinational Companies: Many multinational companies look for highly reputed institutes that can offer consultancy services. Institutes can encourage their faculty members to undertake such consultancy services. The faculty can pay 30% of the income generated from such services. Scan the competitors and their preparedness. Beware of the misinformation propagated by the competitors. They may quote lesser than your bid and win the project. Later, they may try to off load the project on your organization. Don't get into their trap.

Offering Continuing Education Programs: Every engineering college should establish a continuing education center and offer short-term courses to the employees of companies. Such programs will be very useful to both employees and the sponsoring companies. The programs could be planned in consultation with the company executives.

Executive Development Programs: Institutes with postgraduate programs can offer executive development programs with advances in technology. This will enable the executives to get necessary exposure to industry-relevant technologies. These courses can be nationwide.

Part-Time Long-Term Graduate and Postgraduate Programs: Engineering colleges in metropolitan cities can organize part-time evening courses so that many companies can sponsor their employees. This model will provide a win-win solution. The graduates can offer many suggestions to add new courses based on industrial development. The 50% of the fees collected can be shared with the faculty members. These courses will bring better links with each other.

Bidding for Consultancy Projects Under International Development Agencies

Many International Development Agencies like the Asian Development Bank (ADB), the United Nations Development Agency (UNDP), the World Bank, and the United States Agency for International Development (USAID) offer loans for the development of vocational education, technical education, engineering education, etc. As a part of capacity development, they will offer many consultancy projects to reputable engineering institutes. The institutes can bid for capacity development, quality improvement, and efficiency improvement of the project institutes. Many technical universities established Curriculum Development Centers, In-house Faculty Development Centers, Instructional Materials Production Centers Educational Video

Production Centers, Multimedia Learning Package Production Centers, Interdisciplinary Research and Development Centers, etc. can undertake projects under IDAs in overseas countries. Further, the project team has to involve the clients wherever needed. Planning should be based on the terms of reference (TOR).

Formation of Consortium of High-Performing Institutions to Undertake Projects: Many industrial consultancy projects demand a multidisciplinary approach. There are specialists with a lot of success stories in solving complex projects. Hence, it is essential to form a consortium with a set of well-performing universities to bid for industrial projects. The consortium members can prepare joint technical and financial proposals to bid for appropriate projects. Institute-Institute cooperation will help solve complex projects. The project funds can be shared based on the contribution of the institutions.

University's Foundation: This is a professional unit of the university for generating additional funds through donations from industries, alumni, philanthropies, etc. Many philanthropists provide funds for certain development projects like offering skill development programs for out-of-school kids, establishing training programs for physically challenged people, etc. University Foundation can nominate a set of faculty members to prepare project schemes, financial proposals, and letters to philanthropies for granting funds.

Planning and Organizing National and International Conferences: Institutes can plan appropriate conferences in collaboration with professional associations, industries, and national councils/commissions. The institute can fix the fees for the participation of faculty members in the conferences. They can approach various ministries, donors, and industries for granting funds, they can offer space for the exhibition of industrial products, and charge fees. They can also provide space for advertisements in the program brochures. The gains can be utilized for developing corpus funds.

Sports: Universities conduct many sports events at district, state, and national levels. The best sportsmen participate in these sporting events and attract the public. Many universities sell tickets for such events. This fetches a substantial amount of income to the institutions

6.7 Financial Management

Institutes can prepare budget estimates for each year and estimate the deficiencies. From this, they can plan to generate revenue to meet deficits. They can invest some funds in various industry bonds, and government bonds. Many foreign universities invest in international stock exchanges. They need a strong professional who is an expert in investments in foreign exchanges.

Investment Analysis: The institute has to commission experts who can advise on investment analysis in various stock markets. They have to assess the risks, return on investments, and sell at an appropriate time. This is a risky activity.

Break-Even Analysis: In many development programs, the institute has to prepare a break-even analysis. If the project takes a long time to break even, an appropriate decision has to be taken to modify the project.

Return on Investment (ROI): Return on investment will provide clear information on various projects undertaken. There are intangible returns too, like gaining reputation of the institute, programs, and employability for the graduates. When industries are constantly upgrading their design and manufacturing, naturally they will recruit the best graduates who have industry-specific skills. Hence addition to financial returns, one has to focus on intangible returns.

Norms for Sharing the Project Gains: There are many norms, to share the project gains. The sharing varies from 50 to 70% between the institute and all those involved in the project including the head of the Institute, School deans, and faculty members when the institute's resources were used. The highest ratio is 70% for all the concerned and 30% for the institute when no institute resources were used for the project. Many institutes offer 60% to the concerned members and the other staff involved in the project and retain 40% for the institute. Hence, appropriate decisions have to be taken for externally funded projects.

Utilization of Funds Accrued under the Head Contribution to Institute: The institute usually utilizes the funds for creating corpus funds. However, the corpus funds can be utilized for capacity development, purchasing modern equipment, building new departments, institute expansion, etc.

Management of Corpus Fund: There are many rules and norms to manage corpus funds. If the institute has been established by the government, the Integrated Finance of the Ministry of Education may issue guidelines to manage corpus funds. Sometimes the Board of Governors too may issue guidelines for managing the same. It is better to follow the approved process. Corpus funds can't be used for running the institute like payment of salaries to the faculty and staff.

Sustainability of Engineering Institutions: The sustainability of an institution depends on efficient and effective ways of revenue generation and utilization of funds. The generation of funds depends on the educational policies, effective management of the resources, outstanding faculty skills, resources of the institute, and readiness of the faculty members to undertake complex consultancy projects and complete them without any flaw. It also depends on a strong and highly skilled leadership team to steward the allocation of funds and manage all stakeholders involved.

The institutes have to create active linkages with various companies. These depend on the leadership, demands on the highly skilled faculty members, intellectual properties developed, human capital developed, a scaffolding of the high-performing faculty members, the creativity of the faculty team, and the attributes of the graduates. Institutes have to manage vulnerability, uncertainty, complexity, and ambiguity.

7 Critical Success Factors in Internal Revenue Generation

The following factors are to be critically considered for planning and to generate sufficient internal revenue using the resources and the expertise of the high-performing faculty members:

- The institute should have financial autonomy to generate internal revenue and the resources of various departments should be up to date.
- The departments and the outstanding faculty should be empowered to bid for consultancy projects under various government departments, private companies, multinational corporations (MNCs), and International Development Agencies (IDAs).
- There should be approved norms for preparing financial bids and sharing the project gains.
- The project faculty members should be up to date in technologies, critical analysis, and problem-solving skills.
- If there is a need for any external experts or adjunct faculty members. They can be included in the project teams and payments should be included in the financial proposal.
- The project members should be insured against any accidents that may occur during the project.
- The project team should be aware of the risks involved in undertaking the projects and they should develop solutions to resolve them.
- Additional development activities that are not covered in the original contract or agreement should be undertaken after a supplementary contract based on the additional financial proposal approval.
- All the conflicts should be resolved amicably without any loss of time.
- If the clients demand a reduction in the financial proposal, there should be adequate provision to offer around a 5–10% reduction.
- The projects should be completed very well before the deadline to avoid penalties.
- The highest quality should be maintained in the project.
- If taxes are to be paid, the financial proposal should envisage such expenditure and should be included in the project proposal.
- Project completion reports should be prepared in-depth so that future faculty teams can learn project management.
- In the case of projects under IDAs or MNCs, the project team should envisage any change in the value of the local currency against the US Dollar or Euro.
- Every stage of the project has to be carefully completed well before the deadline.
- If there is a need to provide service beyond project completion, the expenditure should be assessed using value analysis. The minutes of the periodical meeting with the client or his project manager should be based on the terms and conditions stated at the time of announcement.
- Bank guarantees to be tendered at the time of award of the contract.

8 Summary

Financial planning and management is one of the most important functions of HEI. It is a tool which will open the doors for the strategic growth and development of HEI. We need to develop this skill set in the faculty so that, as future leaders at department level, functional level, institutional or university level we will have leaders who are equipped with both academic and administrative capabilities. It is not an overnight process; hence steps should be taken to develop everyone concerned by organizing induction and training programs. This will help institutes do a well-organized financial planning and project the expenses early, to provide enough space for planning to manage the resources.

Revenue generation by an engineering college is linked to policy, culture, equity, ethics, integrity, leadership, the scaffolding of high-performing faculty members, linkages with industries, and international development agencies, and delegation of needed authority to bid for consultancy projects. The outstanding faculty needs to be empowered to prepare effective and efficient technical and financial bids to win consultancy projects under various agencies. Revenue generation indicates the skills of the faculty to solve complex programs of companies. This enables the generation of funds to supplement grants-in-aid. Further, this activity reduces the gap between companies and educational institutions. Engineering graduates can also be exposed to analysis-design-prototype development—mass manufacturing—marketing maintenance- and scrapping-design of innovative products based on advanced technologies. By undertaking sponsored projects, the institute gets a reputation. They can contribute to knowledge capital and human capital. Return on investments will increase. The economy of the country will increase by attracting foreign direct investments. Sustenance of industrial growth depends on the needed engineering graduates with adequate skills and attributes. The institutes can utilize the funds generated for corpus funds to supplement grants-in-aid.

References

1. AICTE Guidelines (2000) ISTE News Letter, vol xx, issue no.1
2. Hanover (2020) Alternative revenue generation strategies. https://insites.hanoverresearch.com/hubfs/Alternative-Revenue-GenerationStrategies.pdf
3. Indian Institute of Technology—Madras: Institutional Consultancy. ICSR Guidelines
4. Hearn JC (2013) Diversifying campus revenue streams: opportunities and risks. https://ihe.uga.edu/rps/2013_01
5. Shokouhian M (2019) Revenue generation challenges. Penn State University. https://taim.psu.edu/agenda/sessions?session-4c.aspx
6. Parthenon-EY (2014) Alternate revenues: can institutions of higher education balance their mission and financial goals? https://cdn.ey.com/parthenon/pdf/perspectives/Parthenon-EY_AltrnateRevenue_paper.pdf
7. ProSymmetry.Com.4.famous-project-management Failure Examples. https://www.prosymmetry/resources

8. Sheridan R (2021) General revenue: 47 ideas for raising funds for your college from unusual sources. https://www.academia.edu/6106575
9. Thanikachalam V (1995) Guideline document for internal revenue generation and utilization by technical education institutions in India. Technical Teachers Training Institute, Chennai. https://www.vc.wisc.edu/wpcontents/uploads/site/245/2020/11/Rev-Innovationfull-report.pdf
10. DeZure D, Shaw A, Rojewski J (2014) Cultivating the next generation of academic leaders: implications for administrators and faculty. Change Mag Higher Learn Change 46(6):6–12. https://doi.org/10.1080/00091383.2013.842102
11. Griesel I (2011) The teaching of financial management principles to school leaders in disadvantaged schools. Nelson Mandela Metropolitan University. http://hdl.handle.net/10948/1538
12. Sejdic A (2023) The crucial role of financial literacy in school leadership. Faculty Focus, Higher Education & Learning. https://facultyfocus.com/articles/academic-leadership/cricial-role-of-financialliteracy-in-school-leadership
13. Sorenson RD, Goldsmith LM (2018) The principal's guide to school budgeting, 3rd edn

Leadership Approaches and Change Management in Engineering Education: Case Studies from Four Contexts

Julia M. Williams, Xiangyun Du, Abimbola Windapo, and Beverley Gibbs

Abstract When the term "leadership" is used within higher education contexts, most often we think of individuals who hold formal positions of leadership within institutions, individuals such as the president, provost, dean, or department/unit head or chair. What we have witnessed, however, is the frequency with which individuals who are not in formal positions of power may now be required by circumstance and opportunity to lead. For example, as the demand for innovative teaching has grown over the past twenty years, faculty members have found themselves leading curricular innovation teams in their departments or across their colleges. While not having named formal power per se, these individuals have adapted to their changing circumstances in order to effect the needed change. How do these individuals acquire these necessary leadership skills, knowledge, and abilities in the midst of change efforts? Our four different educational contexts will provide the backdrops as we answer this important question.

Keywords Change · Leadership · Faculty/staff development

J. M. Williams (✉)
Rose-Hulman Institute of Technology, Terre Haute, IN, USA
e-mail: williams@rose-hulman.edu

X. Du
Aalborg University, Aalborg, Denmark

A. Windapo
University of Cape Town, Cape Town, South Africa

B. Gibbs
Dyson Institute of Engineering and Technology, Malmesbury, Wiltshire, UK

R. Kandakatla et al. (eds.), *Academic Leadership in Engineering Education*, Lecture Notes in Networks and Systems 1097, https://doi.org/10.1007/978-3-031-68282-7_5

1 Introduction

1.1 Leadership Approaches and Change Management

As the earlier chapters in this volume suggest, transformation in engineering higher education requires change at multiple levels: in curricula, in academic structures, and in faculty rewards, to name only a few. Likewise, if these aspects of higher education are to change, then those who lead change efforts must themselves adapt to changing circumstances, by acquiring new skills, knowledge, and abilities that make them effective leaders, no matter what their position, institution, or disciplinary experience. In this chapter, we will address various approaches to leadership development that we have implemented in different educational contexts and geographical locations—Denmark, South Africa, United Kingdom, and the United States. Despite our different contexts, we see alignment between successful academic change projects and the leadership approaches and change management strategies associated with them.

Our approaches share important commonalities since many of them are derived from a theoretical foundation in the role of the academy in a changing world. Grant analyses, the critical role that universities play in shaping the future of a society undergoing significant change through the lens of "new power" [1], where power is no longer concentrated and consolidated in formal leadership, but is instead, "open, participatory, and peer-driven" [2]. We see the potentiality of "new power" to revitalize the social contract between universities and society. We realize that higher education founded on these principles requires a substantial review of what we consider to be effective leadership.

When the term "leadership" is used within higher education contexts, most often we think of individuals who hold formal positions of leadership within institutions, individuals such as the president, provost, dean, or department/unit head or chair. The individuals holding these positions certainly employ leadership strategies as they manage in their specific higher education contexts. In order to take on these leadership roles, they often take advantage of leadership training opportunities offered on their own campuses and through other organizations. These programs provide attendees with the tools to lead in higher education from a position already endowed with power.

As authors, what we have witnessed, however, is the frequency with which individuals who are not in formal positions of power may now be required by circumstance and opportunity to lead. For example, as the demand for innovative teaching has grown over the past twenty years, faculty members have found themselves leading curricular innovation teams in their departments or across their colleges. While not having named formal power per se, these individuals have adapted to their changing circumstances in order to effect the needed change.

How do these individuals acquire these necessary leadership skills, knowledge, and abilities in the midst of change efforts?

Our four different educational contexts will provide the backdrops as we answer this important question. Given the breadth and diversity of approaches that are captured in our four case studies, we have determined to give all four cases coherence by focusing on two aspects of educational leadership development: the methods and mechanisms that each of us have employed to prepare others for change work; and the individuals who are then prepared by these methods and mechanisms to lead.

1.2 Relevant Literature

Published literature in the field of leadership development is broad and often focuses on development targeted to individuals in formal leadership roles [3]. In addition, models of leadership are often based in research in business fields [4], rather than in higher education or even academics.

For the purpose of this chapter, we direct readers to the work of Kezar and Lester [5], who study the development of leadership capacity in academic faculty and staff who are not in named leadership roles.

Kezar and Lester provide a series of stories that show how the leadership journeys of these individuals often result in greater equity on campus and create a higher level of engagement with stakeholders. These stories provide important examples of informal leadership development that can serve readers of this chapter. In addition, Williams et al. [6] illustrates the principles of leadership development for faculty involved in a national engineering education change project. It showed that individuals provided with some guidance and opportunities to join a community of practice can adopt strategies that serve them in their academic change projects. Additional literature relevant to leadership development is cited in each case discussed in this chapter.

1.3 Four Cases

Each case presents a different approach to developing educational leadership in diverse contexts.

In case 1, Professor Xiangyun Du, Director of the Aalborg UNESCO Centre in Engineering Science and Sustainability, Aalborg University, Denmark, explains her collaborations with individuals both at her home institution and abroad to improve capacity for change.

Globally and in the past few decades, she has observed ongoing change initiatives in higher education institutions. These were designed to support graduates and prepare them for real world challenges and the demands of their professions and society. She cites, with example, the need for graduates to solve complex problems in interdisciplinary project team, manage stress and conflicts, and address sustainability [7]. Senior leadership often play a decisive role in these change efforts, involving curriculum revision, digitalized support and adopting new teaching methods [8].

In case 2, Professor Abimbola Windapo, Deputy Dean of Postgraduate Studies at the Faculty of Engineering and the Built Environment, University of Cape Town, South Africa, shares her work with faculty at different levels within her institution. Taking a bottom-up approach to dealing with issues in education, Professor Windapo focuses on developing change leadership skills in the service of improving student learning and faculty teaching. As a result of her efforts, she sees that faculty have successfully created a more structured and efficient educational system that promotes student progress toward degree attainment, improves retention and completion rates, and creates vibrant resources and support for training.

In case 3, Professor Beverley Gibbs, Director of the Dyson Institute of Engineering and Technology, Malmesbury, Wiltshire, United Kingdom, draws on her change management experience in two different higher education institutions: a research-intensive large engineering faculty in a multi-faculty university, and new small, specialist institutions conceived and launched to develop 'work-ready' engineering and technology graduates.

She explores collaborative change management strategies that bring together staff, students, and employers to ensure engagement that improves both learning and teaching.

In case 4, Dr. Julia M. Williams, Professor of English, Rose-Hulman Institute of Technology, Terre Haute, Indiana, USA, reports on her work in conjunction with the National Science Foundation's Revolutionizing Engineering Departments (RED) program and the opportunities afforded to impact faculty, graduate students, post-docs, and administrators who are engaged in change work. Prioritizing practical approaches that are research-based, Professor Williams shows how RED projects function as a mechanism for change while enhancing the skills of individuals to improve education on their home campuses.

All four cases in this chapter are focused on improving education and enhancing the learning experience for students. They all collaborate with various stakeholders, including faculty, students, and industry partners, to bring about change. The leaders in each case are actively involved in identifying and addressing the needs and challenges of their respective institutions and recognize the importance of developing leadership skills in others to create lasting change and improve the educational system. However, the leaders in each case have different titles and roles within their institutions, with different foci and responsibilities.

While Case 1 focuses on a global perspective, Case 2 focuses on change management in a specific institution and Case 3 looks at two different types of institutions, a research-intensive and a specialist institution. Case 4 examines change management among multiple institutions who are all funded by the same government program.

2 Case 1: Professor Xiangyun Du, Director of Aalborg UNESCO Problem Based Learning Centre in Engineering Science and Sustainability, Aalborg University, Denmark

Currently Dr. Xiangyun Du serves as the Director of Aalborg UNESCO Problem Based Learning Centre in Engineering Science and Sustainability. It has an overall mission of facilitating universities and other higher education institutions taking an active role in educating engineers and scientists to contribute to the development of sustainable solutions to the present and emerging social, economic, and environmental challenges. To do so, the center she leads organizes and participates in a variety of activities aiming at the dissemination of engineering education research, capacity building and academic staff development, and in collaboration with local, regional, and international partners.

Following the long tradition in Aalborg University over five decades, a problem-based learning approach has been adopted in all the ongoing works as goals and means.

One of the key initiatives made by the university is to support the local development of academic leadership with collaborators. Over the decades, the center's international projects have involved diverse activities working on how change may take place locally with the support of international collaboration.

Among these efforts, one of the often-observed initiative is to organize pedagogical development activities to support educators, who are usually focused on their disciplinary expertise, to implement new teaching related initiatives.

In the past decades, through collaborative work with international colleagues, Dr. Du has been involved in over 30 projects across 20 countries supporting university educators, in particular, engineering educators to develop problem and project-based learning. In general, most of such cross-border projects were initiated by the visionaries of the local universities' senior leadership involving external experts (e.g. colleagues from the center) to organize professional development activities that range from three days to one year in length. These activities, following the principle of learning by doing and experiential learning, have been organized in a project-based learning. The idea was to provide educators a first-hand experience as learners. These activities in turn have produced immediate improvement of these educators' practices by designing and implementing project-based learning in their classrooms [7, 9]. Nevertheless, their evaluation studies have documented a list of challenges for educators in general to negotiate their own change practices. The most reported challenge across all these international projects over decades was related to insufficient institutional support within their departments or study programs [8, 9], This internationally shared challenge is often related to the lack of understanding of the targeted goals demonstrated by the middle leadership and what role they shall play in a change initiative that's focused on teaching and learning advancement [8, 9]. These identified challenges drive Dr. Du's current collaborative work. She has included more actors, other than the frontline of teachers. The mid-level leaders are involved to support their development, so they not only understand the educational situations of their faculties, but also encourage them to take action proactively to make change happen.

Academic middle-level leaders—such as associate deans and department heads—play a vigorous role in the management of higher education institutions by connecting the senior leadership's visions with academic practices involving educators and students [10]. Despite such important roles, it has been often observed that many middle leaders are appointed to their positions because they are experienced academics, but they are seldom provided with formal preparation [11, 12]. In a recent systematic review study, Maddock [13] pointed out that little is known about the role of middle leadership in teaching and learning advancement.

In their collaborative work on academic leadership, Dr. Du and colleagues conceptualize that leadership in higher education should be adaptive, communicative, distributed, shared, and collaborative, instead of being bureaucratic, controlled, and linear [8, 12].

Highlighting the characteristics of interactive and proactive leadership, she has written elsewhere that academic leadership is conceptualized as a complex adaptive system. Her work also highlights the negation of middle leaders within multi-faceted systems connecting multiple intrapersonal, relational, and institutional factors [8].

The following example illustrates how Dr. Du has been involved in targeting change agents' development in Poland [14].

Although it has been undergoing structural changes to address the social and economic demands, the higher education system in Poland is still considered as in need for improvement in the way it isolates academics from real world. Polish institutions of higher education are criticized for their insufficient attention to teaching and learning achievement, emphasizing instead research achievements over teaching, which results in the fact that many academics prioritize research positions over teaching work [14]. In response to the current critique of high rate of unemployment related to unpreparedness for work after higher education, a national level of improving quality of higher education through innovative teaching is being called for by the Ministry of Science and Higher Education [14]. As a result, an agreement was made to organize a study visit for selected middle-level leaders in Polish institutions to universities across Europe who are notable for successful teaching and learning innovation. Aalborg University participated in this partnership with a series of initiatives supporting long-term meaningful change.

One of the first steps was for the UNESCO Problem-Based Learning Centre that Dr. Du leads to organize workshops with 36 middle-level leaders from around 30 institutions across Poland. The purpose of the workshops was to assist these leaders in learning change management to support their efforts to implement problem-based learning on their own campuses. Through these interactive and participatory workshops, middle leaders were invited to share their views of the local challenges, problems, and needs for change, which was followed by designing roadmaps for change to advance teaching and learning. These maps employed different conceptual frameworks provided by the workshop organizers. This encouraged the participants to reflect collaboratively and individually on their leadership roles, anticipated challenges in change implementation, and coping strategies for successful change management. A research-based approach following design-based research [15] principle was employed to thread the workshops.

As Fig. 1 shows, a qualitative study was used to start the workshop, inviting middle leaders to reflect individually on their views on a change roadmap [14]. This activity was followed by discussions with course organizers on theories of adult learning, curriculum development towards problem-based learning and change management in higher education. Following the study outcome, workshop participants were invited to continue their work with others with whom they share collective subjectivity. The result was the creation of consensus regarding feasible and meaningful change road maps for change towards problem-based learning. Through the capture and contrast of individual and collective subjective views on change management, the research outcomes from the workshops provided take-home messages to those participants that capsuled their different but also agreed views regarding sustainable change cross-institutions. For example, they understood the importance of addressing:

(1) creating an institutional culture to support change mindset.
(2) building connectedness and network to motive all change actors to develop a mutual support system.
(3) supporting educators' change readiness individually and collectively in aspects of their shared understanding, motivation, efficacy beliefs, competences, and enactment to actions; and
(4) supporting trust building cross diverse levels of institution, in their programs, departments, colleges, and their universities.

In addition, a consensus was reached regarding collective agency among all actors to engage in pedagogical advancement, which has been historically neglected due to socioeconomic constraints in Poland. Based on these results, such an initiative

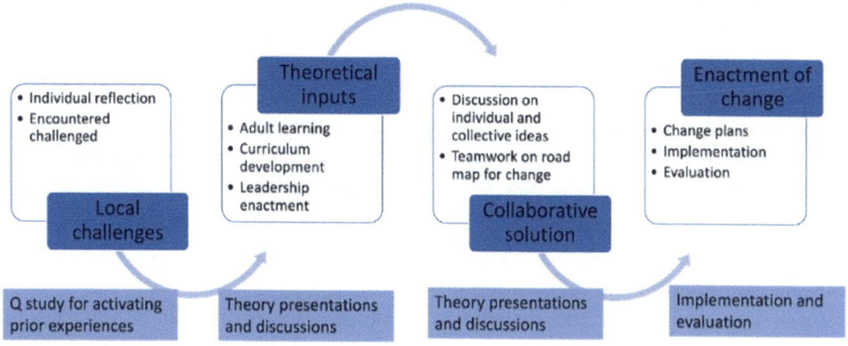

Fig. 1 A design-based research flow of the change management workshop [14]

also provides recommendations that can be used as inspiration across borders and in many societal, cultural contexts regarding the need for systemic and systematic support to be provided to actors to enhance sustainable change. After the workshop the participants were expected to follow up implementing their design and change plan. In the time being, when this chapter was written, Dr. Du and colleagues are in the follow up process with some of the workshop participants.

3 Case 2: Professor Abimbola Windapo, Deputy Dean of Postgraduate Studies at the Faculty of Engineering and the Built Environment, University of Cape Town, South Africa

In her role as Deputy Dean of Postgraduate Studies at the Faculty of Engineering and the Built Environment, University of Cape Town (UCT), Professor Abimbola Windapo promotes academic leadership and drive change in the faculty. She strives to effectively fulfill these responsibilities to ensure the successful implementation of the university's strategic goals and the advancement of postgraduate education in the field of engineering and the built environment.

In the context of the responsibilities of the Deputy Dean of Postgraduate Studies at the Faculty of Engineering and Built Environment (EBE), academic leadership is crucial in supporting the university's strategic goals as outlined in the Vision 2030 of the university. The Deputy Dean contributes to the faculty's strategic goals and provides leadership and direction to advance the postgraduate education agenda. This includes chairing the Postgraduate Planning and Administration Committee (PGPAC) and representing the faculty in various university committees related to postgraduate studies. Additionally, the Deputy Dean collaborates with the Postgraduate Manager to streamline postgraduate processes and ensures effective communication and engagement with postgraduate students, postdocs, and staff. They also serve as a member of important committees such as the Dean's Advisory Committee, EBE Advisory Board, and Faculty Examinations Committee. Furthermore, Professor Windapo's role involves implementing the recommendations of the PGPAC and writing the faculty report for UCT's annual Postgraduate Teaching and Learning report. Professor Windapo also engages regularly with the EBE Postgraduate Student Council and may need to fulfill the duties of the Acting Dean when required by the Dean's Office.

In promoting change leadership development within EBE, Professor Windapo and the staff focused on several specific mechanisms that could help address key challenges, such as high numbers of non-returners, low throughput rates, and problems with the examination and registration processes.

First, they developed and distributed guidelines for postgraduate supervisors. Through an ad hoc committee with inputs from faculty members, comprehensive guidelines for postgraduate supervisors that outline their responsibilities, including setting and enforcing timelines, scoping research projects, providing timely feedback, and improving the communication and collaboration between supervisors and students was developed. This ensures that supervisors have clear expectations and can effectively support students.

Second, the University offers training programmes and workshops for academic supervisors, especially younger ones who lack experience in supervision. These programs guide effective supervision techniques, project scoping, and time management.

Third, the use of a centralized database system, such as the Postgraduate Management Record, helps capture and track relevant information on the postgraduate students. This record highlights student progress, timelines and completion rates and helps supervisors and administrators monitor and manage students' progress more effectively.

Finally, they use communication as a means to ensure transparency. The faculty leadership ensures that relevant information and resources, such as the Engineering and the Built Environment Research-Based Education Handbook, are easily accessible and widely distributed to students and academics. This improves communication and transparency in the postgraduate process.

Through the mechanisms listed above, the following individuals at different levels within the faculty gained skills necessary to take ownership of the change process and drive improvements in the postgraduate space. Academic supervisors, both experienced and less experienced, gained skills in effective supervision, project scoping, and time management through the supervision training programs and workshops. This enabled the academic supervisors to support students more effectively and ensure timely completion of degrees. Faculty administrators were trained to use the Postgraduate Management Record and other administrative processes to track and manage postgraduate students' progress effectively. This improved the efficiency of administrative processes and enabled better student support and transparent records. Additionally, postgraduate students received clear guidelines and expectations from their supervisors and the faculty received clear guidelines regarding timelines and project scoping. This helps them understand their responsibilities and enables them to make timely progress towards their degrees.

Through the bottom-up change leadership approach and by providing training, resources and support, the faculty and university leadership created a more structured and efficient system that promotes student progression, retention and improvement in completion rates.

4 Case 3: Professor Beverley Gibbs, Director of the Dyson Institute of Engineering and Technology, Malmesbury, Wiltshire, United Kingdom

Leaders in UK universities build on the UK's longstanding global reputation in higher education, but are facing continued, and often conflicting, calls for change. These include:

Addressing ongoing expansion of student numbers and calls for fair access and participation.
Pursuit of financial sustainability in a regime of regulated tuition fees that have not kept pace with inflation.
Balancing demanding agendas for research and teaching.
Meeting the evolving needs of employers and economy.
Centralization of 'student voice' and consumerization of the student.
And evolution of a regulatory regime that is diverging across the UK's nation states as it is maturing.

Faculty leadership engaged in engineering education is of strategic importance in a country that has a sustained shortage of graduate engineering graduates and technicians in core roles which is estimated to approach 60,000 per year [16], and yet are subject to sustained criticisms from engineering employers that graduates leaving universities are not "work ready."

Engineering education leaders, therefore, find themselves in a challenging position of having to teach at scale, in a challenging financial context, and introducing the spirit of signature pedagogies such as work-based learning, problem-based learning and authentic assessment that will enhance work-ready, and future-ready engineers.

Leading curricula change in a large, research-intensive university means a number of things practically: building a rationale for change, developing a meaningful vision, empowering different stakeholder communities at the right time and in the right way, making and communicating clear evidence-based decisions, and maintaining a positive tone and momentum.

This case study outlines collaborative leadership strategies across these elements, based on work at the University of Sheffield's Department of Mechanical Engineering in the United Kingdom.

This program of work arose from a concern that the performance of the department's courses in the national subject rankings was falling relative to peer universities. There was a consensus that by building a "bottom-up" mandate for change was critical in coalescing stakeholders in meaningful engagement and generating enough energy to initiate change.

Four months of research with staff, students, employers and sector reports served to generate consensus around a rationale for change; approaches here included exploratory workshops and focus groups, and "blank space design sprints" with students. Comparative analyses of undergraduates' strengths and weaknesses from industrial year-long placement supervisors supported these first-hand observations.

These observations were abled because the employers knew the students first-hand and were able to evaluate them against comparable students from other Universities. Findings here included notable strengths in engineering analysis, and opportunities for development in project conception and management.

This conversation benefitted from the cultivation of an environment where supervisors were able to speak freely and constructively, and this richer approach to industrial involvement was commended in engineering course accreditations at the time.

Existing experience data was mined for insight, and "root cause analysis" workshops were conducted with academic staff [17]. This produced insights that included a description of sometimes passive students who were driven by grades, and difficulties in transition from first- to second-year study. This work was synthesized into compelling change objectives that were based on opportunity rather than critique, specifically: increasing intrinsic motivation in students from 1st Year, improving the Year 1–2 transition, improving employability at end of Years 2 and 3, and transitioning to a large credit structure at Level 7 to reduce assessment loads for students and staff, and to support service teaching to other engineering departments in the Faculty.

A planned stage-gate then secured Teaching Committee consensus on these aims, and 3 options were presented for forward action. Agreement was facilitated, and the case made for the onward resources that would allow the project to move forward with momentum and appropriate contributions. This was the foundation of the project team, comprising 4 thematic curriculum leads (fluids, statics, design and management, dynamics), volunteer students from different years of study, additional curriculum expertise in professional skills and in practical teaching, and Faculty staff who were learning from the process and identifying transferable practice.

The first task of the project team was to develop a meaningful vision in the guise of a learning and teaching ethos [18], each word of which was agreed by the team. It was a novel (but rewarding) step to articulate what was distinctive about the Sheffield University mechanical engineering experience and what the core values were. This approach was shared at the university's annual Learning and Teaching conference in a workshop, with participants later giving the feedback that they adopted the techniques in their own work. Next, to operationalize the ethos, it was synthesized with national accreditation requirements, curriculum design/sequencing reviews by curriculum leads, the pre-existing curriculum practical learning outcome framework, and the University's Graduate Attributes into a unified draft of 20 programme learning outcomes that spanned knowledge, intellectual ability, practical skill, and personal development, each of which are expressed at 4 stages across program years. This was extremely innovative in the sector, took a great deal of negotiation and facilitated prioritization, and was the most challenging part of the entire change process.

This challenge was primarily founded on managing the granularity of these learning outcomes—too coarse, and a founding module may not do the work needed to underpin subsequent learning; too fine, and the learning outcome starts to become a syllabus statement.

The skilled leader at this juncture will be orienting staff towards the learning process and be building alignment between different curriculum areas and experts.

At the University of Sheffield, it was a worthwhile investment given that the Project Team was committed to making this learning outcome framework a fixed point of reference for the programme in subsequent module design and assessment.

Student communities were present in the work in multiple ways [19]. The project approach carried an expectation that students were long-term collaborators in the Project Team and Gibbs worked hard to ensure students remained actively engaged and had a clear voice in the process. She chaired meetings to that effect and supported students in developing their capacity to contribute effectively. This staff-student collaborative work has been published and shared in several different fora across the sector and students co-presented with some members of the project team at the annual Learning and Teaching conference.

This engagement and co-design do not negate or diminish the role of students as partners in the formal governance processes. In fact, the and the Student Union and elected student representatives played an important role in supporting a formal consultation regarding in-session changes.

5 Case 4: Dr. Julia M. Williams, Professor of English, Rose-Hulman Institute of Technology, Terre Haute, Indiana, US

In her academic change work in STEM education, Dr. Julia Williams has found opportunities to work with colleagues at a variety of US colleges and universities who are striving to make change at the course, curriculum, program, department, and college levels. Notable among these endeavors is the Revolutionizing Engineering Departments (RED) Participatory Action Research (PAR) Project, a collaboration with the Center for Research in STEM Equity at the University of Washington, funded by the National Science Foundation.

At the time of this writing, the RED project has funded 26 projects at 23 institutions of engineering higher education in the US with the express purpose of transforming education in the middle years of four-year engineering degree programs. The purpose of REDPAR is to conduct research on academic change with the RED teams and to translate this research into practical, applied strategies that support the work of change agents.

Dr. Williams' work with RED teams has focused on understanding the challenges they face when they develop and implement their change projects. For example, many teams who take on change projects often develop a vision for the change they seek, and this vision is usually the product of the original team that writes the funding proposal. Once the project begins, however, the RED team members frequently neglect to include their vision or the proposed vision of other stakeholders in the creation of a shared vision for change. Using higher education research and emphasizing practical needs, Dr. Williams and her colleagues designed tools that can help teams understand the need for frequent review and revision of their vision in change plans. These tools are captured in REDPAR Tip Sheets that are focused guides to understanding specific change issues and offer strategies for addressing them [REDPAR Tip Sheets, academicchange.org]. The work is also disseminated to the engineering education community through publications in book chapters and research journals [6, 20, 21].

While Tip Sheets provide practical guides to making change, they are not the entire story. RED teams comprise a community of practice that meets regularly online and in person. At these meetings, team members have the opportunity to learn new tools and share their experiences with the challenges of making change. For example, each year the RED community meets in person for a two-day conference. There, teams can share the best practices that emerge from their projects. For example, RED team members from the University of Connecticut Department of Civil Engineering have focused their efforts on supporting neurodivergent students in STEM education. UConn's best practices have been adopted by several other RED teams, thereby supporting both students and faculty who may face similar challenges. Likewise, RED team members from both Clemson University and Montana State University have developed embedded communication practices that can help students improve their writing and speaking in advance of graduation. All of these best practices were shared during sessions at the RED Consortium Meetings. Attendees at the meetings also learn together in a plenary session developed and lead by the REDPAR team. The purpose of the plenary session is to introduce a research-based change tool that they can then practice and apply to their own campus contexts. In 2022, for example, the REDPAR team adapted our Strategic Partnerships Tip Sheet and asked teams to depict potential partners for their projects, using a map metaphor. The results helped teams to understand that there were stakeholders for their projects outside of their departments. These maps also reflected the dominant metaphors of their projects. The RED team from Clemson University used the image of an arch that is part of the visual identity of the civil engineering department and adapted it so they could depict specific partners, like the Student Life staff and their Industrial Advisory Board. By identifying partners and relating them to their RED project, each participating team had the opportunity to consider individuals, departments, and organizations who had not yet been engaged with their project, but who probably should.

6 Reflections by Way of a Conclusion

As leaders approach their own development, we provide the following reflections that can help:

- Consider a systems-thinking approach to support change, ideally as local agents are proactive [8, 22]. Addressing the complexity of change in practice, a systems-thinking approach to change supports change with interdependent strategies, guided by a framework. This includes a collaborative, cross-disciplinary team comprising pedagogical development coordinators and core people from different health disciplines, a coordination team aligning with middle leadership (associate deans, department heads, program leaders, etc.) to balance top-down and bottom-up changes, and a larger group of educators to collaboratively conduct educational research to document and guide their actions.

- The bottom-up change leadership approach described in this chapter suggests that for change to be implemented in higher education institutions in the postgraduate learning space, the university management and leadership have to enable various mechanisms. These mechanisms include developing guidelines for postgraduate supervisors, providing training for academic supervisors, implementing a postgraduate management record, and improving the means of communication and transparency. These are essential in creating a more structured and efficient system and empowering postgraduate supervisors, postgraduate students and administrators to take responsible decisions towards improving student retention, progression and graduation.
- Developing comprehensive guidelines for postgraduate supervisors, ensures that supervisors have clear expectations and can effectively support students. This helps to address the issue of high expectations and provides a framework for adequate supervision. The training programs and workshops for academic supervisors, especially younger ones lacking experience, also contribute to improving supervision techniques and time management skills. Implementing a postgraduate management record helps in tracking relevant information on postgraduate students. This allows supervisors and administrators to monitor and manage students' progress more effectively, ultimately improving retention and completion rates. Also, through adequate communication and transparency that makes relevant information and resources accessible to students and academics, the faculty leadership ensures everyone is on the same page and understands their responsibilities, leading to better collaboration and more timely progress towards degrees.
- Through her work at the University of Sheffield, Professor Gibbs reflects that those principles of the changes introduced held fast throughout her own departure, and through the tumultuous years of COVID-19. On reflection, the process followed showed a good deal of similarity with John Kotter's 8 step model for change implementation [4]. This was not a conscious following of the model, and the intuitive nature of Kotter's 8 steps has been observed elsewhere [3]. The enhancement programme at the University of Sheffield took place in a department that was already successful, which was populated by experienced and long-serving academics who were used to collaborating productively with one another. Gibbs' more recent change projects have taken place in quite different contexts: new UK higher education institutions (less than 10 years old) such as the New Model Institute for Technology and Engineering (NMITE) and the Dyson Institute for Engineering and Technology. In these types of HEIs, curriculum development work is often running in parallel with work to register and/or fully establish the institution in its most formal sense, and with teams who are still learning to work together and who are often in the 'forming' and 'storming' phase [23] of their

own group development. In these contexts, the degree to which the deep collaborative work described in the case study can be replicated undergoes significant challenge. New HEIs have as many features of an entrepreneurial start-up as they do a mature university. They are typified by smaller teams (e.g. an institution of 40–50 staff), high rates of expansion and by definition are learning rapidly and testing their processes. In this context, the effective leader must start out with an honest appraisal of their context and understand where they are on the strategic continuum [24] before consciously choosing a change approach. Failing to take this preparatory step risks tokenistic engagement and wasted precious time and resource.

All of these case studies illustrate the importance of supporting academic staff, administrators, and faculty with the change tools they need in order to effect change in their local contexts. Unfortunately, the work of developing leadership skills occurs, in many cases, simultaneously with the change work. This leads us to reflect that we could do more to prepare individuals for the change leadership roles they will take on. There can be no doubt that higher education in Europe, Africa, Asia, and the United States is undergoing profound change, but disciplinary training takes precedence over leadership development in nearly every graduate program and in every discipline. It is interesting to think how the experiences of individuals could be transformed if some development occurred during their graduate study. In doing so, we could aspire to a more complete preparation of the next generation to assume leadership in the higher education issues that will only proliferate in the coming decades.

References

1. Grant J (2021) The new power university. Pearson Education, Harlow, UK
2. Heimans J, Timms H (2014) Understanding "New Power." Harv Bus Rev 92(12):48–56
3. Brock J, Peak K, Bunch P (2019) Intuitively leading change: completing a kinesiology department-to-school transformation using Kotter's 8-stage change model. J Phys Educ 6:14–24
4. Kotter JP (1996) Leading change. Harvard Business School Press, Boston MA
5. Kezar AJ, Lester J (2011) Enhancing campus capacity for leadership: an examination of grassroots leaders in higher education. Stanford UP, Palo Alto, CA
6. Williams JM, Andrijcic E, Mohan S, Margherio C, Litzler E, Swan AL (2022) Faculty development for academic change in the national science foundation revolutionizing engineering departments (NSF RED) context. In: Linder SM, Lee CM, Stefl SK, High KA (eds) Handbook of STEM faculty development. Information Age Publishing. https://www.infoagepub.com/products/Handbook-of-STEM-Faculty-Development
7. Du XY, Kolmos A, Ahmed MAH, Spliid C, Lyngdorf N, Ruan YJ (2020) Impact of a PBL-based professional learning program in Denmark on the development of the beliefs and practices of Chinese STEM university teachers. Int J Eng Educ 36(3):940–954. https://doi.org/10.1177/17411432231206247
8. Chaaban Y, Qadhi S, Al-Thani H, Floyd A, Du XY (2023) Supports and constraints to middle leadership development in higher education: a Q-methodology study. Educ Manag Admin Leadership 1–22. https://doi.org/10.1177/17411432231174092

9. Du XY, Naji KE, Ebead U, Ma JP (2021) Engineering instructors' professional agency development and identity renegotiation through engaging in pedagogical change toward PBL. Eur J Eng Educ 1–23. https://doi.org/10.1080/03043797.2020.1832444
10. Youngs H (2017) A critical exploration of collaborative and distributed leadership in higher education: developing an alternative ontology through leadership-as-practice. J High Educ Policy Manag 39(2):140–154. https://doi.org/10.1080/1360080X.2017.1276662
11. Floyd A, Qadhi S, Chaaban Y, Al-Thani H, Du X (2023) An integrated systems model for understanding experiences of academic leadership development in Qatar. J High Educ Policy Manag. https://doi.org/10.1080/1360080X.2023.2225148
12. Qadhi S, Du XY, Chaaban Y, Al-Thani H, Floyd A (2023) The role identities of women middle management academic leaders in STEM Higher Education. Eur J Eng Educ 1–17. https://doi.org/10.1080/03043797.2023.2263377
13. Maddock LC (2023) Academic middle leaders, middle leading and middle leadership of university learning and teaching: a systematic review of the higher education literature. J High Educ Policy Manag 1–36. https://doi.org/10.1080/1360080X.2022.2160888
14. Du XY, Guerra A, Chen JB, Lindsay E, Nørgaard B (2023) Exploring academic middle leaders' viewpoints on supporting change in higher education institutions using Q methodology—a case in Poland. Educ Manage Adm Leadersh 1–26. https://doi.org/10.1177/17411432231206247
15. Anderson T, Shattuck J (2012) Design-based research: a decade of progress in education research? Educ Res 41(1):16–25
16. State of Engineering Report (2018) www.engineeringuk.com
17. Gibbs B, Tarnowski K, Wood GC (2018) Poster presentation: integrative projects supporting capability-driven curricula. New approaches to engineering higher education (Engineering Professors' Council/Institution of Engineering and Technology), London
18. O'Neill G (2015) Curriculum design in higher education: theory to practice, 1st edn. In: Chapter 3: Needs analysis and educational philosophy. UCD Learning & Teaching, Dublin, Ireland
19. Wood GC, Gibbs B (2018) Students as partners in the design and practice of engineering education: understanding and enabling development of intellectual abilities. In: Malik M (eds) Proceedings of the engineering education research network annual symposium
20. Margherio C, Doten-Snitker K, Williams J, Litzler E, Andrijcic E, Mohan S (2020) Cultivating strategic partnerships to transform STEM education. In: White K, Beach A, Finkelstein N, Henderson C, Simkins S, Slakey L, Stains M, Weaver G, Whitehead L (eds) Transforming institutions: accelerating systemic change in higher education. Pressbooks. https://openbooks.library.umass.edu/ascnti2020/chapter/margherio-etal/
21. Doten-Snitker K, Margherio C, Litzler E, Williams J, Ingram E (2020) Developing a shared vision for change: moving toward inclusive empowerment. Res High Educ. https://doi.org/10.1007/s11162-020-09594-9
22. Du XY, Guerra A, Nørgaard B, Chaaban Y, Lundberg A, Lyngdorf NE (2022) University teachers' change readiness to implement education for sustainable development through participation in a PBL-based PD program. Sustainability 14:12079. https://doi.org/10.3390/su141912079
23. Tuckman B (1965) Developmental sequence in small groups. Psychol Bull 63(6):384–399
24. Kotter JP, Schlesinger LA (2008) Choosing strategies for change. Harvard Bus Rev 86(Iss. 7)

Creating Quality Learning Experiences

Transformation to Outcomes Based Education

Prakash Tewari, B. Kanmani, David Knight, and Marlene Kanga

The rules of the world are changing. It is time for the rules of teaching and teachers' work to change with them
—Andy Hargreaves

Abstract This chapter explores the transformative journey of engineering education towards Outcomes Based Education (OBE). As the global landscape of education evolves, there is an increasing emphasis on outcomes and student-centric learning. The shift from traditional education models to OBE is a transformative response to the demands of a changing world and an ever-growing need for adaptable, competent graduates. OBE aligns education with industry needs and societal demands, such as—ethical responsibilities, environmental sustainability, safety, and reliability. There are also aspects like economic consideration, innovation and technological advancement, regulatory compliance, globalization and cultural sensitivity, etc. that OBE takes into account so that it aligns with the said societal demands. It also emphasizes holistic development that falls under the category of cultivation of well-rounded skill set and mindset that include components such as technical proficiency, problem-solving, communication skill, leadership and teamwork, ethical awareness and social responsibility, project and finance management, lifelong learning skill, etc. All these includes critical thinking, like the ability to analyze, evaluate and solve complex engineering problems. OBE helps shape educational systems to produce well-prepared graduates for the challenges of the twenty-first century. Here, four case studies, drawn from various educational institutions and settings worldwide are presented. These case studies provide practical insights by looking into the curriculum design, delivery,

P. Tewari (✉)
KLE Technological University, Hubballi, India
e-mail: pg_tewari@kletech.ac.in

B. Kanmani
Department of Electronics and Telecommunication Engineering, BMS College of Engineering, Bengaluru, India

D. Knight
Department of Engineering Education, Virginia Tech, Blacksburg, USA

M. Kanga
Former President World Federation of Engineering Organizations Australia, Paris, France

R. Kandakatla et al. (eds.), *Academic Leadership in Engineering Education*, Lecture Notes in Networks and Systems 1097, https://doi.org/10.1007/978-3-031-68282-7_6

assessment, and attainment. Their alignment for the success of the implementation of OBE, along with examples that connect Outcomes-Based Education to pressures to determine 'What Works.' The chapter also presents innovative approaches, such as how the global student Hackathon could be used to demonstrate the achievement of graduate attributes in engineering programs and key skills that are valued by employers. The evidence of success and challenges in implementation is presented through case studies and equips educators, administrators, and policymakers with valuable tools to navigate the journey towards OBE effectively. The aim is to ensure continuous improvement of educational systems and preparing students for a rapidly changing world.

Keywords OBE · Outcomes · Engineering education · Program assessment · Project based learning · Outcomes assessment · Competencies · Performance indicators · Graduate Attributes

1 Introduction

1.1 Outcomes Based Education

Outcomes-based education (OBE) is a student-centered curriculum design and instruction model that focuses on what students can do after being taught. OBE shifts the focus in education from teaching to learning.

According to Spady [1],

> Outcomes-based education means starting with a clear picture of what is important for students to be able to do, then organizing the curriculum, instruction, and assessment to make sure that this learning ultimately happens.

Academic leadership plays a crucial role in implementing OBE in educational institutions, as it requires vision, strategic planning, resource allocation, stakeholder engagement, and a commitment to continuous improvement. By providing direction, support, and guidance, academic leaders of educational institutes can help create an environment where OBE thrives and contributes to student success and institutional excellence.

Need for OBE

Outcomes-based education, a performance-centric methodology, has emerged as a significant reform framework in the international education scenario.

Nations that seek signatory status at the International Engineering Alliance for multi-lateral mutual recognition of engineering qualifications, i.e., the Washington Accord (WA), Sydney Accord (SA) or Dublin Accord (DA), must implement Outcomes Based Education [2].

This commitment signifies the dedication of the engineering education system to ensure long-term excellence in preparing engineers who are well-equipped for professional practice in the industry. It is mandatory for national accreditation agencies that are signatories to the International Engineering Alliance Accords to demonstrate that their accreditation process is based on outcomes-based practices by their engineering institutions.

Relevance of OBE

Outcomes-Based Education (OBE) is highly relevant in the context of academic leadership in engineering education. OBE ensures that program outcomes align with accreditation standards and that students are adequately prepared for their careers in engineering. OBE promotes transparency in education by clearly articulating the expected outcomes to all stakeholders, including students, faculty, employers, and accrediting bodies. It also incorporates OBE principles that encourages data-driven decision-making which empowers academic leaders to make well-informed choices regarding resource allocation, faculty development, and curriculum adjustments. Finally, OBE can ensure that students have the skills and competencies necessary to work in diverse international settings. Diversity in the context of OBE refers to various dimensions of diversity among students, faculty, and other stakeholders within the educational environment. The diversity could be cultural, demographic, cognitive, and interdisciplinary. OBE seeks to cultivate diverse skills and competencies that empower students to succeed in their academic pursuits, professional careers, and personal lives while promoting inclusivity, equity, and social responsibility.

2 Case Studies

In this section, four case studies are presented.

- The first case study considers Outcomes-Based Education within broader systems.
- The second deals with OBE through curriculum.
- The third is about the examination reforms that are required to ensure the success of OBE implementation.
- And the fourth provides an example of an innovative approach to demonstrate student learning and the effectiveness of Outcomes Based Education through the World Engineering Day Hackathon, an international competition for engineering students.

2.1 Case Study 1: Thinking About Outcomes-Based Education Within Broader Systems

This set of examples considers some of the additional pressures and considerations related to outcomes-based education within the United States higher education context.

As the costs of higher education have increasingly been borne by individual students and their families as opposed to taxpayer dollars, colleges and universities have been asked to provide measurable evidence of the added value for obtaining a postsecondary education credential (e.g., [3]). There has been a renewed focus on asking higher education programs and institutions to find ways to measure its effectiveness and progress (e.g., [3–6]). Thinking about the educational experience as a system is one way to con-nect the outcomes-based education movement with the pressures facing colleges and universities to demonstrate "what works."

Terenzini and Reason's framework [7, 8] brings together an enormous set of literature within higher education research to demonstrate that students' precollege experiences help shape their engagement within their institutions. A wide variety of curricular (e.g., major coursework, general education coursework), classroom (e.g., instructional practices, climate), and out-of-class experiences (e.g., student organizations) all encapsulate students' engagement in their college experience, which directly influence educational outcomes. All of those experiences are shaped by organizational factors (both at the institutional level and program context), which refer to practices, policies, cultures, and structures.

Figure 1 depicts this logic using outcomes associated with a U.S.-based national-scale study [9] focused on learning outcomes of undergraduate engineering programs that were inspired by the National Academy of Engineering's Engineer of 2020 report [10]. However, any set of learning outcomes could be inserted into that right-hand column based on a program, institution, or national body's interests.

Examples Connecting Outcomes-Based Education to Pressures to Determine "What Works"

Using the framework in Fig. 1, there have been many efforts seeking to explain the variety of educational experiences that relate to different educational outcomes. For example, accounting for differences in students' pre-college characteristics, Lattuca et al. [11] focused on the student experiences related to students' interdisciplinary skills.

The researchers analytically linked together different variables in Fig. 1 and found that explicit curricular emphases on interdisciplinary topics and skills as well as certain kinds of out-of-class activities (i.e., study abroad, non-engineering organizations, and humanitarian engineering projects) related to students' interdisciplinary skills most strongly. Importantly, these relationships were moderated by faculty members' views about interdisciplinary education within engineering.

Any outcome could be swapped into this kind of approach to understand an educational system.

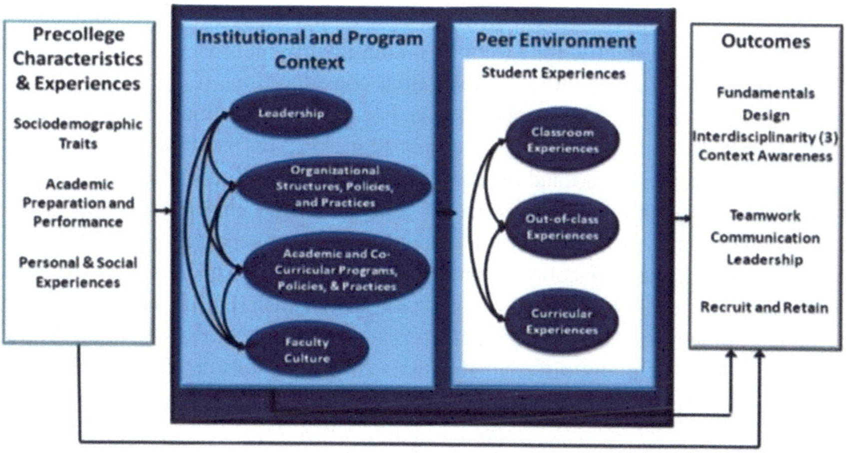

Fig. 1 Example systems-level framework to explain student outcomes [9]

For example, Knight and Novoselich [12] analyzed students' leadership skills and showed that different kinds of curricular emphases (i.e., emphasis on core engineering thinking, professional skills, and broad and systems perspectives) related most strongly to leadership skills.

Palmer et al. [13] focused on contextual competence as an outcome (i.e., students' abilities to consider a range of contextual factors when designing solutions) and showed that students who enrolled in programs offering an entrepreneurship minor as well as students who participated in engineering-related professional organizations, humanitarian engineering projects, and non-engineering organ izations and service work reported higher levels of contextual competence.

These studies and others (e.g., [14, 15]) serve as examples of how the range of in class and out-of-class experiences relate to educational outcomes at the program, institution, or national-scale—in other words, answering the question of "what works" within outcomes-based education discussions.

Thinking about the set of outcomes themselves as a system is another way to conceptualize the right-hand column. Knight [16] created a typology of undergraduate engineering students based on the combination of outcomes listed in Fig. 1. Students who were well-rounded across different outcomes had different educational experiences than the students who were less balanced across the different outcomes (e.g., high in engineering and design skills but less proficient in professional skills).

These examples demonstrate that programs and institutions can consider the set of learning outcomes that aim to help students develop holistically and interdependently as well as discretely.

Maintaining Focus on Learning Outcomes at Different Organizational Levels

Lattuca and Stark's Academic Plan Model [5] emphasizes that academic plans all sit within a broader context, and those influences within an institution, discipline, or the surrounding sociocultural context all affect educational experiences and outcomes—either directly or indirectly, for example, taking faculty members' time in different directions. These kinds of broader contextual considerations (e.g., an institution's emphasis on research versus teaching) have been shown to influence learning environments within undergraduate engineering (e.g., [17, 18]), and so it is critical to maintain a focus on educational outcomes throughout the organization.

In addition to thinking about students' undergraduate engineering experiences and outcomes as a system in the above examples, it is also very important to consider that the undergraduate educational enterprise sits within a broader system (i.e., the institutional environment) that has a wide variety of different aims. For example, the undergraduate education mission at a research-intensive university in the United States is just one of many different functions of the institution. These universities also emphasize research, economic development activities, graduate education, and service and out-reach in communities, to name a few. Finding ways to ensure that outcomes-based undergraduate education is considered alongside other priorities at the college- or university-level is very important, but the level of specificity will look different. Many U.S. colleges of engineering focus on a scorecard or series of metrics within a strategic plan.

As an example, from the College of Engineering at Virginia Tech, some of the undergraduate education outcomes tied to strategic plan monitoring include the following:

One and two-year student retention rates within engineering at the undergraduate level.

Four- and five-year graduation rates within the college of engineering.

Overall undergraduate engineering students' satisfaction with their educational experience at the time of graduation.

Including educational outcomes alongside other metrics, such as college-level research expenditures and scholarly productivity, revenue availability, and faculty and staff hiring and retention data.

These helps academic leaders maintain a view of that part of the mission which they consider for their resource allocations and workloads. The educational outcomes are of a lower specificity than what was previously described, but if aligned well with program- and course-level educational outcomes, should be useful indicators of when greater attention and more resources are needed for undergraduate education. Stepping down one level of the organization, outcomes-based assessment at the academic program level (e.g., mechanical engineering at Virginia Tech) might focus on outcomes more specifically like in Fig. 1 and relate development of those outcomes to the broader range of experiences shown in Fig. 1. This is the kind of approach that we typically use in ABET visits—we adopt a systems-level view to connect experiences to different program outcomes. And then stepping down one more organizational level to a course, a teacher may pay close attention to specific concepts in its outcomes-based assessments. In evaluating a course following the systems logic of Fig. 1, a teacher may be able to link students' experiences in specific lecture sessions to demonstration of different learning outcomes on a final exam. The degree of specificity of the outcome changes based on the level of the organization, but the logic of the systems approach can be applied to each one.

In summary, these examples are aimed at helping readers think about outcomes-based education as being one component of a larger system. In the U.S. context, and in many others, there have been pressures to demonstrate the value-add and effectiveness of higher education—essentially, determining "what works." These examples demonstrate systems approaches to link the outcomes that we care about in engineering to a variety of different student experiences. It is also important to think about how undergraduate engineering sits within a broader system of higher education as well. Depending on the level of the organization (e.g., course, program, college/faculty), the specificity of the educational outcome changes, but the logic remains the same. In the case of the broadest college/faculty level, maintaining some focus on educational outcome is critical to be sure. Sufficient resources can flow toward that mission of the university amidst a wide range of often-competing demands. As academic leaders, we must think about outcomes-based education within broader systems.

2.2 *Case Study 2: OBE Through Curriculum*

Having seen the broad framework of the OBE process, in this case study, we demonstrate an example of implementation of OBE through the Curriculum, as it is an important component for ensuring quality of graduating engineers.

The National Board of Accreditation (NBA) has defined 12 Program Outcomes (POs) for all engineering programs and are directly aligned to the 12 Graduate Attributes (GAs) defined by the Washington Accord (WA) [19, 20]. In addition, programs can define 2–4 Program Specific Outcomes (PSOs). Effective implementation of OBE happens through a well-designed Curriculum, together with suitable pedagogy and relevant assessments. While the POs and PSOs are developed through the curriculum, additional development of skills is through student association with co-curricular and extra-curricular activities on campus. Programs effectively implementing OBE are accredited, by NBA. Hence, the first step in the OBE process is the design of the curriculum, followed by defining Course Outcomes (COs) that map to POs/PSOs.

The All-India Council for Technical Education (AICTE), the apex body for technical education in India, prepared a model curriculum for undergraduate and post graduate engineering programs. This model curriculum includes the Program Specific Criteria (PSC) defined by ABET and the 8 elements of Knowledge Profile of WA [21–23]. In addition, AICTE has defined guidelines for the three-week induction program; guidelines for engaging in internships during semester breaks; thrust areas; and guidelines for active engagement with community service during weekends/holiday through the Activity points [24, 25]. Further, the National Education Policy (NEP 2020), by the Ministry of Education, has ensured the development of expected skills of twenty-first century, leading to holistic development and an

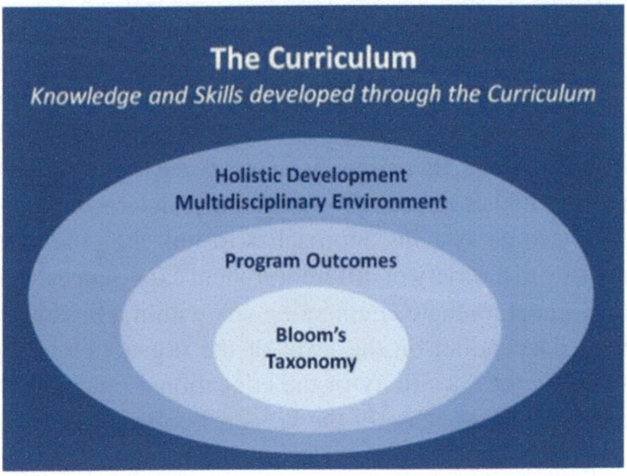

Fig. 2 The Knowledge and Skills to be developed through the Curriculum

ability to perform in multi-disciplinary environment [26]. Hence, programs design curriculum considers these guidelines, as their Vision-Mission, the Program Educational Objectives and the suggestions by the stakeholders; and carefully include courses under various curricular components; to develop the essential knowledge and skills, as indicated in Fig. 2 [27].

The Curriculum: An Example

We now have an example of the Curriculum designed for the Undergraduate Program in Electronics and Telecommunication Engineering, offered to the Batch admitted 2018 [28].

The curriculum includes various curricular components like: Basic Science Courses (BSC), Engineering Science Courses (ESC), Professional Core Courses (PCC), Professional Elective Courses (PEC), Open Elective Courses (OEC), Projects (PRJ), Seminars based on Internships (SMR), Humanities and Social Sciences including Management (HSMC), and Non-Credit Mandatory Courses (NCMC).

The BSC, ESC and PCC develop the required core competency.

The PEC and the OEC provide opportunity for students to select and take up courses based on their passion. One of the strengths of the Curriculum is the inclusion of four projects:

(i) A hardware project to rebuild an existing product/equipment in the V semester.
(ii) Project based on implementing multi-media standards, to address the Program Specific Criteria of ABET [22].
(iii) Project to reproduce an identified research work in the VII semester.
(iv) And the major project in the VIII semester that can be an extension of any of the earlier projects or any other independent project.

Through the COs of the project, all the POs are mapped. The Curriculum requires students to complete 12–16 weeks of internships during semester breaks, which are measured through report and seminar during the semester. Students need to engage in social work during weekends/holidays and earn minimum 50 AICTE Activity points.

To ensure holistic development, the curriculum includes one NCMC every semester, and includes: Kannada Language; English Language; Physical Activity; Cultural Activity; Human Values through Literature; Person ality Development; Communication and Aptitude Skills; two MOOC courses of minimum 4 weeks duration in Engineering/ Management/Science/Liberal Arts.

Pedagogy

There is emphasis on experiential learning of the engineering concept [29]. Accordingly, it is ensured that most courses have a laboratory experience to comprehend the concepts: either through exclusive credits or through pedagogy adopted. Students have been introduced to engineering tools like: C, C++, LabVIEW, Multisim, Matlab, Simulink, Python, Graphical Network Simulator, High Frequency Structure Simulator (HFSS), Keil, Packet Tracer, Hardware Description Language (HDL), and Code Composer Studio. In most laboratory courses, faculty avoid use of in-built functions; and use in-built functions purely to verify the results obtained through the programming tool. The tool is used to understand, apply, design, and analyze concepts of the course.

The assessments become another factor that helps in addressing and measuring the POs/PSOs through the COs of the course. The first three POs can be assessed in the conventional written examination; while it is possible to design laboratory assessments to address and measure PO4 and PO5. The assessments of the laboratory sessions are planned to address the skills; and hence few courses include open-book/ the open-resource examination. The higher POs, PO6 through PO12, are addressed through alternate assessments like course projects; course seminars; programming based assignments; projects; technical seminars; internship based seminars; reproduction of research work; abstract for a technical document; development of a tool-box using the engineering tool and any other.

It is ensured that in every semester, every PO is addressed by at least one course. Hence, the POs and the PSOs are well addressed through the COs of the courses of the curriculum. Since assessments form a key to measure of OBE, the next case study is focused on examination reforms.

Attainments

All courses have equal weightage for the internal assessment and the semester end performance. The COs of the course is mapped to POs/PSOs with a mapping strength: 1/2/3. The method of allocating the mapping strength in the course articulation ma trix is not defined by NBA. In our program, the mapping strength is based on the weightage of the internal assessments in terms of marks allocated, mapping strength of 3 (>20%), 2 (10–20%) and 1 (<10%). This method is measurable, and justifiable, and is not subjective or dependent on the interpretation of the course instructor. The performance of students in various assessments leads to the attainment of the CO. In our program, the attainment depends on the percentage of students securing greater than a set threshold (For example: 60%), multiplied by the CO-PO mapping strength.

The overall attainment of the POs is through the cumulative contribution of attainments of all courses of the Curriculum. The process of computing the attainments is purely towards continuous improvements in pedagogy and assessments; and is not reflected in student Grade cards [30].

Student Performance Through the Curriculum

The Batch admitted in 2018, and graduated in 2022, was offered the above curriculum. The program has an intake of 60 students. Of the 59 students admitted, 90% of the students successfully graduated within the stipulated period of four years. Six students successfully cleared the National level Competitive examination, Graduate Aptitude Test in Engineering (GATE) conducted every year in all major engineering specializations [31]. Two students have secured admission in Georgia Tech College of Engineering and University of Massachusetts Amherst. In addition, students have secured awards/ recognitions like: AWS Educate Cloud Ambassador; Microsoft Learn Student Ambassador; Best Paper Awards; Best Project Awards in contests conducted outside the college. Five student projects have culminated in publications in International Conferences. The placement percentage of the Batch is 83%; with the highest package of Rs. 18.85 Lakhs and average of Rs. 8.5 Lakhs per Annum. Students have secured placement s in organizations like: Samsung Semiconductor India Research, Synopsys, Mathworks, Athena Health Inc, Delloite USI, CISCO, TCS, Infosys, Oracle, IQVIA, Robert Bosch, Tech Mahindra.

The details of achievements of this graduating Batch, together with their experience is published as the Annual magazine [32]. Hence, an effective implementation of OBE, through a well-designed curriculum that ensures all POs/PSOs are addressed, which finally translates to visible student success stories that are made available in public domain through the Annual magazine on the department website.

It is interesting to note that this simple practice has enhanced peer perception; leading to further improvement in the quality of students admitted to the program. Hence, OBE processes can be identified as the strategic plan for imparting quality engineering education.

Success of OBE

The organization structure of the Institution and the department for academic processes leading to collective contribution of all faculty in design and delivery of the curriculum, towards ensuring effective implementation of OBE (ensuring development of POs defined by NBA), has culminated in commendable student performance [32]. The action taken based on attainments of COs/POs has resulted in further improvements in course design/delivery/assessment leading to enhanced student performance. The success of OBE is reflected by the program getting accredited by NBA.

Some Reflections Based on Academic Processes

NBA does not define the method to compute the attainment but emphasizes on the actions taken based on attainments. However, in our program, towards exact computation of attainments, we engaged in micro analysis through mapping of every question in every assessment to the relevant CO, together with performance of every student in every question. This process has proved to be computationally intensive. On retrospection, it is desired to shift the focus to quality assessments, since POs are addressed through the COs, through carefully designed assessments. One can explore simplified methods for computing the attainments, based on assessments and the overall score/letter grade awarded for the course [33].

The program has the practice of defining around six COs for every course. However, on retrospection, it is desired to address four to six POs through every course, together with ensuring every PO is addressed by at least one course every semester. Hence, the emphasis is on the number of POs addressed through the course and not on the number of COs defined for the course.

2.3 *Case Study 3: Examination Reforms*

Reforms in examinations play a pivotal role in enhancing the quality and relevance of international engineering education.

The following case study of reforms in examinations at KLE Technological University explains the context, the process, evidence of success, problems encountered & resources required for its implementation.

Objectives of the Practice

To bring about comprehensive reforms in the assessment of students learning within the evolving landscape of engineering education.

Objectives:

(a) Formulating and adapting Assessment Strategy for Outcomes Based Education
(b) Enhancing the Structure and Quality of Assessment: Incorporating Bloom's Taxonomy Framework
(c) Evolving strategies to assess higher order Abilities & Professional Skills

Adaptation OBE framework requires a different strategy for assessment that ensures realistic measurement of attainment of program outcomes. Further, looking at the needs of the employers, it is imperative that examinations incorporate a suitable emphasis on the assessment of higher-order abilities and professional competencies.

(a) Outcomes based education (OBE)—adapting the OBE framework proves to be a challenging task, particularly in establishing a seamless connection between examination questions and assessment tools with the program's intended outcomes. The lack of clear alignment between program outcomes and assessment tools results in inaccurate and unreliable measurement of students' achievement of outcomes. This disconnect poses a significant obstacle to effectively implementing the Outcome-Based Education (OBE) framework, rendering the entire effort unproductive.

(b) Within the current examination system, the predominant emphasis is placed on memorization. However, it is imperative that the assessment process extends its reach to evaluate higher-order abilities, including the capacity to apply knowledge, solve complex problems, engage in critical analysis and synthesis, as well as design innovative solutions. Moreover, the acquisition of professional skills such as effective communication, teamwork, and the ability to engage in lifelong learning has emerged as pivotal factors in enhancing graduates' employability. Therefore, it is crucial that the examination system accords due importance to assessing these higher-order skills and professional competencies.

The Practice

a. *Formulating and Adapting Assessment Strategy for Outcomes Based Education*

Program Outcomes (POs) are fairly high-level generic goals that cannot be precisely measured. It is a considerable challenge to make these POs observable and measurable at the course level. To bridge the gap between high-level program outcomes, course content & outcomes, and assessment, it is crucial to enhance clarity and specificity in defining program outcomes. This can be achieved through a two-step process that entails identifying Competencies and Performance Indicators (PIs).

i. Identifying Competencies: For each program outcome, establish competencies, which denote the various skills and abilities implied by the program outcome statement. Different competencies often require distinct assessment methods. This step aids in establishing a shared understanding of the competencies we aim for our students to attain.
ii. Defining Performance Indicators: For each identified competency, outline explicit Performance Indicators (PIs). PIs are clear statements that specify what is expected of students in terms of learning outcomes. They serve as measurement tools during assessment, allowing us to know the extent of attainment of outcomes.
iii. Once this process has been completed for the program, the assessment of Course Outcomes for all courses are designed by aligning assessment questions from diverse assessment tools with the established Performance Indicators. Employing this approach, where examination questions are directly correlated with PIs, offers enhanced clarity and a more accurate assessment of COs and POs. The visual representation of this process is depicted in Fig. 3.

b. *Enhancing the Structure and Quality of Assessment: Incorporating Bloom's Taxonomy Framework*

The questions featured in semester-end examinations and internal test papers serve as crucial determinants of the level of learning expected from students in their courses, and consequently, within the overall program.

To consciously align the curriculum and assessment with higher-order cognitive skills, the Bloom's Taxonomy framework has been adopted. Bloom's Taxonomy offers a crucial framework, not just for designing curriculum and instructional strategies, but also for formulate suitable examination questions across different cognitive

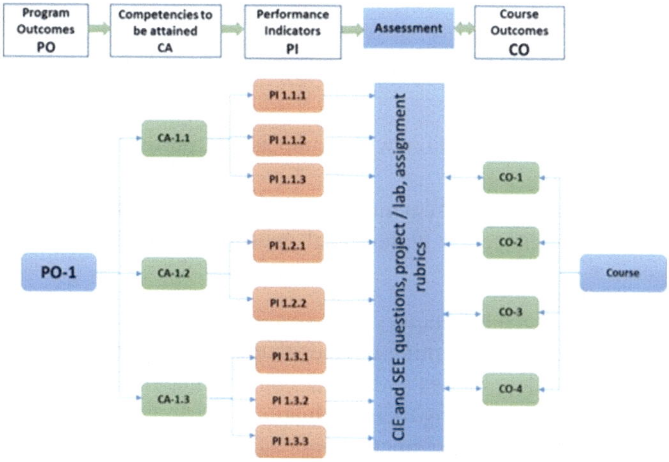

Fig. 3 Connecting program outcomes to assessment [2]

levels. This strategic approach aims to guide educational programs toward nurturing advanced abilities that transcend mere recall and understanding, emphasizing application, analysis, evaluation, and creation.

Conventionally, Continuous Internal Evaluation (CIE) and Semester-End Examinations (SEE) primarily assess the first three cognitive levels of learning (remembering, understanding, and applying), and to a certain extent, the fourth level (analysing). These assessments are typically conducted within time constraints. The evaluation of higher-order abilities such as analysis, evaluation, and creation are better suited for extended coursework or a variety of student projects, including course projects, minor projects, internship experiences, and final-year projects.

c. *Evolving Strategies to Assess Higher Order Abilities & Professional Skills*

The primary challenge in assessing these higher-order abilities and professional skills lies in their difficulty to be learned and evaluated within the confines of our conventional courses and examination system. To confront these challenges, substantial reforms have been initiated in how we structure our curriculum, shape student learning experiences, and assess their outcomes. Examples of these transformative strategies include incorporating projects of core courses, conducting open-ended experiments in laboratory settings, implementing project-based learning approach, embracing blended learning approaches, providing co-curricular learning experiences, encouraging mini/minor projects, fostering final-year capstone projects, and facilitating industry-internship experiences.

These innovative approaches are aimed at enhancing the assessment of higher-order skills and professional competencies, recognizing their vital role in preparing students for real-world challenges and careers.

Evidence of Success

i. Introduction of the new structure of question papers for all program courses with the following:
    - Mapping of each question with competency, performance indicator, and bloom's level
    - Limiting weightage of remembering and understanding level questions to 40% and weightage for higher-level questions to 60% in all the question papers.

ii. Introduction of structured enquiry and open-ended problems in the laboratories.
iii. Enhancing experiential learning courses that create opportunities for students to learn higher-order abilities and professional skills; the student will be doing at least seven projects during their graduation.
iv. The overall increase in assessment weightage for higher-order cognitive skills and professional skills in the program
v. Increase in high-value placements. The average salary increased from Rs. 400,000 in 2016 to Rs. 800,000 in 2022.
vi. Adaptation of these reforms and practices by AICTE, resulting in 'AICTE examination reform policy-2018.'

Problems Encountered and Resources Required

i. Developing clarity about the expectations of POs in the context of program and courses and connecting them to assessment questions.
ii. Bringing clarity and specificity to the PO's so that all the faculty have a common understanding.
iii. Designing of new educational experiences that lead to learning of higher order abilities and professional skills and give an opportunity for assessing them.
iv. Developing rubrics for assessing professional skills.

2.4 *Case Study 4: The World Engineering Day Hackathon: A Case Study for Project Based Learning in Engineering*

Innovative Approaches to Project Based Learning to Enhance Student Engagement and Learning Outcomes

Project Based Learning (PBL) [34] has been increasingly recognized as an effective way to actively engage students in real-world projects and thus enhance their learning experience. PBL is particularly suited to engineering education, where students can be involved in projects that enhance their learning and understanding and expose them to the real-world work environment.

The World Engineering Day[1] (WED) Hackathon [35] is an international competition for engineering students that provides a unique opportunity for project-based learning. The WED Hackathon engages engineering students in the celebration of engineering and provides an opportunity to develop and demonstrate the graduate attributes that have been achieved in their engineering program and key skills that are valued by employers [36]. The student project submissions are judged relative to the recently reviewed International Engineering Alliance (IEA) Graduate Attributes and Professional Competencies (GAPC) Benchmark [37]. It is an opportunity for engineering educators to demonstrate their leadership in implementing the Framework in their programs for graduate outcomes that align with the skills that are increasingly required, such as digital skills and so-called soft skills in a rapidly changing world.

The Benchmark was reviewed between 2020–2021 by IEA signatories, WFEO and its international engineering partners, including the International Federation of Engineering Education Societies (IFEES) and the Global Engineering Deans Council (GEDC), with consultation involving other international networks of consulting engineering companies and women engineers, across 60 countries.

The Benchmark is supported by UNESCO and the World Federation as the pre-eminent model for outcomes-based education. To facilitate understanding and uptake, the Benchmark has been translated into the six UNESCO official languages, English, French, Spanish, Russian, Chinese, and Arabic.

Engineering educators can be confident that the reviewed GAPC Benchmark is a contemporary framework that, when implemented, will ensure that their graduates will have not only the basic engineering skills but also abilities to work in teams, communicate and engage with stakeholders, all requirements from employers.

The GAPC Benchmark requires outcomes where students can demonstrate their ability to think criticality, be innovative, develop solutions to new problems and communicate their solutions effectively. A key requirement is for students to be thoughtful about the impact of their solutions on society, the environment and the economy, particularly in the context of advancing the UN Sustainable Development Goals. Engineering students who participate in the WED Hackathon have the opportunity to demonstrate their skills against these criteria (Fig. 4).

Each of the challenges for the WED Hackathon are aligned with one of the UN Sustainable Development Goals.

In 2023, the overall theme of celebrations was *Engineering innovation for a more resilient world"*. and the challenges related to UN SDG #2 "No Hunger", for sustainable food access., UN SDG # 6 "Clean Water and Sanitation", for a solution for clean water access in scarce and polluted situations, and UN SDG #3—"Good Health and Well-being" for improved city design and infrastructure. In order to replicate the work environment, very limited time is provided for students to plan,

[1] The World Federation of Engineering Organizations (WFEO) has members that are the professional engineering institutions of the world from every continent. It has more than 100 national, and international members. The Federation was founded under the auspices of UNESCO in 1968. Collectively, the Federations represents more than 30 million individual engineers. The Federation proposed the declaration to UNESCO of World Engineering Day for Sustainable Development. Approved in November 2019, this Day is celebrated on 4th March each year.

Fig. 4 Students demonstrate the efficiency of their 3-D printed solution for water purification in areas hit by significant floods and requiring clean water in the WED 2023 Hackathon

research and provide a solution. Students have adapted very well to this environment and have used various digital tools and virtual meetings to collaborate, as shown in Fig. 5. Students collaborate, innovate, and learn quickly. The process is expected to be empowering, build confidence and improve academic performance [38–40].

The competition does not require an extensive report but emphasizes hands-on learn ing and the development of a solution to one of three challenges that relate to a UN Sustainable Development Goal. Engineering educators can expand the benefit of the program with additional discussions and workshops that can reinforce the learning that is achieved. Teams that are shortlisted to Stage 2 are asked to prepare a 5-min video of their solution in any language with English subtitles.

A unique aspect of the competition is that the solutions are judged by young engineers from the World Federation and its partners. The young judges are expected

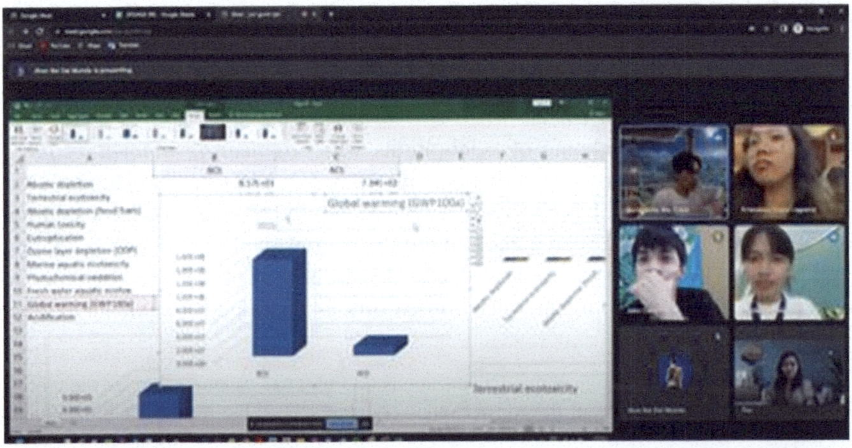

Fig. 5 Students Collaborate to develop their solutions to the WED 2023 Hackathon challenge

to have a better appreciation of the challenges faced by the engineering students and how these have been overcome.

The final stage of the competition is judged by leaders from the World Federation and from its partners from peer international organizations in engineering. These judges provide a high-level global viewpoint to the submissions that are shortlisted. The winners are therefore the highest caliber entries that have been received from around the world. Engineering educators that support their students in participating in the Hackathon can be justifiably proud that their students are finalists or winners.

The winners of the Hackathon achieve global visibility with their' videos shown globally as part of the World Engineering Day global streaming of events over 24 h. The winners are announced on 4th March, by WFEO and UNESCO. The cash prizes enable teams to develop their solutions and possibly take them to market.

The winners of the Hackathon come from developed and developing countries demonstrating the ingenuity of teams that have limited resources. In 2023, Team Sponge from Ateneo Naga University, The Philippines (Fig. 6) who addressed Challenge #3 was the outright winner with an innovation to repurpose waste materials to produce a permeable road surface that can absorb rainwater runoff, reduce flooding, and improve catchment health. Team Aquam Soils, University of British Columbia–Okanagan Canada, achieved second place with a low-cost, portable water treatment device for use in areas hit by natural disasters. Team Aqua-Smart, University of Mauritius, was in third place, developing a low-cost solution for water treatment for remote and rural areas.

Opportunities and Support for Engineering Educators

The WED Hackathon is an innovative approach for engineering educators to enable their students to demonstrate their skills and learning that they have attained in their

Fig. 6 Students from Ateneo Naga University, The Philippines present their prize-winning solution for the WED 2023 Hackathon

engineering programs., especially for those universities that are accredited by IEA signatories.

Engineering educators have an important role in encouraging students to take part in the competition, form their teams and hold discussions and workshops to support teams as they develop their ideas.

In the first two years of the Hackathon, registrations have been received from around the world, from both developed and developing countries.

Interestingly female students have participated in higher proportions relative to the proportion of engineering enrolments globally and are also a high proportion of the winners. More research is being done on the reasons for the participation of female students and their perceptions on the value of the international competition.

The results are inspiring and demonstrate the capacity of students to innovate and develop engineering solutions that can have a positive impact to solve some of the critical challenges that the world is facing today.

3 Leadership Takeaways

- Academic leaders can draw inspiration from these ideas as they continue to position their classes, programs, and institutions to be student-centred, forward-looking, and operating in a mode of continuous improvement.
- This chapter presents a variety of perspectives in terms of scope and scale as well as national and international contexts related to OBE.
- This chapter presents the OBE environment with a student-centred view.
- We have shown how academic leaders can understand how programs can best help students by developing a suite of deliberately defined learning outcomes that align with the workforce needs of the future.

Acknowledgements The World Federation of Engineering Organizations acknowledges the partnership with Engineers Without Borders International in developing the challenges for the World Engineering Day Hackathon in 2022 and 2023. Authors acknowledge the Management of BMS College of Engineering for providing the Professional environment to plan, implement and share Academic Initiatives/perspectives. The authors acknowledge the permission given by the Vice Chancellor of KLE Technological University and Chairman, AICTE Committee on Examination Reforms to present the case study on Examination Reforms conceived and implemented at the KLE Technological University.

References

1. Spady WG (1994) Outcome-based education: critical issues and answers. American Association of School Administrators, Arlington
2. Examination Reform Policy, AICTE, November 2018

3. Shavelson RJ (2010) Measuring college learning responsibility: accountability in a new era. Stanford University Press, Palo Alto, CA
4. Kuh GD, Ikenberry S (2009) More than you think, less than we need: learning outcomes assessment in American higher education. National Institute for Learning Outcomes Assessment
5. Lattuca LR, Stark JS (2009) Shaping the college curriculum: academic plans in context. Wiley
6. Novoselich BJ, Knight DB (2022) Measuring a moving target: techniques for engineering leadership evaluation and assessment. New Directions for Student Leaders 2022(173):63–71
7. Terenzini PT, Reason RD (2005) Parsing the first-year of college: a conceptual framework for studying college impacts. Paper presented at the Annual Conference of the Association for the Study of Higher Education. Philadelphia, PA
8. Terenzini PT, Reason RD (2010) Toward a more comprehensive understanding of college effects on student learning. Paper presented at the Annual Conference of the Consortium of Higher Education Researchers (CHER), Oslo, Norway
9. Lattuca L, Terenzini P, Knight D, Ro HK (2014) 2020 vision: progress in preparing the engineer of the future
10. National Academy of Engineering (2004) The engineer of 2020: Visions of engineering in the new century. National Academies Press, Washington, DC
11. Lattuca LR, Knight DB, Ro HK, Novoselich BJ (2017) Supporting the development of engineers' interdisciplinary competence. J Eng Educ 106(1):71–97
12. Knight DB, Novoselich BJ (2017) Curricular and co-curricular influences on undergraduate engineering student leadership. J Eng Educ 106(1):44–70
13. Palmer E, Terenzini PT, Harper BJ, McKenna A, Merson DS (2011) Design in context: where do the engineers of 2020 learn this skill? In: Proceedings of the American Society for Engineering Education, Vancouver, Canada
14. Lattuca LR, Terenzini PT, Volkwein JF (2006) Engineering change: a study of the impact of EC2000. ABET, Baltimore, MD
15. Sheppard S, Gilmartin S, Chen HL, Donaldson K, Lichtenstein G, Eris Ö, Lande M, Toye G (2010) Exploring the engineering student experience: findings from the academic pathways of people learning engineering survey (APPLES) (TR-10–01). Center for the Advancement for Engineering Education, Seattle, WA
16. Knight DB (2014) Reversing the logic: An outcomes-based student typology for determining "what works" in promoting an array of engineering-related student learning outcomes. Educ Eval Policy Anal 36(2):145–169
17. Brown MG, Knight DB (2014) Engineering practice in the academic plan: external influences, faculty, and their teaching roles. In: 2014 ASEE annual conference & exposition, pp 24–502
18. Knight DB, Cameron IT, Hadgraft RG, Reidsema C (2016) The influence of external forces, institutional forces, and academics' characteristics on the adoption of positive teaching practices across Australian undergraduate engineering. Int J Eng Educ 32(2(A)):695–711
19. National Board of Accreditation (NBA), for promoting International quality standards for Technical Education in India. https://www.nbaind.org/. Accessed 25 Oct 2023
20. International Engineering Alliance, Washington Ac-cord. https://www.ieagreements.org/accords/washington/. Accessed 25 Oct 2023
21. Model Curriculum for Engineering programs. https://www.aicte-india.org/education/model-syllabus. Accessed 25 Oct 2023
22. Accreditation Board for Engineering and Technology (ABET), ISO 9001 non-profit organization that accredits Universities and programs. https://www.abet.org/. Accessed 25 Oct 2023
23. The booklet outlines the history and development of the Washington Accord. https://www.ieagreements.org/assets/Uploads/Documents/History/25YearsWashingtonAccord-A5booklet-FINAL.pdf. Accessed 25 Oct 2023
24. Guidelines for the Induction Program for Undergraduate Engineering students. https://www.aicte-india.org/sites/default/files/Detailed%20Guide%20on%20Student%20Induction%20program.pdf. Accessed 25 Oct 2023

25. Guidelines for Internships during semester breaks. https://aicte-india.org/sites/default/files/AICTE%20Internship%20Policy.pdf. Accessed 25 Oct 2023
26. The National Education Policy 2020, by the Ministry of Education. https://www.mhrd.gov.in/sites/upload_files/mhrd/files/NEP_Final_English_0.pdf. Accessed 25 Oct 2023
27. Kanmani B, Shreenivas B, Gururaj C (2023) Ensuring OBE through curriculum: an example. Indian J Tech Educ 46(4):October–December 2023; 1–14. www.isteonline.in
28. The Curriculum for the Undergraduate Electronics and Telecommunication Program. https://bmsce.ac.in/Syllabus/TE/UG/UG%20Syllabus%202018-22.pdf. Accessed 25 Oct 2023
29. Buddhi K (2023) A pedagogy for engineering concepts focusing on experiential learning. In: Guralnick D, Auer ME, Poce A (eds) Creative approaches to technology-enhanced learning for the workplace and higher education. TLIC 2023. Lecture notes in networks and systems, vol 767. Springer, Cham. https://doi.org/10.1007/978-3-031-41637-8_8
30. Kanmani B (2020) Effective Implementation of OBE leading to accreditation. Indian Society for Technical Education (ISTE), under the Working Professionals Learning Project, an AICTE Project, 10 September 2020. ISBN: 978-81-89731-23-6
31. National Level competitive Examination, Graduate Aptitude Test in Engineering (GATE). https://gate.iitkgp.ac.in/. Accessed 25 Oct 2023
32. The Annual Magazine by the Batch Graduating (2022). https://bmsce.ac.in/reports/TE/MAGAZINE_ETE_2018_2022_4.pdf. Accessed 25 Oct 2023
33. Kanmani B, Gururaj C, Shreenivas B (2024) Simplified method for computing attainments in the OBE framework. Indian J Tech Educ 47(1):January–March 2024. www.isteonline.in
34. De Graff E, Kolmos A (eds) Management of change, implementation of problem-based and project-based learning in engineering problem-based and project-based learning. Sense Publishers.https://www.researchgate.net/publication/227057453_Problem-Based_and_Project-Based_Learning. Accessed 18 Aug 2023
35. The World Engineering Day Hackathon Homepage. https://worldengineering-day.net/hackathon/. Accessed 18 Aug 2023
36. Ratcheva VS, Leopold T (2018) 5 things to know about the future of jobs. World Economic Forum. https://www.weforum.org/agenda/2018/09/future-of-jobs-2018-things-to-know/. Accessed 18 Aug 2023
37. International Engineering Alliance, Graduate Attributes and Professional Competencies (2021). https://www.ieagreements.org/assets/Uploads/Documents/IEA-Graduate-Attributes-and-Professional-Competencies-2021.1-Sept-2021.pdf. Accessed 18 Aug 2023
38. Spence J (2019) 10 things that happen when students engage in project-based learning. https://spencerauthor.com/10-things-happen-students-engage-project-based-learning/. Accessed 18 Aug 2023
39. Smith M (2019) Does project-based learning work? Let's investigate. TES Mag. https://www.tes.com/magazine/archive/does-project-based-learning-work-lets-investigate. Accessed 18 Aug 2023
40. Evans CM (2023) Student outcomes from high-quality project-based learning: a case study for PBLWorks. National Center for the Improvement of Educational Assessment (Center for Assessment), Dover. https://www.pblworks.org/sites/default/files/2020-01/PBLWorks%20HQPBL%20Teacher%20Case%20Study%20Report_FINAL.pdf. Accessed 18 Aug 2023

Creating Student-Centric Learning Environments Through Evidence-Based Pedagogies and Assessments

Juan D. Ortega-Alvarez, Mitra Mohd-Addi, Aida Guerra, Sivakumar Krishnan, and Khairiyah Mohd-Yusof

Abstract Various sectors contributing to global development require transformation due to the changing nature of the world today. This transformation is not only materialistic but also related to human capital. The demand for skilled human resources is increasing with the rapid advancements in engineering and technology. Future engineering jobs will demand highly proficient individuals who can effectively manage technology, work independently, and collaborate with others. Education must adapt to produce highly skilled and ethical engineers, necessitating a shift in teaching practices. Educators must take on the task of changing the roles of learning by establishing student-centric learning environments that are aligned with the intended outcomes and assessments, in response to the call for education reform. This change necessitates a long-term visionary institutional leadership that solidly supports the purposeful development of SoTL-related skills, such as comprehension of the literature on educational research, application and assessment of innovative practices, critical reflection on practice for improvement, and dedication to sharing in a scholarly manner. Educators are increasingly embracing research-based and empirically supported pedagogical practices in their efforts to establish student-centered learning environments. Evidence-based pedagogies emphasize practices that have been shown to improve student learning and outcomes. Examples of these strategies include

J. D. Ortega-Alvarez (✉)
Virginia Tech, Blacksburg, VA 24061, USA
e-mail: jdortegaa@vt.edu

Universidad EAFIT, Medellin, Colombia

M. Mohd-Addi
Universiti Teknologi Malaysia, 81310 Johor, Malaysia

A. Guerra
Aalborg University, 9220 Aalborg, Denmark

S. Krishnan
Shri Vishnu Engineering College for Women, Bhimavaram, Andhra Pradesh 534202, India

Vishnu Educational Development and Innovation Center, Hyderabad, Telangana 500075, India

K. Mohd-Yusof
Purdue University, West Lafayette, IN 47907, USA

R. Kandakatla et al. (eds.), *Academic Leadership in Engineering Education*, Lecture Notes in Networks and Systems 1097, https://doi.org/10.1007/978-3-031-68282-7_7

active learning, presenting students with real-world problems to address, facilitating learning through relevant projects, peer teaching, use of open-ended and probing questions to foster critical thinking and conversation, and the delivery of these strategies tailored to the individual student needs and learning styles. Complementary to these pedagogical shifts are assessments that align with student-centric learning objectives. Authentic and ongoing assessments that occur during the learning process play a pivotal role to provide feedback and guide instruction, helping both teachers and students identify areas for improvement and promote deep learning experiences. These tools offer students opportunities for self-assessment and reflection, encouraging the development of metacognitive skills and a deeper understanding of subject matter. The use of evidence-based pedagogies and assessments aims to improve the quality of education by aligning teaching methods and evaluation practices with research findings, ultimately enhancing student learning and outcomes. Educators adapt and refine these strategies to suit their specific teaching contexts and student needs while remaining informed by the best available evidence. The presented case studies advocate for discussion and incorporation of evidence-based practices to support the creation of student-centric learning environments. The implementation of evidence-based practices can cultivate well-rounded, adaptable learners who will be more equipped to succeed in the ever-changing context of twenty-first century education and beyond. Ultimately, educators, scholars, and institutions contribute to the scholarship of teaching and learning by sharing empirical evidence and insights, while the enactment of leadership at different levels further contributes to enrich and strengthen the broader educational community.

Keywords Student-centered learning · Scholarship of teaching and learning · Evidence-based practice · Reflective practice · Learning communities

1 Introduction

In traditional educational settings, teaching and instructors have been at the center of the teaching and learning process. Quality of education, within this paradigm, could be measured by teachers' experience and mastery of the content [1]. Also within this paradigm, the objectives of a teaching and learning experience (e.g., a course or a workshop) could be defined in terms of what the attendees could expect *the instructor* to cover, discuss, or teach during said experience. Most readers will surely remember those days when "Presenting the typical configurations of heat exchangers" was a perfectly valid objective of a Heat Transfer course, or "Exploring different active-learning strategies" could have been the undisputed outcome of a faculty development workshop.

In the last few decades, educational researchers have gathered enough evidence to challenge the notion that this paradigm, which relies heavily on the *transmission* of information, is the best we can do to educate people. Collectively, educators have come to agree that *student learning* should be the ultimate outcome of education, which shifts the focus from a teacher-centered, transmission-oriented perspective to a student-centered, learning-oriented perspective [1, 2]. In this new paradigm, the quality of education must be measured by what *the students* know, can do, or can be after a teaching and learning experience. Moreover, while teachers' mastery of the content is still important in this paradigm, a new skill becomes particularly relevant: Pedagogical content knowledge [3]. Therefore, designing and implementing learning environments that foster student learning of specific knowledge, skills, and ways of making sense of the world becomes the job description of a teacher today. To create such environments, institutional and departmental leadership as well as faculty need to reflect on and reconsider their roles.

This shift in roles requires a revised vision of the teaching duty of college faculty—a vision that includes the understanding, use, and assessment of subject-matter content and empirical principles to guide the teaching and learning process. To describe such a vision, Boyer [4] aptly coined the term *Scholarship of Teaching*. Boyer's work focused mainly on the recognition of teaching as a scholarly activity of college faculty and a way to reconcile the seemingly different aims of the teaching and research duties of faculty. However, it clearly marked a transition between the two paradigms discussed above:

> As a *scholarly* enterprise, teaching begins with what the teacher knows. Those who teach must, above all, be well informed, and steeped in the knowledge of their fields. [...] Teaching is also a dynamic endeavor involving all the analogies, metaphors, and images that build bridges between the teacher's understanding and the student's learning. Pedagogical procedures must be carefully planned, continuously examined, and relate directly to the subject taught [4].

This revised view of scholarship, Boyers' posited, "will not emerge, however without strong leadership at the top [4]." Therefore, a reconceptualization of teaching and learning as a scholarly activity necessitates change at the individual and the institutional level.

Building upon Boyers' ideas of education and academic change, scholars have increasingly highlighted the importance of the students as a protagonist of the teaching and learning process. This way, the Scholarship of Teaching eventually became the *Scholarship of Teaching and Learning* (SoTL).

In educational research literature, there are multiple definitions and slightly different understandings of what SoTL entails. Research-to-practice and practice-to-research tactics in education like *action research* and *classroom research* are often associated with SoTL. In this chapter, SoTL development is defined along four key components:

1. Familiarity with and ability to identify issues in own teaching and learning related practices, and understand educational literature aimed at both practitioners and researchers.
2. Implementation and assessment of innovative teaching practices, including generating evidence of student learning.
3. Critical reflection that leads to decision making for further improvement.
4. Commitment to sharing results through scholarly conferences and publications.

Going through the SoTL process requires academics to scrutinize their own practices with open minds and focus on issues to be overcome. For instance, a professor unsatisfied with the level of student engagement during her class sessions might want to try flipping her classroom. To that aim, she must first familiarize herself with flipped learning, explore the resources available to implement this technique, and then devise ways to assess the results of such an implementation. Therefore, design and implementation of the intervention to improve the issues requires reviewing related literature and principles and assessing the outcomes. Critical reflection on the implementation and outcomes as well as the literature review help in internalization and meaning making, which can lead to transformation in mindset and practice [5]. While writing and publishing will disseminate the practice, they are also effective in guiding critical reflection as well as consolidating thoughts, knowledge, and practice.

Although SoTL is often seen as confined to classroom practice, Fanghanel et al. extended SoTL in higher education beyond classroom practice (micro-level), to institutional (meso-level) and national/international (macro-level) levels [6]. At the micro level, while the objective is always to improve student learning, all the four components of SoTL lead to self-improvement of the practitioner, setting the stage for a possible meso-level SoTL implementation. Academics can work together to conduct SoTL at the departmental level, be it for multiple sections of the same course, or integration of several different courses. Academics conducting faculty development training and mentoring can also implement SoTL at the institutional (meso) level and national/international (macro) level. In that sense, while engagement in SoTL at the micro-level may result from individual choice, the transition to meso and macro levels is only possible through institutional leadership that enables faculty communities through faculty promotion policies and practices designed to align with SoTL.

2 Background

The following sections present five case studies describing different approaches to engaging with SoTL intertwined with leadership interventions at different levels, as illustrated in Fig. 1. Educators and scholars may find that their experiences resonate with one or more of the cases presented, without any intrinsic order or development direction. While the SoTL journey of some faculty starts individually, others might find themselves in the middle of a community that spurs and facilitates their engagement. The common threads in this space of experiences are the aim of fostering student-centered learning (SCL) by drawing upon the principles of SoTL defined before, and the leadership efforts required for academic change to happen and endure.

The case studies come from different national, cultural, and academic contexts. The first two are from Universiti Teknologi Malaysia (UTM), a research technology-based university. Among the earliest universities in the region to embrace outcome-based education in its curricula, SCL is promoted along with initiatives to build

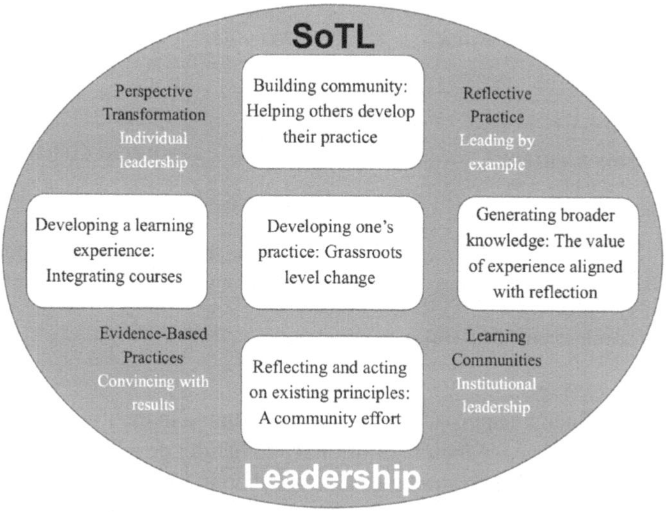

Fig. 1 Five case studies, five SoTL experiences

a community of practice in engineering education. The third case study discusses a set of faculty development initiatives at a privately funded residential center for education and innovation in India. These initiatives were driven by visionary leadership involving four different educational institutions spanning the last six years and have resulted in the formation of a nascent faculty community with a focus on student engagement. The fourth case study comes from Aalborg University (AAU), in Denmark, well known for its problem-based, project-organized learning (PBL) model supported at the institutional leadership level. The final case study is from a first-year engineering (FYE) program at Virginia Tech, a research university in the United States. This FYE program currently ranks in the top ten nationally in terms of student enrollment and is administered by a department of Engineering Education with the dual mission of advancing both engineering pedagogy and scholarship.

A few key concepts are common to the experiences described in the case studies, with *active learning* at the forefront. At its essence, active learning embraces the idea of placing students at the center of the teaching–learning process. This involves designing and implementing pedagogical activities such as Think Aloud Pair Problem Solving (TAPPS) or Small Group Discussions that necessitate students' active and overt involvement with the content, rather than passively listening to a lecture. Active learning encompasses a wide range of strategies, and there is empirical evidence of its effectiveness in promoting deeper student learning compared to traditional teaching [7]. These strategies vary along a continuum from more instructor-led to more student-led activities, as shown in Fig. 2 [8]. Figure 2 also describes the potential for a scaffolded approach that educators can work towards by implementing and experimenting with inquiry-based and active learning practices initially before moving towards implementation of problem-based learning (PBL) and project-based

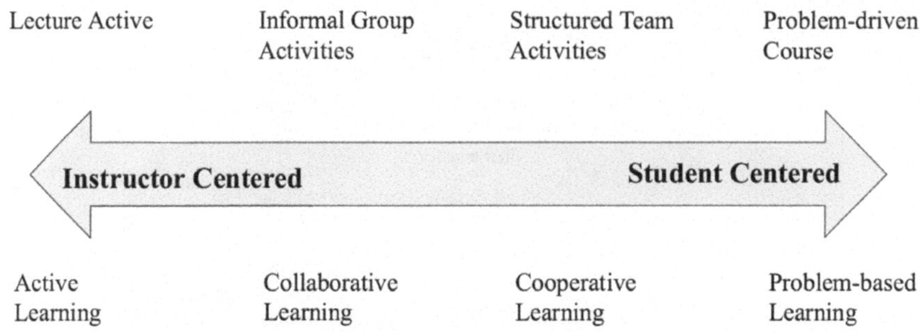

Fig. 2 Active learning continuum (Adapted from Lord et al. [8])

learning (PjBL). Such an approach may be more suitable in institutional ecosystems that are in the early stages of their institutional and faculty development journey. The next paragraphs present some strategies along that continuum that are relevant to the case studies in this chapter.

At the extreme of student-led activities sits problem-based learning (PBL). To understand PBL, picture a group of students who are given a scenario where they have to identify and solve meaningful problems (e.g., how to reduce the carbon footprint of the campus?). Then, they start pulling the information and learning the knowledge they require to solve the problem from the instructor, class materials, and other resources available to them. In general, PBL is an approach where a group of students learn by identifying and solving real and authentic problems. Traditionally, curriculum can be organized around problems in two ways: as cases, which emerged in medical education (see for example, seven steps of Maastricht or the McMaster model), or as projects, which emerged from technical and engineering education (see for example the models of Aalborg University and Roskilde University since the '70s). PBL models tend to comprise a set of fundamental characteristics such as problem orientation, collaboration, self- and participant directedness, exemplarity, interdisciplinarity, and application of theory to practice [9]. When organized around projects, this pedagogical approach is often referred to as project-based learning (PjBL). PBL and PjBL have significant overlap: both start from a meaningful question related to a real-world situation, and both pace the instruction according to students' progress and knowledge required to advance towards a final answer or solution. In PjBL, students work in a task-oriented way, creating a plan and allocating their time and other resources over a sustained period, often the whole term.

Teamwork is usual in PBL and PjBL, but there are different ways to foster student learning through teamwork. One of them is Cooperative Learning (CL), which is a pedagogical approach that fosters learning through interactions among students. There are three main pillars in CL: positive interdependence, individual accountability, and face-to-face interaction [10]. Positive interdependence emphasizes that students work together to achieve common goals, ensuring each member's success contributes to the team's success. Individual accountability requires students to contribute to the group's work and be responsible for their learning. Interacting through face-to-face interaction promotes open communication and exchange of ideas among team members as well as developing other interpersonal skills like conflict resolution and collaborative decision-making. In addition to the three main pillars, regular team assessments help to monitor progress and adapt learning strategies. Group processing is also key in fostering continuous improvement and a sense of shared responsibility among team members. These principles not only enhance the learning experience but also prepare students for teamwork and problem solving in their future engineering careers.

CL principles are grounded in the social interdependence theory, which posits that when an individual perceives that their success is tied to the success of others, they are more likely to collaborate, share resources and help each other [10]. CL also aligns with sociocultural theory by creating an environment where students can scaffold their learning through interaction with peers who may have different levels of expertise. In a CL environment, students engage in discussion, problem solving and peer teaching, allowing them to build their knowledge collaboratively and develop a deeper understanding, which aligns with social constructivism, where learners construct their understanding of concepts through their experience and interactions with others. These theoretical underpinnings support CL as an effective approach in engineering education.

But students are not the only ones experiencing new approaches to learning in this student-centered paradigm. Faculty must develop SoTL-related capacities such as pedagogical content knowledge and evidence-based practices. Embracing SoTL, either as the result of a personal journey or with the help of training and facilitators, is a transformative experience [11, 12]. Faculty approaching SoTL for the first time usually experience a profound transformation of their perspective as they are exposed to the breadth and depth of teaching as a scholarly activity [13]. Continued practice and a shared sense of validation helps cement this perspective transformation. Transformation of the institutional ecosystem requires a leadership that champions sustained and focused efforts over a period of time, which results in the development of self-driven faculty learning communities. Institutional leadership can help address workload inequities and recognize such communities and the work required to nurture them in an effort to expand the new culture. Faculty learning communities can provide the support and sense of belonging that help educators persist in their efforts to engage with SoTL and further advance the student-centeredness of their practice [14, 15]. This, in turn, helps transform student culture and continuously innovate teaching and learning institutional practices.

The case studies that follow should be read and analyzed in light of the background and key concepts presented in this introduction.

3 Student Centered Learning Case Studies Tied to SoTL

3.1 Cooperative Problem-Based Learning (CPBL)

This case study illustrates that those leading changes at the grassroots level should encourage consistent SoTL implementation which can lead to innovation beyond the classroom. Academic leaders should also realize the importance of underpinning innovations on education theories and the literature to avoid reinventing the wheel.

Process Control and Dynamics in UTM was a problematic course for chemical engineering undergraduates in the early 2000's. It was disheartening to see the unfortunate results of the students: 30% of around 200 students divided into 3–4 sections each semester failing and those who passed scoring poorly with only a few getting B or better. It was easy to blame students—but to lead the improvement efforts, taking the scholarly approach is crucial. It initiated identifying of the change required and possible interventions needed.

With the aim of enhancing students' learning and creating a supportive learning community, Cooperative Learning (CL) was implemented in the course in the 2003/2004 academic year. Translating CL principles into the instructional design, students were divided into teams of 3 or 4 according to the chosen criteria. The CL environment transformed the class into a supportive learning community resulting in significant improvement in their scores [16].

Despite the improvement, students still had problems in understanding the context of the concepts in the course. The global challenges of twenty-first century also require various skills that students must prepare for. It was decided that there should be an improvement in the course by implementing PBL and CL. Analyzing various PBL models and considering the constraints, a floating facilitator model, with students divided into small groups, utilizing a typical inductive PBL process [17, 18] was selected. In the inductive PBL model typical in medical schools, learning starts with a problem without first providing a lecture on the topic. There is a wealth of research on this PBL model illustrating how it was translated into practice resulting in positive impacts on students [19–21]. CL principles were embedded in the PBL process to develop functioning teams to support learning. After the first implementation [22], students go through four PBL problems to learn 80% of the course content. To lead the change in all sections of the course, improvements were made each semester. The impact of it was studied following the SoTL process, which led to the development of Cooperative Problem-based Learning (CPBL) framework shown in Fig. 3 [23], where CL principles are explicitly infused into the PBL process. The framework gives an overall picture of the CPBL process while providing a step-by-step guide through the learning process.

It is important for leaders to understand the scholarly aspects of classroom innovations because this will enable adaptation of the implementation into different context, requirements, and constraints.

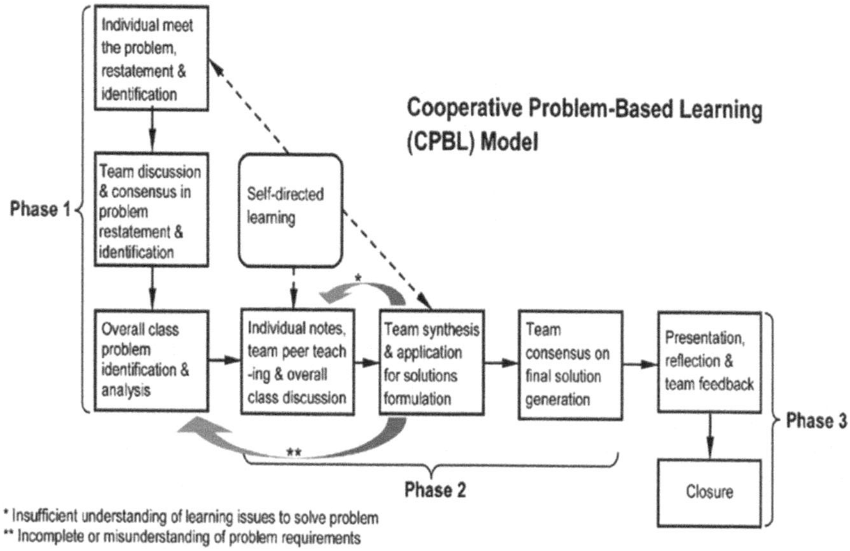

Fig. 3 The CPBL model [23]

When the *Introduction to Engineering* course for chemical engineering students was introduced in 2005, CPBL was chosen so that students can experience engineering problem-solving processes while solving sustainable development related problems.

Problems were crafted based on the theme chosen for the semester, such as alternative energy, river conservation, low carbon society, microplastics, etc. As in Process Control, various studies were conducted that show positive impact on students [24, 25].

When the courses had to be taught online during the COVID-19 pandemic, it was easy to go back to the principles and redesign the suitable learning activities to implement online with the help of technology. In addition, elements of empathy were added to the peer feedback to help students support each other [26]. To increase support for meaningful learning while avoiding increasing workload, several parts of the sustainable development problem were integrated with the *Introduction to Programming* course, enabling the courses to grade different parts relevant to the courses from the same project.

The takeaway pointers of this initiation are:

Following the SoTL pathway will enable academic leaders to drive continuous growth and flexibility, while knowing the effect of the innovations made through assessment and reflective practice.

Underpinning the innovation on strong theories and research enabled systematic course and instructional design while preventing guess work.

Most importantly, overcoming classroom problems were not solely based on personal experience, which helped avoid the reinvention of the wheel.

3.2 Integrated Project Learning Experience

Engineering courses are mostly done in isolation, emphasizing individual course content during learning. In support of SCL, several instructors supplement their efforts by including PBL and PjBL into their respective courses. Since each project requires concentrated time, effort, and thoughtful discussions, the idea of having a project for every course would be extremely daunting for students. However, for learning to be successful and meaningful, students must have a holistic view and establish interconnections between courses. Among the ways to make this happen at the curricular level is to integrate projects between suitable courses taken in the same semester. However, this requires coordination and leadership to convince the suitable course instructors to collaborate.

This case study discusses the implementation of an integrated project designed to span two courses in an electronic engineering program in UTM that interconnects the two courses. The two courses are Physiology & Introduction to Medicine and Medical Instrumentation which are core courses for Medical Electronics majors. As the integrated project is intended to address the learning outcomes of both courses, learning activities supporting the project, as well as assessments, are planned and designed using constructive alignment principles. To ensure coordination between the two courses, the topics covered over the same period are carefully chosen to be closely connected. As an example, the topics on human cardiovascular systems (for the Physiology & Introduction to Medicine course) and electrocardiogram (ECG) (for the Medical Instrumentation) are covered concurrently to provide students with a more engaging learning environment. Although all topics are discussed in separate sessions, students will be able to relate and make connections between the two topics.

Utilizing CL principles for both classes, 3–4 students were grouped into heterogeneous teams, considering various factors including race, gender, cultural background, and academic achievement to promote diversity in thinking and opportunities to adapt with others on mutual concerns [27]. They stay in the same team throughout the semester and in each in-class activity to develop accountability and positive interdependence. When students work together over time, they can engage in critical discussions and provide one another with valuable feedback. This depth of interaction tends to lead to deeper learning and better retention of knowledge.

Then, each team is placed in a scenario as a team of interns attached to the Research and Design department involved in the company's annual bioelectronics or biomedical design project. When tasked with designing an instrumentation tool to address a physiological problem, the teams had to undergo related training to fulfill the requirements. After team formation, students identify issues that they would like to address and current works done to address those issues from literature and putting them in charge of defining their own objectives in a field of interest. Then, each team proposed a conceptual design or improvements to the existing systems they decided to work on, for which they received feedback during the proposal presentation. At the end of the project, students implement the proposed solution as a working prototype. Each team is assessed at different stages of the project through the project proposal, progress report, project demonstration and written report. The deliverables serve as assessments for both courses addressing different learning outcomes. Scaffolding activities were also provided to support the implementation of the integrated design project such as hospital visit, introduction to reference management software, sessions on presentations and proposal preparation.

The presented case study describes the development of one's SoTL journey from the micro level (individual) to another phase towards the meso level (institutional). One of the main challenges to sustain the integrated project is to be able to get the same cohort of students that are enrolled in the same two courses. Implementation of the integrated project requires support and commitment from another instructor who is also keen and understands the needs to integrate both courses—to create a more meaningful learning experience where students get to relate interconnection between the two courses.

In this case study, the role of a mentor plays a crucial role in addition to receiving support and assistance from a fellow colleague. The main instructor had the opportunity to be the mentee of a senior mentor, engaging in CL and PBL practices. The experience of going through training and observing the mentor's CPBL-implementing class provided the inspiration for incorporating CL into the integrated project.

Personal development on the SoTL path requires the support of a community of practice to sustain motivation and efforts in establishing a student-centered learning environment. To benefit from and contribute to such a community, one must take initiative, be an effective team player, and develop strong communication skills—all essential qualities of academic leaders.

3.3 *Building Reflective and Scholarly Practitioners*

In India, two faculty development programs related to teaching and learning are The Wipro Mission 10x (2007–2012) and the International Engineering Educator Certification (IIEECP) (2015–present).

The focus of the Wipro Mission 10X program, which reached 10,000 faculty members from over 700 higher education institutions in India in three years, was to enhance the employability skills of graduating engineers tenfold through teacher training [28]. The initial focus of this program was training on presentation skills and active learning methods, and establishing a teachers' forum for exchange of ideas. A second phase of the program (2010–2012) focused on developing academic leaders and institutionalizing the best practices. Sadineni and Joshi [28] observed significant change in faculty mindsets indicated by perception-related measures in this program as well as the use of innovative teaching methods.

The IIEECP program focused on pedagogical practices including active learning, collaborative learning, course delivery and teaching with technology [29, 30]. One of the authors was involved as a facilitator in the first phase of the IIEECP program.

The IIEECP program was further developed to include institution-level activities and was implemented at the Vishnu Educational Development and Innovation Center

(VEDIC), which is a residential center established in 2016. The center was initiated by visionary leadership that recognized:

(i) the need for institutional support through the empowerment and recognition of faculty for teaching efforts and learning outcome and,
(ii) the need for sharing best practices internally across multiples institutions and disciplines governed by one educational society.

The previous case study described the implementation of SCL concepts into a two-course project. Now, the focus turns to preparing faculty to identify and implement innovative practices aligned with SCL through the building of faculty communities around SoTL. The case study shared here will focus on the creation of a faculty community through leadership and faculty workshops, classroom observations and a faculty teaching colloquium [31].

The workshops included the topics of outcome-based education, session planning, active learning, collaborative learning and learning assessment. It further included faculty submitted assignments and feedback on these assignments. The residential nature of the workshops enabled reflection and discussion during and after the sessions. The programs were deliberately designed to be interdisciplinary, and the faculty were assigned in diverse small groups to enable diversity of thinking and sharing from varied experiences. Further, they were introduced to simple heartfulness relaxation and meditation techniques in the evenings as an approach to self-management based on their interest. A total of 18 batches of 532 faculty and 40 department leaders from 4 institutions attended the workshops and one of these batches was exclusively for the department leadership teams. The latter batches had selected faculty co-facilitators who were practitioners and a second level selection of 127 faculty was made based on the completion of assignments.

The second phase of interventions included classroom observations and confidential feedback. The aim was to improve student engagement and promote higher order thinking in students. The observers chosen from the second level selection mentioned above served as mentors and provided feedback to the instructors in a meeting after the observation. Checklists with positive indicators and rubrics were provided to the faculty observers during an orientation session. The observers were also made aware of a scaffolded approach prior to the observations. The observed parameters included planning content, learning goals, audio and visual presentation, real-life examples, use of resources, and engagement at different levels (overall, individual, questions, and connection).

In the third leg of the interventions, faculty presented their teaching experiments at a monthly colloquium and selected best practices were recognized annually. A group of peer faculty gave feedback to the faculty using a checklist as listed in Table 1 below. An online orientation session was held for the faculty reviewers that included detailed rubrics and positive indicators. The overall goal of the program was to encourage and support the faculty in implementing inquiry-based teaching and active learning strategies.

Table 1 Review parameters for the teaching learning colloquia

Category	Points	Sub-category
Scientific approach	5	Teaching problem
	5	Learning goals
	5	Choice of practice and/or technology
	15	Description of practice
	5	Assessment design
	15	Design for student engagement
Impact	20	Impact on student learning
	5	Reflective report/teaching diary
	10	Publications
Presentation	5	Clarity & focus
	5	Articulation & simplicity
	5	Pace

The observer and reviewer orientation programs enabled faculty reflection and helped promote a culture of student engagement. In summary, these faculty development interventions along with the reviewing and recognition mechanisms resulted in the development of a faculty community of teachers supporting and mentoring each other at the institutional level. For more case studies about building faculty learning communities and recognition as enablers for SoTL and beyond see Kalish and Stockley [14].

The major outcomes included sharing of best practices during the pandemic and afterwards and resulted in several international conference publications. This transition from implementation to dissemination, as mentioned before, is a clear marker of engagement in SoTL.

To demonstrate institutional support and visibility for faculty, institutional recognition began with a certificate of recognition appreciating the specific areas of achievement since 2018.

3.4 Problem-Based Project-Organized Learning

Since 1982, Aalborg University's (Denmark) first year programs comprise a course where the main objective is to '*facilitate the students learning of process competences in connection with the PBL model*', in a PBL course [32]. This is typically achieved by enabling students to analyze, reflect and report their working and learning processes while carrying out a collaborative, problem-based project. The result is a process analysis report, which accompanies the final project report. The process analysis report is an evidence of student learning, which derives from reflections over the actions and experiences lived during the first semester of their studies [33].

Currently, the course is named Problem Based Learning (PBL), and as part of it, students must produce three process analysis reports (PAR1, PAR2 & PAR Final) throughout their first semester. The timeline describing students' and teachers' activities is illustrated in Fig. 4.

The process analysis report is part of student's assessment in the first semester of first year engineering programs. The process analysis report is produced as part of the PBL course [34] and reports how students manage their collaborative, problem-based project work and what they have learned from their experiences by going through reflective processes. These processes are facilitated by the teacher through consultancy and feedback sessions.

In connection with SoTL, the PBL implementation helps students to report systematically by using a framework and probing questions provided by the lecturer. This, in turn, allows the lecturer to generate evidence of student learning. There are explicit criteria for feedback (written and oral) with suggestions for improvement. Students go through three rounds of reflection at different levels and varying their focus, even though always seeking to elaborate on their own learning using the same framework and probing questions. This process generates information for the teachers to reflect and revise their own teaching whilst students report and adjust. The use of reflective reports makes students practice exemplary and their learning explicit.

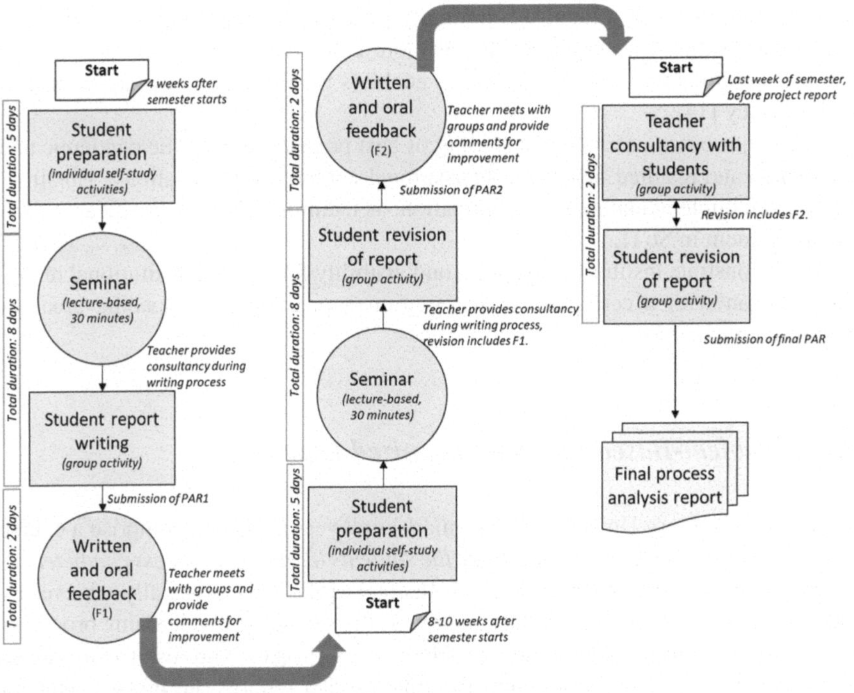

Fig. 4 Representation of timeline and activities to produce the process analysis report (PAR = process analysis report; F = feedback documents)

The above refers to SoTL at the micro-level within the Aalborg University ecosystem. However, the structures and frameworks that regulate and make possible the PBL teaching and learning strategies described are the outcome of leadership at meso- and macro-levels. One example at the macro level of the rectorate leadership is the creation of the Institute of Advanced Studies in PBL (IAS-PBL). IAS-PBL "*functions as a hub for research, knowledge sharing, competence development, collaboration and experiments with PBL*" [35] and it brings together researchers and practitioners from the faculties of engineering and science, medical sciences, social sciences, and humanities. At the meso level, department chairs and study board leaders recognize the importance of educating students and faculty for PBL, its principles, and the importance of educating future professionals beyond the knowledge domain. Therefore, they allocate resources to develop and deliver courses, workshops and events which strengthen students and faculty capabilities to learn and perform in a PBL environment.

The Aalborg University leadership ecosystem makes Aalborg PBL a worldwide reference, especially in the context of engineering education. The motto is "PBL is part of Aalborg DNA", which means it is ingrained in the organization's culture which nurtures initiatives aimed at pedagogical innovations [36].

3.5 Structure Versus Autonomy in Project-Based Learning

The previous case studies have presented approaches to SoTL that rely on individual and community efforts, and leadership support at different levels. This case study discusses an approach to scholarly and reflective teaching led by a team of instructors. Notably, this example unfolds within an environment where institutional leadership actively supports and recognizes engagement with SoTL. The case study highlights the reflection of a team of first-year engineering instructors on the effectiveness and efficiency of their own teaching and learning practice. The context for this case study is the General Engineering program at Virginia Tech, administered by the department of Engineering Education.

Project-Based Learning (PjBL) has proven to be an effective teaching and learning strategy in engineering classrooms [37]. Introductory engineering courses benefit greatly from the introduction and practice of professional skills like problem solving, teamwork, and communication, afforded by the context of a transdisciplinary engineering project [38]. Since the scope of an introductory course is often unrelated to a single engineering discipline, there are a myriad of projects that could support the learning outcomes of such a course. This opens multiple possibilities for the instructors, from presenting students with a challenging problem to solve through a project, to allowing them to select—or even define—the goal of their own project. Meadows et al. (2012) and Shepard (2013) found that providing students with choice and autonomy increased their interest in engineering and promoted engagement, motivation, and overall achievement of learning outcomes [39, 40].

PjBL and motivation through meaningful activities are two examples of evidence-based strategies that have been widely implemented in engineering classrooms. However, at the intersection of PjBL and motivation lies a question worth exploring: ***What is the impact of giving students more or less autonomy in a PjBL class?*** Based on the research mentioned above, self-selected projects are expected to enhance student motivation. However, this advantage may come at the cost of increased instructor time and discomfort, particularly when dealing with diverse projects outside their expertise. This cost can become evident particularly in large classes, which are typical in first-year engineering programs at major colleges and universities. To shed light on the question posed, a team of seasoned instructors and trained educational researchers reflected on the experiences reported by students in multiple sections of the same first-year engineering PjBL course. The course was taught by three different instructors that provided varying levels of autonomy for students to choose their semester-long, team-based project [41]. One instructor assigned a specific project that students could scope in terms of the scale and end users of the final product. The second instructor assigned a specific project that allowed students more scoping autonomy in terms of the actual problem being addressed. The third instructor allowed students to suggest a problem that was meaningful to them and frame the project around designing artifacts that could help solve this problem. The instructor made sure that the scope and complexity of the self-selected projects were adequate for the course. The research team set out to compare the impact of each approach in terms of students' perception of their learning and autonomy. At the end of the semester, the team administered a simple survey with both Likert-scale and open-ended questions to gauge students' perceptions.

The results showed no significant differences in terms of perceived learning across the three approaches. On the other hand, there were small yet significant differences in terms of perceived autonomy commensurable with the face value of each approach, although all approaches yielded scores on the upper end of the scale. The research team was intrigued by the small differences that seemed to defy their expectations. Qualitative analysis allowed them to see that ***students found and valued opportunities to exercise their autonomy well beyond the selection of the project***. For instance, students would often report on their autonomy to select and work on project-related tasks aligned with their interests and strengths. Other students would construe autonomy as their ability to share and discuss their ideas about different aspects of the project with their teams. Insights like these were found in the more structured and more open-ended approaches alike. The bottom line of this work is that, if instructors are not comfortable leveraging motivation by allowing students to work on self-selected problems, they still have plenty of options to build autonomy into a predefined project. This can be done, for instance, by encouraging students to scope the project within a context meaningful to them, allowing them to form sub teams to tackle different parts of the project, and providing guidance and training to foster equitable teamwork [42].

Lessons learned from this effort stem from a common interest in—and autonomy for—improving student learning and teaching efficiency drawing upon the principles of SoTL, within a community of practice supported by administrative leadership.

Instructors take the lead in assessing and shaping their own practice, and institutional leadership provides support and recognition to foster this type of initiative. Dissemination of these lessons at venues for engineering education practitioners and researchers is the required step to close the scholarly loop, which does not mark the end but the start of a new iteration.

4 Conclusion

Designing learning environments is the main role of teachers in the paradigm of student- and learning-centered education, but it is not a trivial task. It necessitates the intentional development of skills associated with SoTL: Understanding educational research literature, implementing and assessing innovative practices, and participating in the broad conversation through educational venues like conferences and journals. SoTL, however, is not equally developed everywhere. Strong communities of practices in certain countries/regions support the continued development of SoTL scholars. Academic and institutional structures and leadership play an important part. Educational research is not equally appreciated and recognized across the globe, and alternate pathways to the professoriate focused on teaching with a significant research component (e.g., collegiate faculty) are not offered by many institutions. Although SoTL is normally practiced at the classroom or micro level, this continuous reflective practice has the potential to grow on a larger scale at the institutional or meso-level, with the appropriate support from the management. Meso-level SoTL also has the potential to lead to national and international or macro-level SoTL, which will have a strong impact on improving the quality of learning, and thus the quality of graduates.

Academic leadership, therefore, takes different shapes at different levels of the educational system, from the individual students and instructors to the top administration of educational institutions. Academic change can only happen and be sustained through bottom-up and top-down approaches combined. From a bottom-up approach, the grassroots level change can be fostered through SoTL. SoTL sparks leadership at this level and adds legitimacy and trustworthiness to move ideas up the chain of leadership. In contrast, a top-down approach could be sparked by the proven benefits of student-centered learning environments, as academic leaders and decision makers aim at promoting the relevance and credibility of the educational model embraced by their institutions.

Echoing Parker Palmer's stages of academic reform (1992), this chapter ends with four simple recommendations for academic leaders aspiring to create and sustain quality learning experiences for students in engineering institutions [43]:

- Academic leaders should never underestimate the transformational power of the grassroots level of SoTL, whether in their own practice or of their colleagues.
- Learning communities and faculty communities of practice are powerful multipliers of the efforts of grassroots leaders.

- Top and middle leadership should continue to create tracks that truly recognize and reward the work done in fostering and nurturing faculty learning communities and SoTL efforts.
- Supporting the dissemination of SoTL work is crucial, as it provides academic leaders with the evidence required to initiate change from the institutional level.

Nurturing a culture of institutional learning is the key to driving sustained academic reform. By promoting continuous improvement and development within educational systems, institutions can become more flexible, adaptable, and responsive to the evolving needs of society.

References

1. Barr RB, Tagg J (1995) From teaching to learning—a new paradigm for undergraduate education. Change: Mag High Learn 27:12–26
2. King A (1993) From sage on the stage to guide on the side. Coll Teach 41:30–35
3. Shulman LS (1986) Those who understand: knowledge growth in teaching. Educ Res 15:4–14
4. Boyer EL (1990) Scholarship reconsidered: priorities of the professoriate. Princeton University Press, Lawrenceville, NJ, p 08648
5. Mezirow J (2003) Transformative learning as discourse. J Transform Educ 1:58–63
6. Fanghanel J, Pritchard J, Potter J, Wisker G (2016) Defining and supporting the Scholarship of Teaching and Learning (SoTL): a sector-wide study. Literature review. HE Academy, York
7. Prince M (2004) Does active learning work? A review of the research. J Eng Educ 93:223–231
8. Stefanou C, Lord SM, Prince MJ, et al (2012) The effect of different active learning environments on student outcomes related to lifelong learning. Int J Eng Educ 28:606–620
9. De Graaf E, Kolmos A (2003) Characteristics of problem-based learning. Int J Eng Educ 19:657–662
10. Johnson DW, Johnson RT, Smith KA (1984) Cooperative learning. Interaction Book Company, New Brighton
11. Mezirow J (1997) Transformative learning: theory to practice. New Direct Adult Continuing Educ 1997:5–12
12. Kitchenham A (2008) The evolution of John Mezirow's transformative learning theory. J Transform Educ 6:104–123
13. Ortega-Alvarez JD, Vieira C, Sanchez-Pena M, Streveler RA (2018) The challenges of assessing transformative learning: lessons learned from an instructional design workshop for Colombian engineering faculty. Int J Eng Educ 34:1604–1614
14. Kalish A, Stockley D (2009) Building scholarly communities: supporting the scholarship of teaching & learning with learning communities. Transformative Dialogues: Teach Learn J 3
15. Cox MD, Richlin L (2004) Building faculty learning communities: new directions for teaching and learning. Wiley
16. Yusof KM, Hassim MH (2003) Cooperative learning in process dynamics and control course for undergraduate chemical engineering students. 7th triennial AEESEAP conference proceedings. Universiti Malaya, Kuala Lumpur, pp 115–121
17. Duch BJ, Groh SE, Allen DE (2001) The power of problem-based learning. Stylus Publishing, Virginia, USA
18. Tan OS (2003) Problem-based learning innovation: using problems to power learning in the 21st century. Thomson Learning, Singapore
19. Dochy F, Segers M, Van den Bossche P, Gijbels D (2003) Effects of problem-based learning: a meta-analysis. Learn Instr 13:533–568

20. Polanco R, Calderon P, Delgado F (2001) Problem-based learning in engineering students: Its effects on academic and attitudinal outcomes. In: Little P, Kandlbinder P (eds) The power of problem-based learning. PROBLARC, Newcastle, pp 111–125
21. Helmi SA, Mohd-Yusof K, Phang FA (2016) Enhancement of team-based problem-solving skills in engineering students through cooperative problem-based learning. Int J Eng Educ 32:2401–2414
22. Yusof KM, Tasir Z, Harun J, Helmi SA (2005) Promoting problem-based learning (PBL) in engineering courses at the Universiti Teknologi Malaysia global journal of engineering education. Glob J Eng Educ 9:175–184
23. Mohd-Yusof K, Helmi SA, Jamaludin MZ, Harun NF (2011) Cooperative problem-based learning (CPBL): a practical pbl model for a typical course. Int J Emerg Technol Learn 6:12–20
24. Sadikin AN, Mohd-Yusof K, Phang FA, Abdul Aziz A (2019) The introduction to engineering course: a case study from Universiti Teknologi Malaysia. Educ Chem Eng 28:45–53
25. Samsuri S, Mohd-Yusof K, Phang FA, Helmi SA (2019) Do teaching and learning environments influence students' conative domain development? In: Research in engineering education symposium 2019 (REES 2019). Cape Town, South Africa
26. Busu TNZ, Mohd-Yusof K, Abdul Rahman NF (2020) Empathy enhancement among engineering students through cooperative problem-based learning. In: Proceedings IEEE TALE 2020
27. Valls R, Kyriakides L (2013) The power of Interactive Groups: how diversity of adults volunteering in classroom groups can promote inclusion and success for children of vulnerable minority ethnic populations. Camb J Educ 43(1):17–33. https://doi.org/10.1080/0305764X.2012.749213
28. Sadineni N, Joshi RK (2012) Mission10X transforming teachers for enhancing employability skills of engineering graduates. In: 2012 ASEE international forum. San Antonio, TX, pp 17.36.1–17.36.10
29. Kurundawade SM, Kavale SM (2018) IIEECP intervention in teaching-learning process: an experience. J Eng Educ Transform 31:192–199
30. Krishnan S (2018) Observations from faculty development workshops in the current Indian context. J Eng Educ Transform 31 (Special Issue)
31. Krishnan S (2019) A phased faculty development program to improve teaching and learning. In: 2019 ASEE annual conference & exposition. Tampa, Florida
32. Holgaard JE, Guerra A, Kolmos A, Petersen LS (2017) Getting a hold on the problem in a problem-based learning environment. Int J Eng Educ 33:10701085
33. Holgaard JE, Dahms ML, Kolmos A, Guerra A (2017) Empowering students to co-construct the PBL environment. In: 6th international research symposium on PBL: PBL, social progress and sustainability. Aalborg University, Aalborg, Denmark, pp 386–398
34. Aalborg University (2023) AAU modules. Problem based learning. https://moduler.aau.dk/course/2023-2024/TECENGPBLE20?lang=en-GB. Accessed 25 Jan 2024
35. Institute for Advanced Studies in PBL Aalborg University. https://www.iaspbl.aau.dk/. Accessed 25 Jan 2024
36. Bertel LB, Winther M, Routhe HW, Kolmos A (2022) Framing and facilitating complex problem-solving competences in interdisciplinary megaprojects: an institutional strategy to educate for sustainable development. Int J Sustain High Educ 23:1173–1191
37. Prince MJ, Felder RM (2006) Inductive teaching and learning methods: definitions, comparisons, and research bases. J Eng Educ 95:123–138
38. Davis LM, Caldwell BS (2023) Utilizing transdisciplinary project-based learning in undergraduate engineering education. Prod: Manag Dev 20
39. Meadows LA, Fowler R, Hildinger ES (2012) Empowering students with choice in the first year. In: 2012 ASEE annual conference & exposition
40. Shepard T (2013) Implementing first-year design projects with the power of choice. In: 2013 ASEE annual conference
41. Chowdhury T, Ortega JD, Twyman C et al (2022) Understanding student experiences in a first-year engineering online project-based learning (OPjBL) course. In: 2022 ASEE annual conference & exposition

42. Stoddard EL, Pfeifer G (2018) Working toward more equitable team dynamics: mapping student assets to minimize stereotyping and task assignment bias. In: 2018 CoNECD-The collaborative network for engineering and computing diversity conference
43. Palmer PJ (1992) Divided no more: a movement approach to educational reform change. Mag High Lear 24(2):10–17

Inclusion of Inter- and Multi-disciplinary Experiences Through the Adoption of Experiential Learning in Undergraduate Engineering

Juhi Bansal, Jamie Gurganus, Clarie Lucas, and Marcello Nitz

Abstract In the rapidly evolving landscape of engineering education, leaders grapple with the intersection of technological advancements, global complexities, and evolving student expectations. The imperative to reevaluate conventional practices and embrace innovative pedagogies has never been more pronounced. Engineering, once confined to specific disciplines, now operates at the convergence of technology, science, arts, and social sciences. Graduates are increasingly expected to navigate this interdisciplinary terrain and address multifaceted challenges. To meet this demand, engineering education must transform, integrating authentic multidisciplinary experiences into the curriculum. This chapter tackles a paramount concern for engineering education leaders: the provision of experiential learning opportunities that enhance students' capabilities to tackle intricate problems affecting the planet and humanity. The chapter presents four case studies from North America, South America, Europe, and Asia, offering insights and best practices from global educational leaders. These cases showcase successful initiatives and methodologies that have significantly impacted engineering education, providing a blueprint for navigating the evolving demands of the field.

Keywords Experiential learning · Multidisciplinary · Interdisciplinary · Curricula change · Change agents · Life-long learning · Undergraduate students · Recommendations for implementation

---

J. Bansal
Ahmedabad University, Ahmedabad 380015, Gujarat, India
e-mail: juhi.bansal@ahduni.edu.in

J. Gurganus (✉)
University of Maryland, Baltimore County, Baltimore, MD 21250, USA
e-mail: jgurganus@umbc.edu

C. Lucas
King's College London, Strand, London WC2R 2ND, UK
e-mail: claire.1.lucas@kcl.ac.uk

M. Nitz
Mauá Institute of Technology, Praça Mauá, 1, São Caetano do sul, São Paulo 09580-900, Brazil
e-mail: nitz@maua.br

R. Kandakatla et al. (eds.), *Academic Leadership in Engineering Education*, Lecture Notes in Networks and Systems 1097, https://doi.org/10.1007/978-3-031-68282-7_8

1 Introduction

1.1 A Subsection Sample

Leaders in engineering education find themselves at a crossroads. The world, marked by rapid technological advancements, complex global challenges, and shifting student expectations, demands a new approach to engineering education. The need to review the practices and introduce innovative and holistic pedagogical methods has never been more critical.

Engineering has therefore transcended its traditional boundaries. No longer confined to the silos of specific disciplines, it now operates at the intersection of technology, science, arts, and social sciences. The demand for graduates who can navigate this complex terrain and solve multifaceted problems is rising. To meet this demand, engineering education must evolve, incorporating multidisciplinary authentic experiences into the curriculum.

This chapter addresses one of the pressing concerns and challenges facing deans, administrators, and leaders in engineering education: providing multidisciplinary authentic experiences through experiential learning situations and aiming to advance students' competencies to face complex problems that threaten the Earth and humanity. Four different case studies are presented representing countries from North America, South America, Europe, and Asia sharing best practices from educational leaders around the globe. It showcases successful initiatives, case studies, and methodologies that have made a tangible difference in engineering education.

Highlight of four case studies The first case is from **Ahmedabad University, India**, a private non-profit research university, which focuses on interdisciplinary learning and research, recognizing the need for problem solvers who can bridge gaps between various fields like engineering, sciences, humanities, and management. They offer diverse, flexible, and interdisciplinary programs encouraging cross-disciplinary engagement based on a unique proprietary pedagogy called ENABLE (Engagement and Application Based Learning and Education).

Led by Juhi Bansal at the Centre for Learning Futures, they have introduced experiential learning through the Independent Study Period, allowing students to work on real-world challenges spanning multiple disciplines. This approach aims to prepare students with the ability to view a problem as a whole and not in silos of singular disciplines and also develops skills like collaboration, critical thinking, and adaptability, essential for addressing complex global issues. The university has faced challenges in convincing faculty to embrace these changes but has shown progress through various initiatives and support systems that promote interdisciplinary and experiential teaching. These endeavours have helped design a one of its kind interdisciplinary studio-based common core called the Foundation Programme at Ahmedabad University.

The second case is from **King's College in London**, where one of UK's oldest Engineering departments was re-launched in 2019. They embarked on a journey to reimagine engineering education, challenging traditional practices and norms.

Through a process of deconstruction, they questioned the purpose of engineering, ultimately defining it as making things better for society. They distinguished between disciplines and do-mains, focusing on the latter to address real-world challenges. Their curriculum encouraged students to create bespoke domain specialisms and adopted a design-centric approach, with design at the core of their education. They emphasized non-functional learning outcomes to develop engineers who are socially responsible, innovative, and globally aware throughout their academic journey.

The third case is from the USA, **University of Maryland Baltimore County**, where Jamie Gurganus has dedicated around 15 years to promoting, researching, and establishing improved engineering education, with a focus on inclusivity regardless of students' backgrounds. She highlights their commitment to innovative teaching and learning practices at the UMBC, emphasizing the university's mission to redefine excellence through inclusive culture, innovative teaching, and civic engagement. They show their efforts to integrate experiential learning and multidisciplinary experiences into the curriculum. They also delve into specific course examples, including an introductory engineering project, service learning, and a senior capstone design program. Additionally, they reinforce the importance of ABET accreditation in maintaining quality engineering education in the United States.

Last case is from the **Mauá Institute of Technology**, in Brazil, where Marcello Nitz facilitated the adoption of the Grand Challenges Scholars Program (GCSP), emphasizing alignment with its own goals and vision. The GCSP was initiated by the U.S. National Academy of Engineering in 2009, which is an undergraduate initiative that encourages students to address complex global issues through multidisciplinary approaches. The program has encouraged students to engage in research, interdisciplinary activities, international experiences, leadership, and service projects. This participation in the GCSP extends the impact beyond the number of students involved, fostering a culture of collective activism for the greater good and a more inclusive approach to engineering education. Ultimately, the GCSP is a driver to circular changes and attract more students to the field by highlighting the societal impact and its potential to create positive change.

Insights from Authors to the Readers: Through these shared experiences, we aim to inspire, motivate, and empower leaders to catalyse positive change within their institutions. We as authors wish to convey to our readers that these are only a sampling of approaches to multidisciplinary authentic experiences. With that in mind, we hope our reading community gains the following from this chapter.

One-Size-Fits-All Approach: Although there are unmistakable and wide-ranging variations in culture, ethnicity, and religion among different institutions, it's important to recognize that the challenges faced within the realm of engineering education bear striking similarities.

Leadership Across the Board: We all hold the potential to be leaders within our institutions. Nonetheless, the realization of this potential hinges on the active engagement of administrators, faculty, and staff at an institutional level. Without their collective commitment, these transformative experiences may not thrive as intended.

2 Case Studies

2.1 Ahmedabad University

Ahmedabad University is a private, non-profit university that offers students a liberal education focused on interdisciplinary learning and research thinking. Our liberal arts education molds independent thinkers and compassionate leaders who go on to engage innovatively with the complex challenges of our societies. Established in 2009, Ahmedabad University is a comprehensive university. We offer undergraduate, graduate and doctoral studies in areas such as engineering, humanities, management, natural sciences, and social sciences.

As our world confronts problems of diverse nature, that cannot be compartmentalized within the confines of a single discipline— Industry and society alike now seek problem solvers who can bridge the gap between technology and humanities, engineering and arts, biology, and mathematics and so on.

The monumental challenges posed by climate change, healthcare advancements, urbanization, and rapidly growing artificial intelligence, are multifaceted issues. It demands students who can collaborate across disciplines, engage in holistic problem-solving, and understand the societal, economic, and environmental dimensions of their work. Traditional silos of knowledge are no longer sufficient to tackle such complex, interconnected challenges. And that's where interdisciplinary and experiential learning become the need of the hour.

While redesigning our undergraduate curriculum in 2016 we recognized the demands of the fast-changing world and thus the need to update the curriculum urgently. The fact that we are a liberal education, multi-disciplinary university helped our cause greatly. We were able to build a curriculum that was:

- *Diverse*: We designed academic programmes and courses that span various disciplines like engineering, sciences, humanities, and management. This diversity enables engineering students to explore subjects beyond their core curriculum, fostering a broader perspective and encouraging cross-disciplinary engagement.
- *Flexible*: Students are encouraged to take courses from other disciplines. There are GERs in varied disciplines like performing arts, mathematics and data, biology etc.
- *Interdisciplinary*: Whether it's our majors, programs or even research opportunities, students are explicitly encouraged to blend knowledge from different fields. These programs provide a structured framework for students to engage in cross-disciplinary learning. Students can also collaborate with faculty and peers from diverse backgrounds on research projects that address complex, real-world problems, helping them gain practical experience in multidisciplinary teamwork.
- *Faculty collaboration at course level*: What better way to get students to appreciate interdisciplinary learning but by getting two or more faculty from different disciplines to collaborate on a course that is based around a problem. For example—faculty in 'Music and Materials' come together to understand why a given material

is preferred over the other in a particular instrument and how does sound/ bass change when you change the material.

Ahmedabad University positions itself as a liberal education-driven university that plans to usher in very close interdisciplinary linkages between areas. Our programs are designed to offer the students unprecedented locus of choice and flexibility in crafting their career interests. The university has introduced a new pedagogy called Engagement and Application Based Learning and Education (ENABLE). It is a novel concept of designing learning of the theoretical underpinnings of their field of study in the process of coming up with real-time solutions for a problem. For example, a course on Economics, Sociology and Anthropology called the Sociology of Bazaar had students study different traditional market set ups (both formal and informal) in detail and draw from them to design an ideal "market" set up.

To bring experiential learning into the forefront of teaching and curriculum design, we came up with a 7-step process for course design called ENABLE. Faculty are encouraged to design new courses and redesign their existing courses using this pedagogy. More and more courses at the University use ENABLE pedagogy now.

One of the first steps we took in this direction was designing the Independent Study Period in 2016. The idea was simple—During the winter break (month of December between the Monsoon and the Winter semesters) students will be required to take a studio-based course for credit. The course will be designed and taught by experts, practitioners, and pioneers in various fields from across the world including faculty at the University, and has to meet the following criteria:

- It should be designed around a passion or interest or should endeavour to solve a real-life problem. Courses that are based on "grand challenges of the society" or a real-life problem that students solve over the duration of the course. These challenges can be based on problems that we see around us on a daily basis: water, food production, health, climate, international relations etc.
- It should surpass disciplinary boundaries (Blend of 2 or more disciplines).
- Driven by theory—The courses should have theoretical depth. The academic concepts taught through the course should be enumerated clearly.
- Course should be hands-on and use a "learning by doing" methodology.
- There should be a tangible physical output at the end of the course which will be displayed at the ISP Expo every year.
- The course is typically spread over 13–14 working days with 40–45 h of faculty-student engagement for 3 credit courses. Additionally, students work (in groups or individually) for about 5–6 h every day on the physical output or solution to the grand challenge.

Some very interesting courses that have been taught over the years are:

Thermodynamics in Historical Wars—This course covers several historical events and relates them with the concept of thermodynamics. The course includes the stories of Napoleonic wars, Jin-Song wars, Anglo-Mysore wars and World wars. Different concepts of thermodynamics such as entropy, heat of reaction, first law of thermodynamics and phase equilibrium are linked with specific historical events. The course contains many activities, namely making rockets, fire lance, chemo-car, heat engines by experimenting with different materials at different temperatures.

Science of making Musical Instruments—A course taught by an Indian classical musician who is also an engineer by training. It became a way for students to not just pursue their interest in music but also fabricate and handle materials, thus developing a deeper understanding of concepts of Physics and the science behind sound, besides collaborative learning and organisational skills.

Innovating for Impact: Using Technology to Create Change—The course is divided into three main sections: Ideation, Development, and Testing. In the Ideation phase, students will learn how to identify problems and opportunities, and how to generate and evaluate innovative ideas using a variety of techniques such as brainstorming, design thinking, and market research [RC9] [J10].

Some problems identified by students included developing a wearable device using Arduino and sensors to monitor the health and wellness of elderly individuals. The device may track vital signs, activity levels, and medication schedules and alert caregivers or family members in case of emergencies. They also created a robot that assists with household chores and developed an automated system for reducing food waste.

Once students have identified a problem and generated a viable solution, they move on to the development phase of the course. This involves introduction to the different technologies and tools available for building solutions, including AI & ML based tools (like CHAT GPT, writesonic, Scribe etc.), 3D Printing, Basic Robotics, and automation, as well as best practices for project management and team collaboration. They also learn about key considerations such as scalability, usability, and data privacy.

Different technologies will be taught from the ground up with no prior technical knowledge as a required skill set.

In the testing phase, students work with beta testers to assess the effectiveness of their innovation. This is a critical phase in the innovation process, as it enables students to gather feedback and make necessary improvements.

Throughout the course, students work in teams to identify a problem and develop a minimum viable product (MVP) to solve that problem. Important life skills like being a team player, leadership skills and communication skills gets honed during this team activity.

Independent study period is how we envision education at Ahmedabad University. Choice-based, liberal, interdisciplinary, hands-on yet deeply rooted in theory. Students are encouraged to go beyond their own domains and pursue courses that are based around their passions or that solve real-life problems.

Courses range from studying thermodynamics in historical wars to building underwater drones to biomimicry. Classrooms turn into studios and thinking outside the box is replaced with reimagining the box altogether.

From building neurocomputers to musical instruments to robots, the Independent Study Period is a time when students take hands-on learning in an academic studio setting to a whole new level.

Benefits of Interdisciplinary and Experiential Learning. Interdisciplinary and experiential education equips students with a diverse skill set, a global perspective, and the ability to address complex challenges.

Students acquire skills like collaboration, adaptability, critical thinking, problem solving and creativity—skills that are invaluable in the real world, thus preparing them for their professional endeavors.

This type of teaching ensures students learn to connect their classroom learnings to the real world and thus solve complex problems like climate change, healthcare, cybersecurity etc. which require interdisciplinary thinking. While this looks easy in theory, there are two major challenges we faced while trying to make big changes.

Challenge 1: Overcoming Faculty Resistance to Change—Convincing faculty to depart from time-tested teaching methods and course designs that have served them well for decades can be a formidable obstacle. Faculty may perceive these approaches as ineffective and hesitate to alter what they see as a successful formula.

Challenge 2: Fostering an Interdisciplinary Mindset Among Faculty—Convincing faculty to embrace interdisciplinarity can be challenging, as it may be perceived as a threat to established disciplinary boundaries. Concerns about the potential dilution of their expertise or added complexity in teaching and research can hinder their enthusiasm for collaboration across disciplines.

Getting faculty to work collaboratively is an ongoing process but over the last 7 years we have found increasing success. Here are a few ways which have helped us inch closer to our goal.

- Demonstrate Impact: Over the years faculty have seen concrete evidence of the positive impact of interdisciplinary and experiential learning on student outcomes.
- Learning Coalition: To facilitate the transition, we instituted something called the Learning Coalition where we conduct periodic peer mentoring sessions and share best practices through workshops and seminars. Through these workshops we endeavor to emphasize how interdisciplinarity can amplify, rather than diminish, the value of their expertise.
- Showcase Best Practices: We regularly highlight successful examples, both within and outside the institution through our newsletters, where faculty members have successfully embraced interdisciplinary and experiential teaching.
- Institutional Support: We always make sure to prove our commitment and support towards interdisciplinary collaborations. These encompass resources for course redesign, technological assistance, and recognition for innovative teaching practices.
- Supportive Culture: We have carefully cultivated a university culture that encourages and celebrates collaboration and interdisciplinary exchange. Faculty are encouraged to set clear objectives and outcomes for interdisciplinary projects.

Looking to the future—We continue to believe that interdisciplinary problem-based learning is the way forward. Through our experiments in the Independent Study Period, we were able to launch our extremely ambitious and much talked about common core Foundation Program. The program is a distinctive pedagogical experiment offering transformative, transdisciplinary, and contextual learning by integrating six domains of knowledge—Constitution and Civilisation, Behaviour, Biology and Life, Materials, Data, and Communication.

2.2 *King's College London*

Would you like to come and lead a new General Engineering course in London and help us to reimagine Engineering Education?

That was the offer given to me in March 2020, a few days before UK went into the first Covid-19 lockdown and the world changed forever. I had already been part of a team running a General Engineering course elsewhere and had been recognised for innovative and inclusive curricula but was straining against the confines of a more traditional University setting. So, when I was offered this opportunity, I couldn't resist the chance to rethink the core of engineering education and purposefully consider how our actions influence the development of the engineers who will ultimately shape the profession.

Within this chapter, I delineate my strategy for revisioning the emerging Engineering Education pathway at King's College.

Deconstructing Engineering. It was clear in those first few months that we were going to need time for deconstruction so that we can re-evaluate traditional engineering practices, methods and systems and challenge established norms. We asked questions as basic as 'what is the purpose of a lab' and 'why do we learn thermodynamics' holding in-depth roundtable discussions on topics such as 'Who is Engineering for'? 'How does Engineering impact the world?', 'How do other disciplines solve problems?'. By immersing ourselves in these questions and in external disciplinary approaches we were able to rediscover Engineering in a new way.

Defining Purpose. Through these explorations we landed on a shared principle: Engineers aspire to:

> Better make things, make better things, and make things better.

This encompassed a comprehensive approach to engineering including consideration of how things are made (the materials used, their environmental impact, the ethics of the supply chain and of automation in manufacturing). It extended to designing and creating higher-quality solutions that function more effectively. Ultimately engineers are driven to enhance the world and society.

Disciplines versus Domains. We distinguish between disciplines and domains. ***Disciplines*** traditionally refer to well-defined fields of study such as mechanical or electrical engineering. These offer depth and expertise in specific areas.

Domains, though, transcend disciplinary boundaries and encompass broader problem spaces such as energy, cities, environment, transport, and infrastructure. Domains also include the space of people, law, politics, and finance. We saw that in traditional curricula students end up far separated from one another as they specialise in their respective domain. It's much like a tree, where students branch off and lose sight of the branches and leaves on the other side, even though they may share some common techniques and approaches.

We wanted a curriculum of possibility (rather than specialty) and identified ways that classical modules might be reimagined and co-created from a domain-perspective.

We came up with a 'knit your own' approach where students can pull together threads from different disciplines and combine these synthesising to create bespoke domain specialisms which juxtaposed traditional ideas and challenge conventional degrees. For example—students may knit together electronics, materials, electrical and thermal fluids modules with an individual dissertation on energy storage to develop a domain specialism which combines traditional disciplines.

Design-centric. At the heart of the curricula is design—taking up 25% of the credits. We deconstructed traditional pedagogy rejecting the blooms triangle where creativity is reserved at the top. Instead, we flipped the structure on its side and backwards that resulted in our 'create-first' approach. Our courses are horizontal, and the approach is such that it motivates students to create a solution that requires evaluation of the situation, understand the needs, and the current approach. Students analyse options and information, analyse themselves and then try out and apply things they might have learned in their concurrent modules. In the process students gain deeper understanding and remember what worked and what didn't. The knowledge gained becomes cemented in their memory by being used.

This whole process iterates each semester.

These projects take students on several learning journey themes from ideation in the first semester through various foci on integration, imitation, inspiration, iteration, and innovation. We know that people are shaped by culture and so it was important for us to establish a learning culture that positively shapes students so that they design better solutions.

We are not only multicultural in our staff and student body but in our exploration of different business cultures. Our authenticity is therefore not based on replicating the real world or real organisations since these vary significantly but on creating our own thriving culture.

Developing Engineers. In developing our programme-level learning outcomes, we introduced the concept of 'non-functional' learning outcomes. Much like non-functional requirements in systems or software engineering, these aren't about what a student should be able to do but instead define how they should evolve as engineers throughout their academic journey. For instance, in year 1 a student should be able to apply methods to solve broadly defined problems, but they should also be more self-aware, culturally competent and embody the traits of a design thinker. By establishing these goals for becoming we are more intentional about the transformative impact that experiential education has on students.

By the end of	Engineering standard, students should	Students will be
Year 1	Apply methods to solve broadly defined problems	Self-aware, culturally competent, creative **design thinkers**
Year 2	Apply knowledge to solve and analyse broadly defined problems using established principles and techniques	Socially, ethically and environmentally aware, innovative and resourceful **problem solvers**
Year 3	Select and apply techniques to solve and analyse complex problems informed by current developments	Commercially & professionally aware, globally and socially responsible **technology innovators**
Year 4	Select from state-of-the art knowledge and understanding to frame, analyse and solve complex problems	Critical, research and state of the art aware **systems thinkers**

2.3 *University of Maryland, Baltimore County*

For around 15 years, I have spent my efforts helping to promote, research and establish the need for better engineering learning. It hasn't been without challenges and working to break cultural barriers to ensure that students, no matter the background, race, religion, or culture have access to become an engineer was an uphill task. With my personal motivation always at the forefront, I take great pride in the ongoing evolution of my institution's commitment to embracing the highest standards in scholarly practices, thereby enhancing the traditional engineering classroom experience.

In this particular case study, I present a concise overview of the mission and vision of the University of Maryland Baltimore County (UMBC), as well as some of the initiatives and projects I've spearheaded. These endeavors revolve around the integration of experiential learning and the promotion of multidisciplinary educational experiences.

UMBC and its teaching innovation. As a minority-serving institution, our former President, Dr. Freeman Hrabowski, and his administration dedicated their tenure to emphasizing the significance of teaching, including offering internal opportunities for faculty to foster innovation. The core of our success in championing innovative teaching and learning practices lies in the institution's commitment to valuing and actively investing in the advancement of these concepts. US News has routinely highlighted our undergraduate teaching programs, ranking #12 nationally. UMBC also maintains its longstanding position as one of the nation's most innovative universities, ranking #15 this year. These earnings are reflected and deeply founded in our mission and vision statement:-

Our UMBC community redefines excellence in higher education through an inclusive culture that connects innovative teaching and learning, research across disciplines, and civic engagement. We will advance knowledge, economic prosperity, and social justice by welcoming and inspiring inquisitive minds from all backgrounds [1].

As a community, our goal was to guarantee that all our students, especially those from marginalized and underrepresented groups, received excellent support and retention. In 2012, the Hrabowski Innovation Fund was officially established and launched, serving as a platform for faculty to genuinely explore, experiment, and implement fresh approaches to learning. This was especially fruitful to non-tenured faculty who had wanted the time and space to try new methodologies.

The College of Engineering and Information Technology (COEIT) has worked to represent this mission and promoted the philosophy of "weaving in" students by focusing on providing accessible pathways to all students irrespective of their backgrounds. This includes implementing more authentic learning into our engineering classes and meeting the students' needs regardless of their prior training. In 2019, COEIT devised a unit, Engineering and Computing Education Program (ECEP), to focus on advancing the engineering and computing professions by establishing multidisciplinary pathways, model and advocate for scholarly, evidence-based teaching within our college, integrate pedagogical discovery, practice, and dissemination through leading edge engineering and computing education research and serve as an incubator for new ideas and programs that shape the future of engineering and computing education.

ABET Accreditation. An important key facet of engineering programs in the United States is meeting our nationally (and internationally, currently serving 40 countries) set standards for quality engineering education. Through the Accreditation Board for Engineering and Technology (ABET), this non-profit non-governmental organization, provides assurance that a college or university program meets the quality standards of the profession for which that program prepares graduates [2]. Engineering programs undergo a rigorous six-year review process, during which they must meet precise student learning outcome criteria. Graduates of an ABET-accredited engineering program gain eligibility for specific licensures, making them eligible for a wide range of federal student loans, grants, and scholarships. This accreditation also enhances their qualifications for various job opportunities.

ABET has identified seven learning outcomes which engineering students must achieve that will prepare them to enter the professional practice of engineering. Among these desired outcomes, there is a consistent emphasis on engineering students gaining a deep understanding of engineering within an authentic and meaningful learning environment.

For example:

Learning outcome 2 explains that engineering students should have the *"ability to apply engineering design to produce solutions that meet specified needs with consideration of public health, safety, and welfare, as well as global, cultural, social, environmental, and economic factors."*
Learning outcome 4 wants students to *"recognize ethical and professional responsibilities in engineering situations to make informed judgements…"*.

To genuinely fulfill these outcomes, our program had to embrace meaningful experiential learning opportunities that offer students a genuine and authentic space for their educational growth.

Class examples. Key highlights of our program include infusing these experiences in earlier programming, service learning and the community, authentic practices—connecting to real challenges, research and clients, and internationalization.

Engineering 101 with Multidisciplinary—UROS Engineering 101 is positioned as an introduction to engineering that covers "thinking like an engineer;" including professional practice, data analysis and curve fitting, estimation, engineering units and dimensional analysis, and the engineering design process. Students are strategically placed into interdisciplinary teams (computer, mechanical, chemical engineering) on a design project, which includes design, construction, evaluation, testing, modeling, and presentation. The purpose of the project is to provide students the opportunity to work in multidisciplinary teams in order to design, build, test, and operate a representative engineering project subject to realistic cost, technology, and time constraints.

UROS is a translation of the Academic Robotics on the Water (AROW) project developed by Captain Jonathan Russell at the United States Coast Guard Academy, and subsequently modified by the engineering 101 instructor team.

Service learning in 1st and 2nd year engineering courses—Service learning in engineering involves the joining of both academic coursework and community service with key features including reciprocity, reflection, and community-expressed needs. Previous studies have shown positive effects of service-learning on a wide variety of cognitive and affective measures, many of which match the criteria of ABET (for example, those dealing with interdisciplinary teams, ethical responsibility, impact of engineering in a global and societal context, and effective communication).

Examples of service learning include Engineering Statics course and Introduction to Engineering Design—In my classes, I have embraced using active learning, problem-based learning, and peer teaching.

As experience and research dictate, application of concepts is one of the most effective ways for a student engineer to learn [3]. In 2012, I read an article from ASEE that discussed integrating service learning into the classroom. I decided to adopt this into my classroom, engaging the students in the engineering design process. For my freshman statics students, this involves working on a project in the greater Baltimore area, designing, and eventually, installing a handicap accessible ramp for a local home in need. For my sophomore design class, this manifests through the semester group design project. The project itself varies, but one recent example included working with a Baltimore City school for children with special needs. Engineering students designed toys for the children to enjoy and aid in their physical therapy.

Senior Capstone Design & Multidisciplinary Capstone with the support of many clients -The Multidisciplinary Engineering Capstone Program is specifically structured to enable students who are pursuing curricula in Computer Science, Mechanical Engineering, and Computer Engineering to participate in a comprehensive system design experience. The primary objective of this course is to instill a strong foundation in team leadership and project management, equipping graduates with the skills needed to make purposeful contributions and positively address the challenges they will encounter in their future careers.

This course also places a strong emphasis on entrepreneurial skills that are essential for functioning effectively in any organization, regardless of its size. These skills encompass innovation and creativity, teamwork, effective communication, and the ability to create value by translating user needs into a fully functional system. Engineers in the industry are often tasked with solving problems that involve juggling budgetary, time, technical, and sometimes even social, ethical, and environmental constraints. In this course, students immerse themselves in an experience that closely mirrors this real-world environment. An Example project included creating a UAV system for a well-known company that is 3D printed. Using a system engineering approach, this ‘super group’ consisted of sub-divided groups of computer and mechanical engineers that tackled aspects of this system in Control and Autonomy, Airframe, Navigation and Communication, and Propulsion.

Students are organized into four-person project teams with the goal of completing a system design project of their own choosing. Additionally, opportunities may exist for students to select projects sponsored by industry partners, nonprofit organizations, or UMBC faculty/organizations. The assignment of project teams and their corresponding responsibilities closely mimics the dynamics of actual engineering organizations. Throughout this process, system engineering concepts serve as the foundation of the development process. Furthermore, teamwork evaluations and assignments focusing on professional ethics reinforce the multifaceted nature of professional engineering practice.

Global Engineering, development of global competencies using COIL methodologies—Global Engineering was designed as a 15-week elective for both senior mechanical engineering undergraduate and graduate students. It was established as an inter-, multi, and trans- disciplinary study and design of engineering technology, entrepreneurship, and the influence of global society and culture on problem-solving. Students use explicit technological examples to explore how different cultures meet the engineering objective (both their paths and final solutions may differ). Topics include the global scope of engineering; differences in engineering around the world; cultural, environmental, and political factors; conducting oneself in a foreign environment; and preparing for an international work/study experience.

Guest speakers present lessons learned in real global engineering cases, where both technical and cultural factors are integral. Students build on these cases by practicing the appropriate application of engineering resources in diverse scenarios.

Using Collaborative Online International Learning (COIL) methodologies, the students at US and Portuguese institutions form multi-cultural teams to design and present their engineering solutions. By working remotely with colleagues at a foreign university, students gain direct experience in cross-cultural collaboration, including joint research, writing, and giving professional and technical presentations.

2.4 Mauá Institute of Technology

As an educational leader at the Mauá Institute of Technology (IMT), my decision to facilitate the Grand Challenge Scholars Program (GCSP) was driven by a deep commitment to nurturing the next generation of engineers and problem solvers equipped to tackle the world's most pressing challenges. The GCSP represents a transformative initiative that aligns with IMT's vision and goals, focusing on instilling in students not only technical expertise but also a profound sense of social consciousness and a global, multicultural perspective. The influence of the GCSP extends beyond the program's participant count, leaving a lasting impact on our academic community and the professional trajectories of our students, reaffirming the significance of this educational endeavor.

The Grand Challenges Scholars Program (GCSP) is an educational initiative for undergraduate students to promote a multidisciplinary approach to solving complex global problems. Launched by the U.S. National Academy of Engineering (NAE) in 2009, the program is inspired by the Global Grand Challenges for Engineering in the 21st Century, announced by NAE in 2008.

These challenges fall into four crossing-cut themes: sustainability, health, security, and joy of living [4]. The main goal of the GCSP is to prepare students to tackle the world's most urgent challenges, addressing critical and complex issues.

Students who complete the GCSP program are expected to develop these five key competencies:

- Viable Business/Entrepreneurship: Understanding the necessity of a viable business model for solution implementation.
- Multidisciplinary: Understanding multidisciplinary engineering systems solutions developed through personal engagement.
- Multicultural: Understanding the necessity of a viable business model for solution implementation.
- Talent: Mentored research/creative project experience on a Grand Challenge-like topic.
- Social Consciousness: Understanding that engineering solutions should primarily serve people and society reflecting social consciousness.
- Participating students are encouraged to engage in learning and research activities during their undergraduate education, such as the following:
- Conduct research or engineering projects related to the identified Grand Challenges.

- Participate in interdisciplinary activities, focusing on integrating various academic disciplines.
- Gain international experience through studying or working abroad or even collaborating with peers around the globe, with an emphasis on global challenges.
- Engage in leadership activities to develop teamwork and other essential skills.
- Get involved in service activities or social entrepreneurship.

These activities are opportunities to gain diverse mindsets and skills. To be accepted in the GCSP network, the higher education institute must give evidence that the program will offer the opportunities and necessary resources and will support, advise, and assess the participant students along their journey.

The Mauá Institute of Technology (IMT) in São Paulo is one of Brazil's most reputable engineering schools. The IMT applied for the GCSP program in 2019 and joined it in 2020. It was not a mere adoption of a detached model. Instead, it stemmed from a shared vision between IMT and the GCSP-NAE goals and reflection on optimising this alignment within the institute's context, vision, and existing activities. The GCSP-IMT was thus conceived with a broader perspective, considering not only the national curricular guidelines for engineering, which already include practices closely identified with IMT, but also emphasising the institute's overarching goal of nurturing professionals capable of innovating through the integration of diverse knowledge domains.

Joining the GCSP network was a strategic decision. Although the number of faculty and students directly involved in the GCSP program is relatively low—40 students out of 2500 undergraduates in 2023—its impact extends throughout the entire community so that the program's influence has reached beyond its participant count.

IMT is actively revising its learning outcomes and learning activities, fostering an environment that nurture graduates capable of addressing complex problems. More transdisciplinary and intricate challenges are being proposed, reflecting the programs' aim to engage students in tackling multifaceted global issues. This extension is paralleled by increased opportunities for international collaboration, enabling students to broaden their perspectives and solutions on a global scale.

Among the active students in the current program, a wide range of projects cover the major areas organised by the challenges of GCSP. The student body is also quite diverse, currently encompassing various engineering disciplines and the Design Bachelor's program offered at IMT. Most of these students have had the opportunity for exchange programs, both within the internal community and at international program seminars, as well as other opportunities provided by the institution, such as presentations to companies and hosting foreign delegations visiting the campus. Understanding some of the projects will contribute to grasping this diversity and how they can be linked to competency assessments.

One of the projects that has gained significant visibility at GCSP seminars worldwide, including the latest 2023 GCSP annual conference, is the project led by a student from the civil engineering program. In his project, the student established direct contact with the Foundation in charge of the Arvoredos' Island, off the coast of São Paulo State. The project's goal was to develop, through Building Information Modeling (BIM), the digital twin of the island. A digital twin is a virtual product created from the concept of BIM and data produced by sensors and measurement instruments. It can reflect the life cycle process of an equivalent physical structure. The main purpose is operational and maintenance management, as it enables a wide range of simulation possibilities. In this specific case, the digital twin of the island will streamline the entire management and decision-making process regarding operational issues for over 90 engineering and sustainability on-going projects taking place there.

As students strive to evolve as system thinkers and adept problem solvers, the curriculum is also being harmonised to integrate the technical and social dimensions of engineering seamlessly. This synthesis is fundamental to the GCSP's mission and underscores the institute's overarching goal of cultivating innovative professionals by effectively integrating diverse knowledge domains. Moving toward impact-centred education necessitates creating and sustaining conditions for thinking and doing within a values-driven culture of collective activism for the greater good thereby moving away from an emphasis on individualism [5], This paradigm shift will demand inclusive innovation and design that values diverse identities, perspectives, and lived experiences.

The approach will embody a culture where everyone assumes the roles of empathetic learners and educators, all committed to an equity ethic and a purpose that transcends self-serving motives to serve humanity [6].

The social perspective of engineering, emphasised within the GCSP framework, also promises to attract more students to this career and reduce the underrepresentation of women. By showcasing the tangible and positive societal impact that engineering can have, more students are likely to be drawn to the field, recognising its potential to drive meaningful change. After all, engineering is for and about people and about making the world a better place.

As the inaugural batch of GCSP-IMT graduated in December 2023, the program's positive influence on the academic community is palpable. The anticipated lasting impact on their professional trajectories underscores the transformative nature of these changes.

3 Key Takeaways

Leaders should be aware of some of the challenges they might face in instigating such change and the various sources of inertia. To successfully implement multidisciplinary and experiential education, it's crucial to identify the points of paralysis, moments where students and faculty might become discouraged. Strategies and support mechanisms need to be in place to help them navigate through these challenges and keep moving forward in their learning and teaching journeys.

Some key takeaways from our experience are shared below to help you to navigate these challenges.

- *Necessity for Curriculum Redesign as a driving factor*: Over the course of writing this chapter, we have united from four very distinct contexts and diverse cultural backgrounds, each with its unique governance of education. Yet we have shared a common problem, goal and solution—that the conventional engineering curriculum, primarily focused on technical subjects, must evolve to foster a broader range of competencies. We believe that experiential and multidisciplinary learning are the best way that we can conceive of delivering engineering education. By integrating inter-disciplinary collaboration and experiential learning, students will be better equipped to address the multifaceted challenges that lie ahead.
- *Inclusive Environment for Change*: To successfully transition from mandate to mindset and to sustain change requires an inclusive environment for multiple stakeholders—faculty, administrators, leadership and students. All need to feel included and actively involved in the process. This requires active listening from all parties to understand all perspectives. The environment for change should be safe for all and each group should have a clearly defined role.
- *Identity and role*: The transformation of Engineering Education involves significant shift in the identity and role of faculty members and students. We move from deeply entrenched disciplinary identity and role as 'sage on the stage' to facilitators of knowledge co-creation. This transition is profound and can expose insecurities. It is necessary to support and address the rebuilding of identity.
- *Resistance and challenge*: We are acutely aware of the inherent inertia in the system, particularly from faculty accustomed to conventional curricula and pedagogy, but also from students. Over-coming this resistance involves many of the above strategies about safety, belonging and identity and keeping the bigger vision and picture in mind.
- *Student adaptivity to new learning modes*: We recognise that students often have a background in rote learning in which they have previously been very successful. Such students may struggle to move towards the more uncomfortable position of embracing complexity and uncertainty and will also need guidance through the transition. This requires a significant amount of unlearning as part of adaption to dealing with wicked problems. This support extends beyond academic learning to include nurturing of adaptability, resilience, and confidence.
- *Local and Global engagement*: Working on community challenges prepares students for global issues and helps students to feel like they belong to something bigger. Community challenges can reinforce the engineer's role in meaningful change in a local context preparing students for less familiar issues later. Hearing about successful projects and working with experienced engineers can further inspire students to believe in their ability to design solutions for a better future.
- *Fostering hope*: Project-based, Multi and inter-disciplinary education immerses students and faculty in complex global issues, building confidence and problem-solving skills, but also raising awareness of daunting problems and repeatedly exposing students to situations requiring large amounts of empathy for people around the world. We need to incorporate hopeful pedagogies into our programmes identifying examples of positive impact which inspire optimism and belief in the

possibility of positive change. We can identify and instill peer-mentoring and coaching, foster creativity and show examples of positive engineering innovation.

4 Conclusions

In this chapter we have explored the motivation for change and described four case-studies from around the world.

In concluding we have reflected on the journey and identified common complexities and challenges as well as some of our observations. In each institution we required a profound cultural shift, a re-evaluation of pedagogical roles, and the courage to embrace change. However, we strongly believe that these changes are needed. The future of education lies in preparing students for the complex, interconnected world they will face, and interdisciplinary and experiential learning is the bridge to that future. We must ensure that our educational institutions shape the leaders and problem solvers of tomorrow and we must acknowledge that our traditional education is no longer fit for purpose. If you are just starting this journey, then we extend not only our advice but also our encouragement for perseverance. Each of our institutions, irrespective of its modernity or tradition, funding, or local context, is dedicated to fostering global change with supportive and invested leadership. Our various mission statements comprise of specific foci on just social development, meaningful service and participation. Engineering should be at the heart of achieving these goals and as Engineering education leaders we share the following values:

- *Passion and Purpose*—We demonstrate genuine passion, enthusiasm for engineering, communities and students and hold a strong belief in purpose.
- *Resilience*—We are open about our own challenges and failures and how we overcame them, sharing that setbacks are a part of learning.
- *Adaptability*—By continuing to challenge ourselves in new experiences and embracing change (for example technology and AI) we build empathy with students doing this for the first time.
- *Interdisciplinarity*—We stretch ourselves beyond the boundaries of our own discipline seeking out others to challenge our approaches and lenses.
- *Advocacy*—We are champions for sustainable, ethical and responsible engineering practice holding these as values rather than constraints on the practice of engineering.

The path may not be frictionless, but you will be making a difference and paving the way for a future in which Engineers are the beacons of hope for the world.

References

1. University of Maryland, Baltimore County. "UMBC Home Page." UMBC. https://www.umbc.edu/. Accessed 15 Nov 2023
2. ABET. "Accreditation." ABET, Inc.https://www.abet.org/accreditation/. Accessed 10 Nov 2023
3. Oakes W, Zoltowski CB, Huff J (2014) Engineering service-learning: a model for preparing students for engineering practice while meeting needs of the underserved. J Eng Educ Transform 27(4):46–56
4. Light J (2012) Case study incorporating service-learning in a statics and dynamics course: the wheelchair ramp design/build. In: 2012 ASEE annual conference & exposition
5. "Engineering Challenges," National Academy of Engineering. http://www.engineeringchallenges.org/challenges.aspx. Accessed 22 Jan 2024
6. Barabino GA (2022) An impact-centered approach to higher education: engineering as a model. Not Alone: Leaders in Focus, July 13, 2022. https://www.elsevier.com/connect/an-impact-centered-approach-to-higher-education-engineering-as-a-model

Reimagining Engineering Education Through Technology

Camilo Vieira, Alejandra J. Magana, Chandan Dasgupta, and Shaheen Hassan

Abstract This chapter presents five case studies that explore various affordances for engineering educators in reimagining education through technology. The first case study delves into the integration of teamwork and technology, addressing challenges in online learning and proposing innovative solutions for effective engagement in teamwork within large-sized classrooms. The second case study explores the integration of computational science and engineering (CSE) and computational thinking (CT) into existing engineering courses, recognizing CSE as a crucial pillar in the digital era. The third case study discusses personalized digital learning through adaptive technologies to enhance participation in engineering education. The fourth case study highlights the nuanced art of managing uncertainty in makerspaces, emphasizing the role of educators in guiding students through productive uncertainty management. The final case study explores the application of learning analytics in engineering design, offering insights into student experimentation strategies. Through these case studies, the chapter aims to provide academic leaders with valuable insights and strategies to redefine engineering education through technology.

Keywords Technology · Engineering education · Teamwork · Learning analytics · Personalized learning · Uncertainty management · Computational thinking

C. Vieira (✉)
Universidad del Norte, 080003 Barranquilla, Colombia
e-mail: cvieira@uninorte.edu.co

A. J. Magana
Purdue University, West Lafayette, IN 47907, USA

C. Dasgupta
University of Twente, 7522 Enschede, Netherlands

S. Hassan
University of Warwick, Coventry CV4 7AL, UK

R. Kandakatla et al. (eds.), *Academic Leadership in Engineering Education*, Lecture Notes in Networks and Systems 1097, https://doi.org/10.1007/978-3-031-68282-7_9

1 Introduction and Relevance for Academic Leadership in Engineering Education

In the rapidly evolving landscape of engineering education and educational technologies, the role of academic leadership becomes key in navigating the transformative potential of technology. As information and communication technologies and computational technologies continue to evolve at a rapid pace, they enable and empower engineering educators in reimaging how teaching and learning occur in the classroom and beyond physical spaces. This chapter delves into five different case studies describing potential affordances that engineering educators may take advantage of in reimagining engineering education through technology.

The first case study revolves around the nexus of teamwork and technology. In today's workplace, engineering graduates require a blend of technical and teamwork skills to navigate the interdisciplinary landscape. In response, instructors integrate teamwork into teaching. But online learning presents challenges. Some of the challenges recorded are the assumptions about social interaction, comfort level with physical proximity and psychosocial aspects. The case study showcases an educational innovation that integrates pedagogical methods and technological solutions to enhance effective engagement in teamwork within the context of large-sized classrooms in higher education. One of the central objectives of teamwork engagement is to also direct focus on the above mentioned psychological aspects so that graduates can seamlessly transit from the educational campus to a professional campus.

The second case study explore the concepts of computational science and engineering (CSE) and computational thinking (CT). In this digital era, engineering students need to be prepared with the knowledge and skills to participate in the computational world they live in, but integrating CSE into existing disciplinary engineering is not trivial. This case study elaborates different approaches to integrating CSE concepts and skills into existing courses.

The third case study discusses the potential for personalized digital learning through adaptive technologies. The purpose is to widen participation in engineering education in order to realize future developments needed to improve global standards of living and the global economy.

Makerspaces, uncertainty management, and materials take center stage in the fourth case study. This exploration delves into the nuanced art of managing uncertainty, which is an inherent and pivotal aspect of collaborative endeavors within disciplines like makerspaces. As engineers and students grapple with various forms of uncertainties—be them being rooted in the nature of the tasks, social interactions, or the availability of resources—the narrative unfolds, highlighting the multifaceted nature of this challenge.

The final case study unfolds in the realm of learning analytics and engineering design. This section describes a case study using learning analytics to characterize student experimentation strategies in design, and how students often conduct limited experiments (if any) to inform design decisions.

In traversing these five case studies, this chapter endeavors to empower academic leaders with insights and strategies to redefine the landscape of engineering education through technology.

2 Case Studies

2.1 Teamwork and Technology

The interdisciplinary nature of 21st-century workplaces requires engineering graduates to possess a fusion of technical and professional skills, such as communication, teamwork, and leadership, among others [1, 2]. Furthermore, as society is becoming increasingly globalized, engineering graduates will work with individuals in different geographical locations who will possess different cultures and backgrounds [3]. Consequently, organizations and governments expect higher education institutions to prepare students to engage as productive members and leaders of global teams [4].

As a response, instructors in higher education have integrated teamwork into their teaching practice [i.e., Chen et al. 5]. Teaching teamwork and collaboration skills in an in-person classroom is substantially different from teaching the same skills when students are instructed asynchronously, remotely, and without in-person interactions. Factors impeding effective online collaboration include: (a) assuming that participants will socially interact merely because the environment makes it possible, and (b) neglecting the psychosocial dimension of the desired social interaction [6]. Online modality requires new forms of presentation and interaction [7, 8] that can address the factors that impede effective online teamwork [9]. Virtual teams need constructive communication and interaction to succeed [10].

Context and Educational Innovation. The case study presents an educational innovation that combines pedagogical approaches and technological solutions that can promote effective teamwork engagement in the context of large-size classrooms in higher education. The case study takes place in the context of a software engineering class with a class size ranging from 100 to 150 students. The course introduces theory and practice related to systems analysis and design, and students have an opportunity to practice those as part of a semester-long project where, as a team, students have to document the functional requirements of a system. The course also introduces project management practices, such as Scrum [11], for delivering the functional prototype. Students work in teams of four or five members throughout the semester. The course has been delivered in three different modalities: fully in person in the classroom, hybrid with a lecture in person and a lecture online, and HyFlex during the COVID-19 pandemic [12, 13]. Regardless of the modality, a weekly session was often devoted to teamwork time.

Cooperative learning. For this, the course followed the five elements of cooperative learning for orchestrating the semester-long project [14]. Accordingly, *to promote positive interdependence*, the semester-long project was organized into four deliverables organized into milestones, and one final project. *Individual and group accountability* was established by having elements of the project delivered in teams and individual contributions. Implementing scrum roles and values promoted *interpersonal and small group skills* where students were expected to take leadership roles, communicate and coordinate each milestone, and manage conflict. The *face-to-face promotive interaction* was enabled by devoting class time to work on the project and facilitating teamwork sessions in person and online, using a collaborative platform. Finally, *group processing* was promoted by the scrum practice of writing team retrospectives within each project submission.

Transformative pedagogy. One aspect leading to unsuccessful teamwork experience could be attributed to cultural differences [15]. To promote students' self-awareness of their cultural backgrounds and help them articulate how that would play out in the teamwork experience was formulated. Then a joint reflection and action approach grounded in transformative pedagogy was introduced [16, 17]. Transformative pedagogy encourages students to critically examine their assumptions, grapple with social issues, and engage in social action [18]. Weekly reflections were introduced during class participation. The students were shown a video presenting a case that dealt with the influence of cultural backgrounds on teamwork [19].

Conflict resolution training. Students were also provided with a training session to empower them with techniques to address conflict [20], and techniques to prevent conflict escalation [21]. The training started with a brief introduction to conflict and conflict negotiation. The expert in negotiation then introduced a three-step approach for solving conflicts [22]:

(1) preparing for resolution,
(2) understanding the situation, and
(3) reaching an agreement.

The second topic of the training consisted of describing the role of the mediator and techniques to serve as a good mediator [21].

Technological workspace and collaboration platform. Technological communication and collaboration platforms such as Microsoft Teams and Slack offer capabilities that can support teamwork productivity such as providing shared workspaces, chat, videoconferencing, file storage and application integration. The individual teams' collaboration spaces are permanent and provide instances of all capabilities such as individual chats per team, and individual sharing spaces. Regardless of the course delivery mode, fully in person, hybrid, or HyFlex, students were provided with a virtual workspace.

Evidence of Student Learning. Evidence of the learning has continuously been revised and disseminated through multiple manuscripts [i.e., 12, 13, 23–26]. Specifically, for this case study, we compared student evidence of their learning before (Fall

2019) and during (Fall 2020) the COVID-19 pandemic. The Fall 2020 offering of the course was HyFlex, with an in-person and an online section of the course.

Team Academic Performance. The academic performance measure consisted of students' grades on the team-based project deliverables. Students from all sections obtained the highest performance on Milestones 2 and 4, with average scores over 90%. Overall performance on Milestone 1 was proficient, with average scores over 80%. However, there were differences in student performance, particularly on Milestone 3. While the average score for the in-person section in the Fall 2019 semester was over 90%, but for the sections in the Fall 2020 semester, the average scores were between 76 and 87%. There were no significant differences between the traditional Fall 2019 environment and the HyFlex environment in Fall 2020 in Milestone 1 and Milestone 4. However, there were significant differences observed in Milestone 2 and Milestone 3.

Individual Academic Performance. The term exam was used as the measure for individual academic performance. Exam scores showed that, on average, students from all sections performed comparably at a 'proficient' level with a mean of 88%. Students in the traditional Fall 2019 semester, the HyFlex in-person sessions in the Fall 2020 semester, and the HyFlex fully online session in the Fall 2020 semester all performed at 'proficient' levels.

Team Self and Peer Assessment. At the end of the semester, students responded to a survey that evaluated theirs and their team performance. The overall average of team performance for the in-person sessions (Fall 2019, 2020) were 'high.' Overall, students identified themselves and their team members as dependable in attending group meetings, contributing positively to group discussions, being a valuable member of the team, and helping others with their work when needed. Students also reported completing work on time or making alternative arrangements, doing work accurately and completely, and contributing a fair share of the overall workload. In contrast, the online session scores were 'moderate' for team processes and team outcomes. Meaning that students in the HyFlex fully online session identified themselves and their team members as somewhat dependable in attending group meetings, contributing positively to group discussions, and being a valuable member of the team. Students in the HyFlex fully online session also identified that they somewhat completed work on time or made alternative arrangements, delivered work somewhat accurately and completely, and somewhat contributed a fair share of the overall workload.

Implications and Contribution. This case study brings together evidence-based practices to promote effective teamwork, specifically, cooperative learning [14], transformative pedagogy [16], and conflict resolution training [21]; and the integration of technology to facilitate different forms of delivery. Research suggests that it is not enough to ask students to work in teams and let team behaviors emerge [6]. That is, for students to develop teamwork values and skills, those need to be taught, facilitated, and practiced in the context of projects [27, 28]. In addition to contributing to new knowledge and evidence-based practices, this case study also provides practical contributions that facilitate teamwork skills (a) in a large-scale classroom with over 100 students and (b) that can be delivered via face-to-face, hybrid and HyFlex modes.

2.2 Computational Science and Engineering

Computational Science and Engineering (CSE) is an interdisciplinary field at the intersection of computer science, mathematics, and engineering.

It leverages advanced computational techniques to solve intricate scientific and engineering problems, utilizing algorithms, simulations, and data analysis to model complex systems and optimize designs [29]. Computational skills are indispensable for engineering students and engineering professionals in all disciplines [30]. They support complex problem-solving, foster adaptability in a rapidly evolving technological landscape, and empower students to automate tasks, analyze data, and optimize designs. Proficiency in CSE not only enhances efficiency in engineering processes but also equips students with the tools to make informed decisions, collaborate across disciplines, and drive innovation in research and industry.

However, preparing engineering students with the knowledge and skills to take advantage of CSE practices offers important challenges [31]. While integrating computing into disciplinary courses is important for engineering students, this may overwhelm students. Learning computer programming is itself a complex learning process [32]. This process requires that students understand the problem that they are trying to solve, how a computer works, the programming language syntax, structure, and semantics, etc. Adding mathematical modeling and disciplinary knowledge makes it even more complex for novice learners [33]. Hence, integrating CSE into the engineering curricula should consider scaffolding strategies to support student learning.

Context and Educational Innovation. This case study explores a curricular transformation of a Materials Science and Engineering program, which included adding a course called "Computation and Programming for Materials Scientists and Engineers" (CPMSE). Students learned about computation in the context of Materials Science and Engineering (MSE), and a set of computational modules integrated into core MSE courses to represent disciplinary concepts [30].

The learning outcomes of this course included:

(i) Write MATLAB programs to execute well-defined algorithms;
(ii) Design algorithms to solve engineering problems by breaking these into small tractable parts; and
(iii) Model physical and biological systems by applying linear systems and ordinary and partial differential equations.

Here, we describe how we supported student learning in this course, and how this course contributed to student learning in other courses.

Results. The CPMSE course adopted an inverted classroom approach and provided different forms of support for students to develop their agency in selecting scaffolding strategies.

Over three iterations, we interviewed students to identify how they took advantage of the following resources [31]:

Video lectures with the slides available.
Worked examples with self-explanation activities as in-code comments.
Teaching assistant and course instructor.
Peer-scaffolding.
Code snippets and,
Test cases.

Pedagogically integrated scaffolds, such as self-explaining worked examples [34], and their alignment with students' problem-solving needs influence the effectiveness of student agency. The course structure, allowing students to watch video lectures at their own pace and access slides for specific syntax issues was valued for flexibility and independence. Worked examples, accompanied by video explanations, saw increased usage in later iterations, when the self-explanation activities were integrated. Low performers favored video explanations while high performers engaged directly with the code.

Expert scaffolding (i.e., TA and Instructor) demonstrated a significant difference between high and low performers, particularly in the third iteration, where all low performers used it for troubleshooting compared to only one high performer who employed it for a specific purpose.

Peer scaffolding exhibited variability, with high performers using it for validation and planning, while low performers mainly utilized it for troubleshooting code. Conversely, the provision of code snippets (i.e., structure of the code in terms of functions and parameters) and test cases emerged as crucial scaffolds for all students, aiding them in finding a starting point to understanding the problem and validating solutions. Challenges arose during the validation stage, which resulted in highlighting the need to explore strategies for gradually reducing reliance on test cases, fostering students' ability to create their validation processes. Students' reliance on provided test cases indicated a potential need for exploration on fading test cases and encouraging students to develop their own validation processes, possibly hindered by a lack of disciplinary knowledge at this level.

The research team also explored the effects of this curricular innovation on student beliefs and disciplinary learning [30]. Our findings suggest that students, exposed to the CPMSE course, consistently reported increased self-efficacy to do computer programming. These increases exceeded those reported by students with prior experiences in general-purpose computing courses. Students who participated in the CPMSE experience reported higher self-efficacy compared to those exposed solely to computational modules integrated into disciplinary courses, especially if the latter group had no prior exposure to computational courses.

Regarding academic performance, students generally enhanced their understanding of disciplinary concepts when using modules implemented through a configuring approach. However, students previously exposed to the CPMSE course demonstrated reliable learning not only from modules with a programming approach but also from those with a configuring approach. Notably, conceptual learning gains were observed mainly in students who had prior 'authentic' computing preparation via the CPMSE course when dealing with modules requiring code implementation or alteration.

2.3 Personalized Digital Learning with Adaptive Technologies (Intelligent Tutors) X Participation

Widening participation in engineering at higher education is a key global marker for the advancement in infrastructure, construction and manufacturing that is required to meet the challenges of the twenty-first century. Without sufficient numbers of engineers from different cultures, backgrounds and socio-economic groups, growth is limited [35] both in tangible outputs, and in humanitarian developments. Using technology to create accessible and individualized engineering education pathways, has significant potential to democratize the engineering education landscape.

In this context, the UK has introduced engineering degree apprenticeships to boost numbers of engineers, required for infrastructure development plans of government [36]. Degree apprenticeships require part-time study and are more technical and workplace-focused in nature than corresponding full-time engineering degrees. One of the aims of introducing engineering degree apprenticeships in the UK is to widen participation in engineering education at higher education institutions, to cohorts typically under-represented in this field. Although the numbers anticipated for socio-economic progress have not been fully realized, they are certainly a step in the right direction.

Developments in artificial intelligence in education (AIED) points to making significant strides towards better representation in higher education and the engineering workforce. Accessible, non-judgmental learning in engineering coupled with effective support at the point where it is needed for that learner.

Context: This case study presents pedagogic factors considered in the design of a digital learning resource for first year engineering degree apprentices, that adapts to individuals' current learning. It provides relevant scaffolding required to navigate successfully through the curriculum at an appropriate pace. The automation of individualized pathways through a subset of threshold concepts [37–41] that is nuanced enough to provide support where required and point to next steps in learning when ready is made efficient using technology. Hinge questions [42] are designed to drive each learner's individualized path through their engineering mathematics module according to their needs.

The evolution of this resource has been from videos with hyperlinked resources and guidance of possible pathways, to the more streamlined current version that relies on learner's responses to multiple choice hinge questions to deduce conceptual understanding or misconceptions.

Evaluation of intervention. Using the three-part model of different areas of the brain from the Chimp Paradox [43] with the Learning Zone model [44] and the hand model of the brain [45] can provide a useful framework for understanding the different elements of support required for learners as they move through different zones. Learning takes place when the level of challenge is stretching but without the level of cognitive overload that leads to negative emotions and actions. If the challenge is not demanding enough, then it may feel tedious unless a valued purpose is provided. Intelligence can be applied to large numbers of data points from the current resource to provide more accurate nuance and personalization for learners. This includes the resource checking for the type of learning required based on whether the learner is situated in the chimp, human or computer brain.

The chimp brain. There are some learners at risk of dropping out. The chimp part of the brain here is attempting an initiation to flight. Intelligent use of past data-points and trends can help pinpoint stress points in learning or predict that if one part of the pathway is not completed satisfactorily, then it might be prudent to check for supporting emotional regulation. It can then aid in developing skills in changing blood flow towards the human brain where learning can resume.

The human brain. While blood flows to this part of the brain, conceptual understanding can take place and feel good. It is important to ensure that the correct level of challenge is presented so the learner stays in a state of 'flow' throughout the learning episode. Too much challenge will engage emotional thinking whereas too little challenge will create boredom since the human brain needs to feel fulfilled by its goal.

The computer brain. This is the part of the brain where routines are executed on autopilot since, in everyday situations one must move from working from first principles to a level of mastery requiring some fluency and acceleration. Learning in this part of the brain is more like working within your comfort zone—so there is a risk of boredom setting in. Hence, it is always helpful to create this kind of learning task with a view to enhancing autonomy and self-regulation while providing pauses for reflection.

Adapting to the different parts of the brain in real time. When designing an adaptive system, the use of intelligence via a large data bank can provide useful common pathways through the materials and resources. These can be used to make offers of suitable pauses in content sharing for resetting such as for meta-analysis, reminders of emotional regulation skills and reflection journals. One of the useful ways of exercising the chimp part of the brain is to allow it to express itself freely of all its frustrations, fears and neuroses. At the next stage i.e., 'boxing the chimp' [43], the application of logic and nuance provide the context needed to make sense of and apply perspective to the chimp's point of view. In doing this, learners realize they have the power and choice to deal with their learning differently. Hence, the intelligent learning system can offer options if it notices the learner struggling or detects particular emotions.

The use of AI to emulate what a good teacher does in real time, i.e., respond to the learner with an appropriate level of challenge at the time it is needed, is at the heart of intelligent tutoring systems [46].

The intelligent tutoring system can use data points and trends to make predictions about what are the suitable next steps for a learner. This can be done at speed for numerous students with no restrictions on time, place or pace.

Current discussion of AIED highlights the important missing component of human empathy, since this is difficult to emulate with machine learning systems that is connected to mathematical algorithms. The point to be taken into consideration is that any system will not only produce, but very likely magnify the bias contained in its programmes [47].

Performing algorithms without context is a good approximation to the opposite of human empathy, and it is clear that there is a need for routines to consider appropriate contexts. For example, UK secondary schools were provided predictions of what students should achieve at GCSE based on various factors ranging from prior attainment in national tests to the students' socio-economic background and heritage. It was upsetting for students when they were expected to be doing worse than their actual attainment due to factors the algorithm used that they felt were unfair. It is currently difficult to achieve this with anywhere near the nuance a human can have since algorithms can only follow defined situations and frameworks and not think through new information and judge for itself its significance. Even if we successfully teach intelligent tutoring systems to check for certain contexts, this can only ever be as good as the human creating the code. The human responsible for the code may well have unconscious biases, hence these are difficult to spot and eradicate. Machine learning routines have the potential to magnify the bias by a large factor, hence vigilance and revision should be factored into the design. For instance, Amazon had to stop using AI informed hiring decisions that showed a gender bias when it became clear the used data favored male applicants.

The large number of benefits of AI in education are largely based on the speed at which routines can be executed. It is far superior to humans—thus reducing time and boredom from repetitive tasks especially for large number of learners, such as: marking or providing targeted feedback or next steps. Automating these creates the opportunity for teachers to spend more time on areas requiring human empathy and judgement, while results of automation are checked for fairness and fitness for purpose. At the same time, providing learners in engineering with individualized routes through learning can allow them to make progress at a rate suited to them at that given moment and provide the appropriate scaffolding needed to fill gaps in knowledge or skills. Providing objective personalization in learning for many hundreds of learners is something that is within the reach of AIED although care must be taken to eliminate biases.

2.4 *Makerspaces, Uncertainty Management and Materials*

Managing uncertainty is an essential part of our daily interactions in the real world. Engineers and engineering students frequently work collaboratively and navigate through various types of uncertainties while solving problems at their workplace and classroom [48]. These uncertainties can emerge due to a variety of reasons such as the nature of the task, social interactions during teamwork, prior knowledge of the team members, their self-efficacy, uncertainty orientation, client interactions, availability of resources, and cultural and social norms of the space. Thus, as educators, it is imperative to understand what uncertainty in the context of engineering education means, what goes into managing it productively in classrooms and workplaces, the role of technology, and implications of productive uncertainty management.

Uncertainty is defined as the subjective feeling of doubt, confusion, or wonderment about a situation or interaction [49]. It is subjective because people have different uncertainty orientations [50]. Uncertainty orientation theory suggests that uncertainty experienced by people is influenced by individual characteristics, degree of uncertainty or certainty present in the situation, and goals of the individual (e.g., achievement-related) [50]. While it is experienced at an individual level, it is mediated by social interactions [51]. Hence, it is important to identify and scaffold at the individual and social level.

Uncertainty can be divided into various categories [52–55]:

Perceptual—where there is difficulty in recognizing the meaning of signs/symbols used for conveying the information.
Cognitive—due to inadequate information, there is limited info processing capacity, ambiguities in the problem or solution space, and
Social—ambiguity in roles/positioning of team members or mentors.

Further, these uncertainties could be rooted in various factors such as the nature of tasks, social context, and individual characteristics [56].

Prior research shows that these uncertainties trigger negative responses like stress, anxiety, fear, and decision paralysis. This in turn triggers high extraneous cognitive load and thus disrupt the learning process [52]. Further, there is a natural bias towards processing certain information and neglecting uncertain information. In an academic context and preparing students for real workplaces, uncertainty needs to be nurtured and proactively engaged with in productive ways [57].

Uncertainty in Makerspaces. Makerspaces are becoming ubiquitous in the engineering context. These spaces afford students to create, tinker, invent and learn by making. During these processes the makers or engineering students may face many types of uncertainties mentioned previously. For instance, uncertainties arising from the nature of materials could foster creativity during making process resulting in better design process and higher quality product [58]. In fact, uncertainty faced by the students as a result of the socio-material exchanges mediated the dynamic meaning making that happen amongst collaborating groups as they generated new ideas, refined and tested them to solve design problem [59]. These indicate that such makerspaces in engineering settings need to be designed carefully keeping in mind the types of materials available and the way the groups are formed.

Productive Uncertainty Management. Productiveness can be defined as an iterative process which helps the student/team make progress towards their goal. Research suggests that uncertainty can be managed in various ways. Chen [55] suggests a pathway where students raise, maintain, and reduce the extent of uncertainty. This pathway is facilitated by the teacher who uses the productive disciplinary engagement framework [60, 61] to problematize the content and raise uncertainty. He then gives students authority to participate in a dialogic problem-solving process, make them accountable for their intellectual work, and give access to relevant resources to establish certainty of knowledge [55]. Students engage in peer-critiquing and co-constructing knowledge as they iteratively proceed towards establishing social certainty or consensus [55].

Jordan and McDaniel [49] investigated the role of peer influence in the context of robotics engineering activity and identified socially supportive and unsupportive peer response patterns. Students used Lego Mindstorms robotics kits to build required structures and programmed the components to achieve various functionality.

Supportive responses included instances of peers sharing uncertainty, jointly addressing it, inquiring about relevant uncertainties, and adapting individual and group actions to facilitate collaborative idea generation and problem solving [49].

Unsupportive responses included peers dismissing uncertainty by viewing them as unreasonable and hindrances to solving the problem and withdrawing engagement [49].

Productive uncertainty management is a goal-oriented behavior driven by students' epistemological framing and positional framing and the interaction between these framings [62].

These framings need to be aligned to each other such that knowledge sharing, and group dynamics can happen effectively. Interviews with expert engineers and designers suggest at least six possible strategies for fostering productive alignment of these two framings and thus productive uncertainty management. This includes acknowledging and proactively responding to uncertainty, maintaining uncertainty for a sufficient period by increasing it before converging/reducing, and seeking feedback from reliable sources. This process helps reduce the uncertainty by clarifying doubts and confusions. In continuation, more uncertainties are introduced that would further guide the knowledge seeking pathway, leveraging conceptual and technical tools/techniques to resolve uncertainties during different engineering phases, monitoring micro and macro progress, and maintaining a positive attitude [51].

This prior body of work suggests that engineering students must engage in self-regulation, co-regulation, and socially shared regulation processes to regulate their own and their group members' uncertainty management. These are all opportunities for using technology for engaging students in deep disciplinary learning and fostering twenty-first century skills. For instance, in a computer-supported collaborative learning (CSCL) context, it was found that groups of students that struggled to navigate through the uncertainties posed by the given ill-structured problem also demonstrated compromised socially shared regulation of learning (SSRL) strategies where they hardly engaged in task and content understanding and planning [63].

2.5 *Learning Analytics and Design*

Learning analytics (LA) serves as a method for assessing open-ended learning processes [64]. Using LA to understand complex learning procedures, such as engineering design, presents an opportunity for engineering educators to enhance their practices and expand existing body of knowledge [65]. In comparison to conventional methods for studying student learning, LA approaches offer advantages by enhancing personalization, accessibility, and efficiency [66]. As students interact with technology, they generate a record of their technological interactions and the outcomes of utilizing such technology (i.e., learning data) [67]. By merging computational techniques with the learning sciences, LA expands research capabilities, enabling a deeper understanding of intricate learning processes and distinctions among students.

In this case study [68, 69], we explore how we can use LA to understand specific strategies of the engineering design process. The primary benefit of the design process is its support for solving real-world problems, which are complex issues with no single solution or approach to arrive at the solution. A solution works for a specific context, with given a set of criteria and constraints. However, when examining students' learning processes, the openness inherent in design poses a significant challenge for assessment. This section describes a model for characterizing and assessing students' strategies in conducting experiments as part of the engineering design process using LA.

Context and Educational Innovation. Energy3D is a free and open-source CAD tool that allows creating 3D constructions and simulating their energy consumption [70].

The CAD tool includes energy sensors and simulators for different seasons that provide information to students for design decision-making. Students completed a challenge to design a house located in Indianapolis that would produce enough energy in a year to be self-sufficient, given a set of design constraints and quality criteria.

Testing hypotheses through controlled experiments is an essential part of the design process. Designers often create prototypes of their ideas and evaluate them through controlled experiments to iteratively refine prototypes until finding a solution that meets the constraints and quality criteria of their solution [71, 72]. Learning by design, relies on students' ability to conduct experiments to generate conceptual changes. Thus, conducting controlled experiments is a skill that all students should develop.

Vieira and colleagues [69] proposed a continuum of experiments implemented by designers with different levels of experience. Novice designers do not generate hypotheses or implement experiments to test their hypotheses [72]. When they start conducting experiments, these tend to be poorly controlled experiments in which they mix many variables, preventing them from drawing relevant conclusions. Advanced designers conduct controlled experiments, isolating a variable and testing its effect on a given design, while others may carry out more complex experimental designs, such as factorial experiments, to study interactions between variables.

This case study shows how to use LA to characterize students' experimentation strategies during a learning-by-design activity in Energy3D. One of the opportunities offered by a simulation tool like Energy3D is that it captures all participant interactions as they work on solving the design challenge. This allows developing evaluation methods for this type of open-ended tasks that focus on the process and not just the product.

The research team aimed to compare these student interaction data with Energy3D with data from a think-aloud protocol where students verbalize their design process. Since many students start conducting poorly controlled experiments, an intervention was designed for students to provide recommendations to a researcher about the need to control variables to draw useful conclusions from an experiment. The case corresponded to a researcher looking to identify the best combination of light, water, and fertilizer for plant growth. She did not control the variables at play, that prevented her from drawing useful conclusions. But after reflecting on the kind of recommendations to give to this researcher, students resumed the challenge, and it was expected that they could integrate these recommendations into their own design process.

Results. In this context, a controlled experiment is defined as the modification of only one variable between two data collection processes. For example, if a student installs a tree in front of a house window and runs the energy consumption simulation to see its effect, they can draw a conclusion about the tree's effect based on the simulation results. However, if this student not only installs a tree but also a solar panel and changes the materials of the house, the results will not tell them much about the effect of the tree itself. Likewise, a factorial experiment implies that students seek interactions between variables. Therefore, it is essential to collect information multiple times, modifying these variables in different ways to check for interactions.

Figure 1 shows the number of different actions a student took per minute. The first three lines show the actions the student took with trees, windows, and solar panels, common strategies students could use to reduce energy consumption. The fourth line shows the number of data collection activities on energy consumption throughout the activity. The blue vertical line represents the intervention to promote the development of controlled experiments. This student only collected information about the performance of their construction once before the intervention. In fact, while building the house, they mentioned incorrect strategies for reducing energy

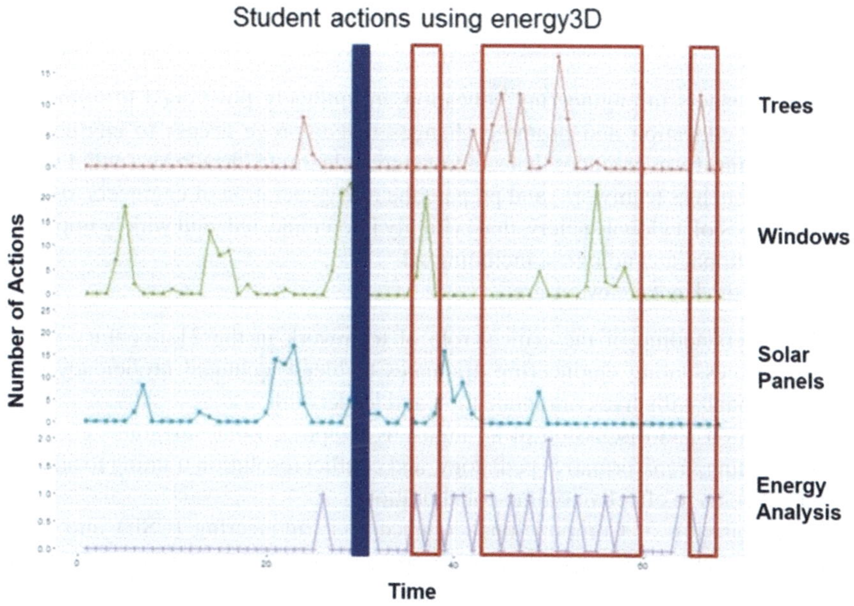

Fig. 1 Number of actions over time per type of actions for a single student. The blue bar represents the time when the intervention took place

consumption: "*I'm going to add another window so that in the summer it can release heat more easily*"—A window without a tree to shade it in the summer can actually increase energy consumption both during the winter and summer. However, the figure shows that the student started conducting controlled experiments after the intervention, where each change was followed by a new test. What the student was saying now supports our hypothesis: "*After adding more windows, energy consumption increased. I'm going to remove some*".

It is essential to clarify that not all participating students managed to draw the correct conclusion about the intervention to change their experimentation practices. While the student we just described reflected on the intervention as "*She didn't control the variables. There are two variables in the experiment,*" another student analyzed the intervention by talking about the importance of sunlight: "*[the researcher] should put the plant by the window...*" However, this section is not about the intervention but about the opportunities offered by learning analytics to understand open-ended activities such as engineering design. Learning analytics take center stage here given their potential to assess the learning process beyond a final product. In this case, learning analytics help identify how students conduct experimentation practices during the design process.

3 Key Takeaways

Academic leaders in engineering education are uniquely positioned to reimagine engineering education and promote changes that improve access to engineering education, transform practices that support student learning, develop essential workplace skills such as teamwork, and prepare the next generation of engineers to solve the challenges of a global society, that is highly interconnected, and widely impacted by computational technologies, including AI.

Some of the key takeaways are:

The understanding of the critical role of teamwork in the 21st-century workplace, necessitating engineering graduates to blend technical proficiency with professional skills like communication and leadership.

Introduction and discussion of an innovative approach that integrates cooperative learning, transformative pedagogy, and conflict resolution training to address impediments to effective online collaboration.

The integration of Computational Science and Engineering (CSE) into engineering education, inverted-classroom approach and diverse support mechanisms, including video lectures, peer scaffolding, and code snippets.

Leveraging Adaptive Technologies and Intelligent Tutors driven by artificial intelligence, is advocated as a strategy to democratize engineering education, accommodating individuals from varied backgrounds.

Automation of learning pathways for first year engineering apprentices, emphasizing the importance of aligning challenges with cognitive load and the potential for further modernization with AI/ML technology.

Highlight on the critical aspect of managing uncertainty in engineering education and makerspaces. It defines uncertainty as a subjective feeling and explores its categories.

Finally, the last case study underscores the significance of Learning analytics (LA) as a powerful method for assessing complex learning processes in dynamic environments, particularly in domains like engineering design.

The evolving dynamism to the study of engineering have put high focus on the need for an equally dynamic and effective leadership team that is ready to think globally, and act quickly. The focus these days are way beyond academics. Innovations, research, and strategies for the betterment of the study of engineering are being floated every day. Cultural exchange happening on a global scale and AI changing the very basic of the rules. It is all very wonderful and invigorating. But to be able to make sure that all factors act at its supreme potential, the leadership team must be equipped enough to step up and act from the forefront.

References

1. Chamorro-Premuzic T, Arteche A, Bremner AJ, Greven C, Furnham A (2010) Soft skills in higher education: Importance and improvement ratings as a function of individual differences and academic performance
2. Vogler JS, Thompson P, Davis DW, Mayfield BE, Finley PM, Yasseri D (2018) The hard work of soft skills: augmenting the project-based learning experience with interdisciplinary teamwork
3. Fruchter R, Townsend A (2003) Multi-cultural dimensions and multi-modal communication in distributed, cross-disciplinary teamwork
4. Lowden K, Hall S, Elliot D, Lewin J (2011) Employers' perceptions of the employability skills of new graduates
5. Chen J, Kolmos A, Du X (2021) Forms of implementation and challenges of PBL in engineering education: a review of literature
6. Kreijns K, Kirschner PA, Jochems W (2003) Identifying the pitfalls for social interaction in computer-supported collaborative learning environments: a review of the research
7. Wallace RM (2003) Online learning in higher education: a review of research on interactions among teachers and students
8. Smyth R (2011) Enhancing learner–learner interaction using video communications in higher education: implications from theorising about a new model
9. Gillett-Swan J (2017) The challenges of online learning. Supporting and engaging the isolated learner
10. Curşeu PL, Schalk R, Wessel I (2008) How do virtual teams process information? A literature review and implications for management
11. Sutherland J, Schwaber K (2007) The scrum papers
12. Magana AJ, Amuah T, Aggrawal S, Patel DA (2023) Teamwork dynamics in the context of large-size software development courses
13. Magana AJ, Karabiyik T, Thomas P, Jaiswal A, Perera V, Dworkin J (2022) Teamwork facilitation and conflict resolution training in a HyFlex course during the COVID-19 pandemic
14. Johnson DW, Johnson RT, Smith KA (2014) Cooperative learning: improving university instruction by basing practice on validated theory
15. Ale Ebrahim N, Ahmed S, Taha Z (2009) Virtual teams: a literature review
16. Nagda BA, Gurin P, Lopez GE (2003) Transformative pedagogy for democracy and social justice
17. Jaiswal A, Patel D, Zhu Y, Lee JS, Magana A, A reflection on action approach to teamwork facilitation
18. Meyers SA (2008) Using transformative pedagogy when teaching online
19. Bahrami P, Kim Y, Jaiswal A, Patel D, Aggrawal S, Magana AJ (2023) Information technology undergraduate students' intercultural value orientations and their beliefs about the influence of such orientations on teamwork interactions
20. Thompson LL (2020) The mind and heart of the negotiator
21. Lewicki RJ, Saunders DM, Barry B (2021) Essentials of negotiation
22. Deutsch M, Coleman PT, Marcus EC (2011) The handbook of conflict resolution: theory and practice
23. Jaiswal A, Karabiyik T, Thomas P, Magana AJ (2021) Characterizing team orientations and academic performance in cooperative project-based learning environments
24. Karabiyik T, Jaiswal A, Thomas P, Magana AJ (2020) Understanding the interactions between the scrum master and the development team: a game-theoretic approach
25. Magana AJ, Seah YY, Thomas P (2018) Fostering cooperative learning with Scrum in a semi-capstone systems analysis and design course
26. Thomas PJ, Patel D, Magana AJ (2021) Characterizing student proficiency in software modeling in terms of functions, structures, and behaviors
27. Salas E, DiazGranados D, Klein C, Burke CS, Stagl KC, Goodwin GF, Halpin SM (2008) Does team training improve team performance? A meta-analysis

28. McEwan D, Ruissen GR, Eys MA, Zumbo BD, Beauchamp MR (2017) The effectiveness of teamwork training on teamwork behaviors and team performance: a systematic review and meta-analysis of controlled interventions
29. Turner PR, Shiflet AB, Cunningham S, Stewart K, Phillips AT, Vakalis IE (2002) Undergraduate computational science and engineering programs and courses. ACM SIGCSE Bull 34:96–97
30. Magana AJ, Falk ML, Vieira C, Reese MJ Jr (2016) A case study of undergraduate engineering students' computational literacy and self-beliefs about computing in the context of authentic practices. Comput Hum Behav 61:427–442
31. Vieira C, Magana AJ, Roy A, Falk M (2021) Providing students with agency to self-scaffold in a computational science and engineering course. J Comput High Educ 33:328–366. https://doi.org/10.1007/s12528-020-09267-7
32. Mselle LJ, Twaakyondo H (2012) The impact of Memory Transfer Language (MTL) on reducing misconceptions in teaching programming to novices. Int J Mach Learn App 1:6
33. Vieira C, Magana AJ, Falk ML, Garcia RE (2017) Writing in-code comments to self-explain in computational science and engineering education. ACM Trans Comput Educ (TOCE) 17:1–21
34. Vieira C, Magana AJ, Roy A, Falk ML (2019) Student explanations in the context of computational science and engineering education. Cogn Instr 37:201–231
35. Campaign for Science and Engineering (CaSE) (2014) Improving diversity in STEM. https://www.sciencecampaign.org.uk/app/uploads/2023/03/Improving-Diversity-in-STEM-2014.pdf
36. The Institution of Engineering and Technology (IET) (2015) Skills & demand in industry. 2015 Survey. https://www.theiet.org/media/8792/2015-skills-survey.pdf
37. Cousin G (2006) An introduction to threshold concepts
38. Meyer J, Land R (2006) Overcoming barriers to student understanding: threshold concepts and troublesome knowledge. Routledge, Abingdon, Oxon; New York, NY
39. Rowbottom DP (2007) Demystifying threshold concepts. J Philos Educ 41:263–270
40. Salwén, H.: Threshold concepts, obstacles or scientific dead ends? Teach High Educ 26(1)
41. Shanahan M, Meyer JH (2006) The troublesome nature of a threshold concept in economics. In: Overcoming barriers to student understanding. Routledge, pp 100–114
42. Black P, Wiliam D (2010) Inside the black box: Raising standards through classroom assessment. Phi delta kappan. 92:81–90
43. Peters S (2020) The chimp paradox. Vermilion
44. Senninger T (2000) Abenteuer leiten-in Abenteuern lernen: Methodenset zur Planung und Leitung kooperativer Lerngemeinschaften für Training und Teamentwicklung in Schule. Ökotopia Verlag, Jugendarbeit und Betrieb
45. Neborsky R (2010) Mindsight: the new science of personal transformation by Daniel, J. Siegel MD. New York: Bantam Books.2010, 336 pp $27.00. In J Group Psychotherapy 60:605–609. https://doi.org/10.1521/ijgp.2010.60.4.605
46. Kuleto V, Ilić M, Dumangiu M, Ranković M, Martins OM, Păun D, Mihoreanu L (2021) Exploring opportunities and challenges of artificial intelligence and machine learning in higher education institutions. Sustainability 13:10424
47. Hwang G-J, Tu Y-F (2021) Roles and research trends of artificial intelligence in mathematics education: a bibliometric mapping analysis and systematic review. Mathematics 9:584
48. Dym CL, Agogino AM, Eris O, Frey DD, Leifer LJ (2005) Engineering design thinking, teaching, and learning. J Eng Educ 94:103–120
49. Jordan ME, McDaniel RR (2014) Managing uncertainty during collaborative problem solving in elementary school teams: the role of peer influence in robotics engineering activity. J Learn Sci 23:490–536. https://doi.org/10.1080/10508406.2014.896254
50. Szeto AC, Sorrentino RM, Yasunaga S, Kouhara S, Lin L (2011) Motivation and performance: uncertainty regulation in Canada and Japan. Motiv Emot 35:338–350
51. Kaur N, Dasgupta C (2022) Investigating experts' and middle school learners' uncertainty management during collaborative engineering design tasks. In: 16th international conference of the learning sciences (ICLS)
52. Hirsh JB, Mar RA, Peterson JB (2012) Psychological entropy: a framework for understanding uncertainty-related anxiety. Psychol Rev 119:304

53. Babrow AS, Kasch CR, Ford LA (1998) The many meanings of uncertainty in illness: toward a systematic accounting. Health Commun 10:1–23
54. Brashers DE (2001) Communication and uncertainty management. J Commun 51:477–497. https://doi.org/10.1093/joc/51.3.477
55. Chen Y-C (2020) Dialogic pathways to manage uncertainty for productive engagement in scientific argumentation. Sci Educ 29(2):331–375
56. Daalhuizen J, Badke-Schaub P, Batill S (2009) Dealing with uncertainty in design practice: issues for designer-centered methodology. In: ICED 09—the 17th international conference on engineering design. Human behavior in design, vol 9. Design Society, pp 147–158
57. Bradac JJ (2001) Theory comparison: uncertainty reduction, problematic integration, uncertainty management, and other curious constructs. J Commun 51:456–476
58. Alekh V, Dasgupta C (2020) Knowing together with materials in collaborative creative making. In: Proceedings of the 17th international conference of the learning sciences—ICLS 2023, pp 1070–1073
59. Alekh V, Dasgupta C (2022) Distributed creativity mediated by social and material interactions in a collaborative pop-up makerspace in India. In: Proceedings of the 16th international conference of the learning sciences—ICLS 2022, pp 1301–1304
60. Engle RA (2012) The productive disciplinary engagement framework: origins, key concepts, and developments. In: Design research on learning and thinking in educational settings. Routledge, pp 170–209
61. Engle RA, Conant FR (2002) Guiding principles for fostering productive disciplinary engagement: explaining an emergent argument in a community of learners classroom. Cogn Instr 20:399–483
62. Kaur N, Dasgupta C (2024) Investigating the interplay of epistemological and positional framing during collaborative uncertainty management. J Learn Sci 33(1)
63. Badhe V, Priyadarshini R, Dasgupta C (2022) Understanding SSRL strategies and its impact on group performance in a CSCL environment: a case study from a project-based learning course in India. In: Proceedings of the 15th international conference on computer-supported collaborative learning—CSCL 2022, pp 187–194
64. Dasgupta C, Magana AJ, Vieira C (2019) Investigating the affordances of a CAD enabled learning environment for promoting integrated STEM learning. Comput Educ 129:122–142
65. Goldstein MH, Purzer Ş, Mejia CV, Zielinski M, Douglas KA (2015) Assessing idea fluency through the student design process. In: 2015 IEEE frontiers in education conference (FIE). IEEE, pp 1–5
66. Thomas S (2016) Future ready learning: reimagining the role of technology in education. 2016 National Education Technology Plan. Office of Educational Technology, US Department of Education
67. Xie C, Zhang Z, Nourian S, Pallant A, Hazzard E (2014) A time series analysis method for assessing engineering design processes using a CAD tool. Int J Eng Educ 30:218–230
68. Vieira C, Goldstein MH, Purzer Ş, Magana AJ (2016) Using learning analytics to characterize student experimentation strategies in the context of engineering design. J Learn Anal 3:291–317
69. Vieira C, Seah YY, Magana AJ (2018) Students' experimentation strategies in design: Is process data enough? Comput Appl Eng Educ 26:1903–1914
70. Adams RS, Goldstein M, Purzer Ş, Chao J, Xie C, Nourian S (2017) Traversing the barriers to using big data in understating how high school students design. In: Design computing and cognition'16. Springer, pp 613–631
71. Zacharia ZC (2007) Comparing and combining real and virtual experimentation: an effort to enhance students' conceptual understanding of electric circuits. J Comput Assist Learn 23:120–132
72. Crismond DP, Adams RS (2012) The informed design teaching and learning matrix. J Eng Educ 101:738–797. https://doi.org/10.1002/j.2168-9830.2012.tb01127.x

Industry Academic Partnerships Leading to Better Preparation of Graduates for Employment

Clare Wood, Bongekile J. Matsenjwa, Jun Dongol Hada, Samuel Rolland, Vasilios Samaras, Jason W. Jones, Gavin Bunting, and Ian Hobson

Abstract Universities play a pivotal role worldwide in preparing graduates for industry and there are proven benefits to the positive outcomes of effective industry–academia partnerships. As we explore in this Chapter, industry-academia partnerships can take various forms and impact at multiple stages of education, with the potential for tremendously positive outcomes in many contexts. Industry-academia partnerships are not only a critical tool in ensuring that the modern workforce remains up to date and readily adaptable to changes in industry practices, but also foster the authentic development of leadership, innovation, and entrepreneurship skills.

Keywords Industry · Graduates · Academia · Internship · Year in industry

1 Introduction

Higher Education is undertaken by students with the expectation of a degree and ultimately a career. Within Engineering, it has been long established that companies are seeking graduates who have critical thinking and problem-solving skills, who can develop creative and innovative strategies and communicate eloquently with confidence. However, the knowledge, experience and skills required by students to find employment opportunities after graduation are evolving and becoming more complex, as industry itself is constantly and rapidly evolving in response to factors such as technology, sustainability, legislation, and the global economy. Indeed, a recent study carried out across industries by Practera [1] found that employers are looking for graduates with skills far beyond the core technical role that such a graduate

C. Wood (✉) · S. Rolland · V. Samaras · J. W. Jones · G. Bunting · I. Hobson
Faculty of Science & Engineering, Swansea University, Swansea, UK
e-mail: c.wood@swansea.ac.uk

B. J. Matsenjwa
University of Cape Town, Cape Town, Rondebosch 7700, South Africa

J. D. Hada
Swiss Agency for Development and Cooperation in Nepal, Kathmandu, Nepal

R. Kandakatla et al. (eds.), *Academic Leadership in Engineering Education*, Lecture Notes in Networks and Systems 1097, https://doi.org/10.1007/978-3-031-68282-7_10

might fulfil, including communication, leadership, conflict resolution, teamwork, self-management, problem-solving, people management, digital and social media skills and service orientation.

It is suggested that students learn most by actively deploying their knowledge via employment, with Khan et al. [2] demonstrating this to be the case across a broad range of skills. It is common in European engineering education that a mandatory placement is included, such as the *stage de fin d'études* in the French engineering training, or even a grounding placement before studies, *Grund-praktikum*, and a technical placement at the end, *F-Praktikum*, as in the German engineering training. Several modern studies have looked to quantify the impact of placements on academic performance and students' qualities and skills, including those by Inceoglu et al. [3], Tanaka and Carlson [4], Duignan [5, 6], and Rolland et al. [7].

In this Chapter we present a portfolio of Case Studies from around the world, centred on Industry-Academia partnerships leading to better preparation of graduates for employment. We present contributions from the United Kingdom (Swansea University) considering both undergraduate and postgraduate Higher Education, as well as documenting the findings of two programmes strategically designed to stimulate the uptake of STEM careers by females in schools and at the graduate stage, based in the Kingdom of Eswatini in Southern Africa and Nepal respectively.

2 Women in Engineering (WomEng) Eswatini – Documented by Dr Bongekile J Matsenjwa, University of Cape Town, South Africa

2.1 Introducing WomEng Eswatini

According to the Global Gender Gap Report 2023 [8] women make up just 29% of professionals in Science, Technology, Engineering and Mathematics (STEM).

In Eswatini, the number of female STEM professionals is less than 10%. Anecdotal evidence from educators in the region indicates that more than 60% of the students who struggle with mathematics or science at high school level later opt for careers that require little or no mathematics or science. Further on, stereotypical views, such as mathematics and science are hard, further hinders the uptake of STEM by females.

Women in Engineering (WomEng) Eswatini is one organization that has done a lot of work in this area. It is a non-profit social enterprise that helps develop high-skilled girls and women, encouraging them into STEM careers. WomEng Eswatini seeks to help close the gender gap in STEM by helping develop girls from primary school right up until they become STEM professionals, through targeted schemes such as GirlEng, LaunchPad and a dedicated Fellowship program. WomEng Eswatini uses a combination of in-person experiential learning and virtual learning to connect, collaborate and seek new opportunities for girls in STEM. Through its proactive, needs-driven

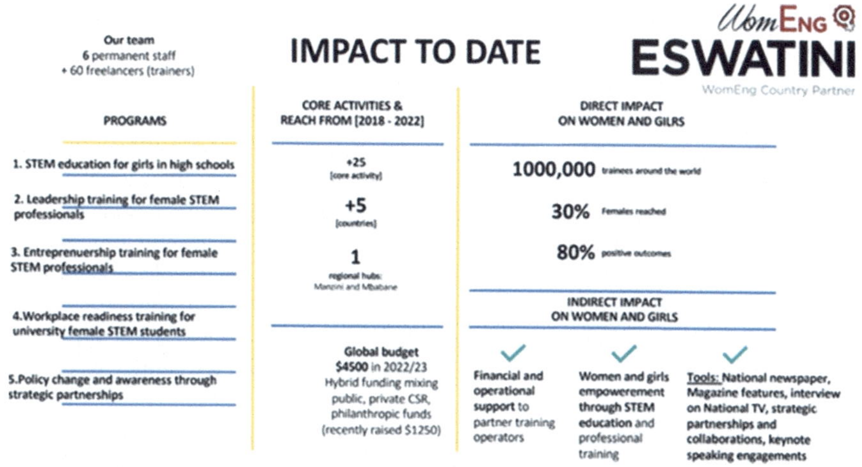

Fig. 1 WomEng Eswatini impact to date. (*Source* WomEng Eswatini, 2023)

bottom-up approach, WomEng Eswatini has been instrumental in developing strong female STEM talent pipelines. At the heart of the organisation is its core philosophy, which provides the framework for all its activities. These include:

1. Promoting STEM education at multiple stages of school education
2. Creating greater awareness and raising the profile of the STEM industry
3. Attraction and retention of females into the STEM sector
4. Supported development of STEM skills, problem solving and leadership
5. Mentorship

The mission of WomEng Eswatini is to develop a thriving, socially and economically engaged female STEM workforce in the Kingdom of Eswatini, through education, industry engagement and mentorship. We aim to do this by inspiring and enabling young women through education and play, raising awareness and breaking STEM stereotypes. The endeavour is to make it challenging, transforming and present new opportunities for STEM leaders. To date, WomEng Eswatini has done tremendous work in Eswatini to attract, develop and retain the pipeline of female STEM leaders through a series of robust programs and interventions at every stage of the pipeline (see Fig. 1).

2.2 *The LaunchPad, GirlEng and Other STEM Education Programs*

The Kingdom of Eswatini has a population of around 1.2 million, with 50.4% of that being female [9]. The employment-to-population ratio for females is 35%. Eswatini uses the Eswatini General Certificate Secondary Examination (EGCSE), a syllabus that is used in more than 70% of the primary and high schools in the country. The completion rate for primary education for both sexes is 65.7% and 30.3% for high schools.

There are many factors which can result in girls dropping out from school, but undoubtedly the absence of female role models to look up to for motivation, poor career guidance at school and the lack of mentorship through education can be identified as key factors. Female students that do complete high school may be influenced to avoid the STEM sector as their career option because of stereotypes, such as the belief that STEM is difficult, the lack of visible female role models, lack of onward career guidance or a lack of access to further STEM education.

The standard curriculum provided in schools is limited in its STEM provision and so students often suffer from a lack of meaningful exposure to initial aspects of STEM education.

The LaunchPad program, shown in Fig. 2, addresses these challenges by connecting professional women in STEM with girls in primary and high schools. These STEM professionals serve as role models to the girls and provide motivation for them to strive towards STEM careers, enhancing the school curriculum and providing career guidance. To date, 3000 girls have benefited and more than 60 schools in Eswatini have been reached through these workshops.

The GirlEng Program targets girls who have indicated an interest in STEM careers. They are then exposed to workshops where they develop practical STEM-related projects which focuses on solving challenges within their communities. This gives them a deeper understanding of STEM in the world. These girls are then connected with female STEM professionals in their areas of interest for coaching and mentorship.

To date, 1500 girls have benefited from these workshops, and WomEng Eswatini has collaborated with 10 schools in Eswatini, establishing five GirlEng Clubs.

WomEng Eswatini has identified that there is a great need to enhance the spectrum of STEM subjects that students are exposed to within high school. Currently, students are only taught chemistry, biology, and physics, with around 40% of schools lacking the appropriate facilities to carry out their practical experiments. By collaborating with the Ministry of Education in Eswatini, Eswatini College of Technology and

Fig. 2 **Left**: WomEng Eswatini LaunchPad. **Right:** WomEng Eswatini GirlEng program

the Taiwan Technical Mission, WomEng Eswatini provides free practical training on basic and enhanced electrical engineering and automation, benefitting around 200 high school girls.

WomEng Eswatini has also established an annual STEM career fair in collaboration with the Royal Science and Technology Park (RSTP), enabling students from primary and high schools to participate in STEM themed competitions and make contact with STEM companies and professionals. Around 10,000 students, both male and female, have benefitted from this program.

WomEng Eswatini career development workshops, run in partnership with STEM companies, professional mentorship schemes and networking events have all contributed directly to an increased selection of STEM careers by girls in high school, with 10 female engineering graduates known to have directly benefitted by being employed by engineering companies. WomEng Eswatini is also working with local and government organizations and policy makers to address the gender stereotypes affecting women in STEM in the workplace.

2.3 *The Fellowship Program*

The WomEng Fellowship program, run in partnership with The Royal Academy of Engineering and AESAP (the registration Council of Architects, Engineers, Surveyors and Allied Professionals), is a technical entrepreneurial challenge and employability programme for the best and brightest female engineers. It is a WomEng's flagship programme for developing leadership, employability, and entrepreneurship skills ~~for~~ in female engineering students. At the heart of the Fellowship is the entrepreneurial innovation challenge, in which Fellows pursue business solutions to global challenges, whilst preparing for the industry. The outcomes include Fellows with increased entrepreneurial and leadership skills, enhanced networks between high-skilled graduates and employers, business mentors and seed capital. Another outcome of this program is an increased awareness within industry and employers to the huge potential of employable high caliber female engineering graduates.

Disciplines benefitting from this scheme include mechanical, civil, chemical, electrical, electronic, industrial, aeronautical, construction studies, mechatronics, telecommunications, mining, and metallurgy. The qualifications awarded ranges from national diploma, higher national diploma, BSc, BEng, BTech, Master's and PhD. 48 Fellows participated in the program from across the Kingdom of Eswatini.

3 The Impact of a Year in Industry on Academic Outcomes in the Context of UK Undergraduate Higher Education In Engineering—Presented by Dr S. A. Rolland, Dr J.W. Jones, and Prof. G. Bunting of Swansea University, UK

3.1 Introduction

Year in industry (YII) schemes provide an opportunity for students to develop new or enhanced skills beyond the academic environment. It also helps them explore further studies. In a recent research study carried out at Swansea University, we aimed to quantify the impact of the year in industry placement scheme on academic outcomes for undergraduate engineering students. The work can be read in full in [7]. The most significant shortcoming of the existing material on this subject in literature is the variation in the sizes of the cohorts included and inconsistency in the approach used to quantify the benefits. Hence, our new study aimed to quantify the academic impact of the year in industry placement for engineering students, with statistically significant cohort sizes, as well as to provide an insight into how grades are affected according to other factors, such as the point within the degree scheme that the placement is undertaken.

3.2 Context and Analysis Method

Our study was set in the context of placement schemes as structured in a typical degree scheme in the United Kingdom. Placements are not a mandatory requirement of the undergraduate training cycle for the accreditation of Engineering Council [10] approved Higher Education degree. However, they can be proposed to undergraduate students as an option to enhance skills and employability. This takes the form of a "sandwich year", where the placement year is added between two academic years. The placement must be at least of equivalent duration to an academic year (40 weeks). Although it is not prescribed that the total duration of the placement is undertaken with a single host organisation, it is consistently the case in practice. The time spent on placement by the student is academically recognised in the curriculum but cannot be used in lieu of any taught component of the degree and is a non-scored component of the degree programme. Students build a portfolio over the course of their placement, with a process of quarterly reporting which aims to provide evidence of the students' development against the UK-SPEC framework set by The Engineering Council [11].

There are two possible routes to undertake a placement, depending on the duration of the students' degree programme—3-year Batchelor of Engineering BEng or 4-year Master of Engineering MEng, as illustrated in Fig. 3.

Whilst this can lead to difficulties in the later analysis of student outcomes, the fact that students may undertake a placement at separate points of the curriculum may be

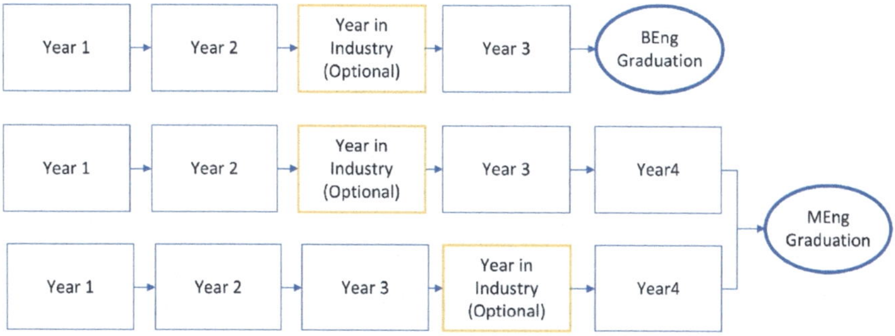

Fig. 3 Degree pathways available for the undergraduate BEng and MEng schemes

used to identify whether the placement has an impact, or whether the students who were going to have better outcomes are simply the students who secure a placement. This present study offers a review of the impact of a year-long placement based on a large data set of student information and records of academic performance collected in the period 1999 to 2021. The data includes all students eligible for the year in industry programmes who completed their studies, a total of 21,279 student-years, representing 6124 distinct students. To be eligible for placement, students are required to maintain an average grade of 55% or more at the end of their second year of study.

3.3 Findings—Batchelor's Degree Routes

In Fig. 4 it can be seen that the students who undertake a YII placement and go on to graduate with an engineering degree with YII outperform marginally their peers from the start of their studies; 0.5% in year 1 and 1.2% in year 2, prior to the YII placement, and outperform significantly their peers by 5.7% in year 3, after the YII placement.

Fig. 4 Average grades per year for YII and non YII cohorts; 5.7% YII improvement in Year 3

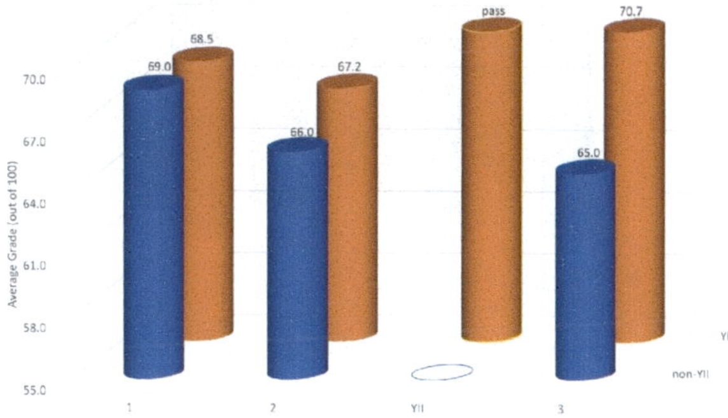

The similarity in trends across groups prior to placement, and the divergence in grade evolution after placement supports a correlation between the year in industry and the improvement of grades. Detailed analysis of the data via t-test (detailed in full in [7]) proves that the non-YII and YII groups are statistically distinct and therefore BEng students who undertake a year in industry achieve higher final-year grades than would have been expected if they did not.

3.4 Findings—Undergraduate Master's Degree Routes

A separate analysis of the Engineering Master's degree (MEng) cohorts is of interest because students can (and do) undertake their placement at two distinct points in the curriculum—between years 2 and 3 or between years 3 and 4. We can therefore determine into whether any effect is by causality or correlation. The comparison of student performance, as shown in Fig. 5, for non-YII routes YII with placement between years 2 and 3, and YII between years 3 and 4 shows that the uplift in academic performance is strongly linked to having undertaken a YII placement. Students who undertake a placement between years 2 and 3 then experience and end of year 3 uplift of 4.1% with respect to the non YII route cohort and 3.5% uplift with respect to those who do later go on to undertake a placement between years 3 and 4.

By the end of year 4, the difference between YII cohorts is reduced; YII students who undertook the placement between year 2 and 3 still maintain an uplift of 2.6% with respect to non-YII Master's students, whilst YII students who undertake the placement between years 3 and 4 experience and uplift of 2.7% with respect to non-YII Master's students.

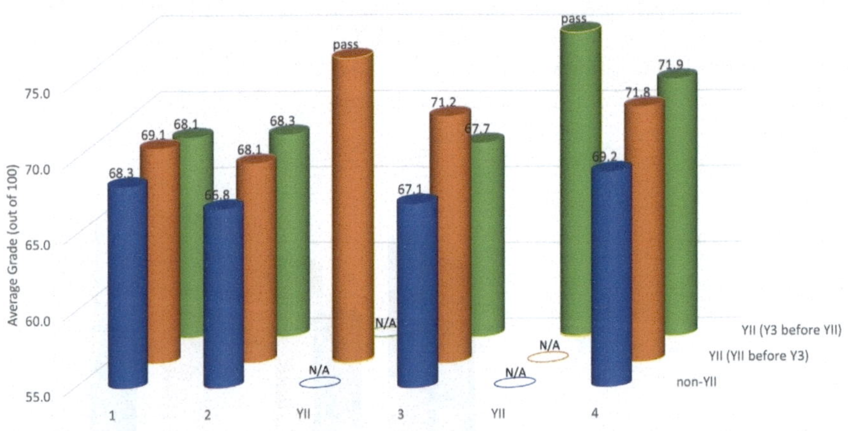

Fig. 5 Average Grade per Year in Master's degree schemes

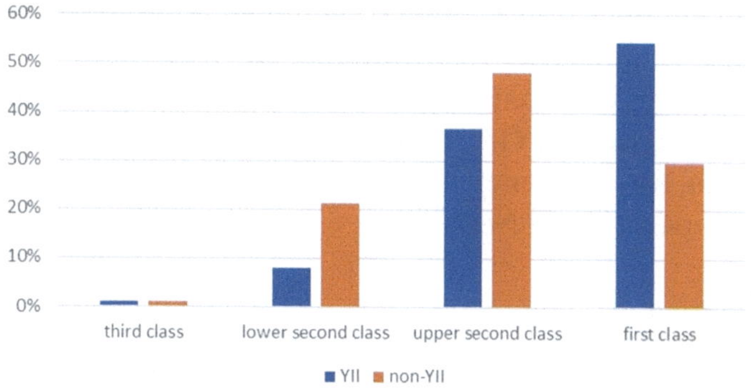

Fig. 6 Comparison of degree outcomes for YII eligible students. First class: 70% +, upper second class: 60–69.99%, lower second class: 50%-59.99%, third class: 40–49.99%

Very extensive analysis of the data, presented in full in [9], provides evidence that the students who do undertake a placement are not statistically different from their peers who do not (but who are eligible to do so) and that the uplift following the year in industry is a statistically verified outcome. Based on the immediacy of the uplift upon return from year in industry shown between MEng cohorts, it is credible that it may be acquired chiefly by experience, knowledge, and skills rather than just acquired maturity.

Ultimately, the year in industry is a significant contributor in the achievement of good degrees. Figure 6 demonstrates very clearly the significantly altered distribution of degree classification between students who graduated with a year in industry degree and those who were eligible, yet did not go on the year in industry.

3.5 Key Takeaways from the Study

1. BEng students achieve significantly better grades in Year 3 after a YII placement.
2. MEng students also achieve better grades in the years of study following a YII.
3. All students' uplifts in grades upon return from placement can be shown to be outside the expected grade evolution based on the non-YII control group.
4. Students in the lower eligible grade categories prior to placement experience a greater grade uplift in the years of study following the YII.

4 Internship for Fresh Female Graduates—Collaboration Between Engineering Institutes and Development Programmes in Nepal—Presented by Jun Dongol Hada, Senior Programme Officer, Rural Infrastructure Portfolio, Swiss Agency for Development and Cooperation in Nepal

4.1 Introduction

The Swiss Agency for Development and Cooperation (SDC) is providing Technical Assistance (TA) to the Government of Nepal (GoN) through the Department of Local Infrastructure (DoLI) under the Ministry of Federal Affairs and General Administration (MoFAGA) in the implementation of Rural Roads (RR), Trail Bridge (TB) and Motorable Bridge (MB) programmes in Nepal since last several decades. While the SDC tries to promote workforce diversity, it has struggled to meet the number of female engineers in these programmes as this profession is primarily and historically dominated by men. There were only handful female engineers in the programmes. Therefore, the internship programme was started to train the young fresh female graduates in the rural roads programme initially in the year 2008, with six female engineering graduates deputed in six operating different districts of the programme. With its success, it continued in the trail and motorable bridge programmes from 2011 onwards until the current time (2024).

So far, approximately 140 fresh female civil engineering graduates have been trained in Nepal and are working in different industries within Nepal and abroad. This includes the government systems and outside within the private sector contributing their knowledge in project design, implementation, management, and supervision. This is an excellent example of how collaboration between development programmes (as industry in this context), engineering institutes (academic in this context) and female graduates bridges the inclusion gaps and ulfils the need of qualified and trained human resources in the engineering industry within the government, non-government, and the private sector.

4.2 Aims of the Internship Programme

This internship programme is designed with a broader objective of transforming fresh graduates into competent technical professionals in the fields of roads, trail and motorable bridge building, so that they are available for the much-needed economic development of Nepal. More concretely, the main objectives are:

- To equip fresh graduates (Intern Engineers) with knowledge and skills in roads, trail and motorable bridge designs and construction.

- To help fresh graduates gain experience in these sectors by giving them space to work in a team of national and international bridge professionals.
- To produce an inclusive human resource in the bridge sector, who are limited but in high demand in the job market.
- To build interest/awareness of road and bridge construction as a field of work to a growing population of female engineers.

4.3 How Was the Collaborative Approach Set up?

Every year, all the Government and private engineering institutes that produce civil engineering graduates are identified and contacted by the Programmes, in collaboration with the engineering institutes, to encourage potential female graduates to apply for the internship opportunities. Numbers of applications are generally in the hundreds, therefore a written test and general interview for aptitude are conducted. Based on the written and aptitude tests, and with consideration of ethnic backgrounds (with those who have been historically discriminated against being given preference), around 20 female graduates are chosen from all over the country to take part in the internship programmes.

In total the training, coaching, and mentoring of the programme lasts between nine months to one year, with each intern assigned a dedicated mentor and coach to provide one-to-one support, as well as being provided with a stipend and other benefits as per the policies and provisions of the GoN's Labor law and standards. At the start of the programme, interns undertake an intensive 35-day theoretical training on trail and motorable bridge design and construction. After this, interns are divided into small groups of two or three and deputed to the respective provincial offices of their choices or their residential provinces, where they are deputed to the trail bridge programme for the first five to six months and then the motorable bridge programme for 5–6 months. In the Provincial offices they learn and experience the real world of work with the trail and motorable bridge building processes alongside the staff and professionals of these programmes.

4.4 What Were the Outcomes?

To date, the internship programme has coached around 140 female graduates. Most have continued to work with the GoN, which is a major success story for the internship programme. Some have opted to pursue higher studies, including travelling abroad, before returning to continue to work in Nepal.

Ms. Sanju Thapa Shrestha, one of the first intern graduates. *"I worked as an intern Engineer in District Roads Support Programme (DRSP) in 2010–11 and then continued as a District Roads Engineer until 2013. This internship gave me a great platform for my professional growth, the opportunity to explore the country and be* *familiar with the real Nepal and help its people. They used to say that "kasto fuchhi engineer le pani kasto ramro kam gareko—what a small engineer has done such a good work". I also won a scholarship to pursue my Master's degree at the Asian Institute of Technology (AIT) in Thailand. Currently, I am working as an Infrastructure Development Expert in the Provincial and Local Governance Support Programme of United National Development Programme, supported by various development partners and the GoN* (Fig. 7).

Ms. Poonam Pandey, graduate intern in 2019. *"Bridge internship was the launchpad of my professional career. It gave me the opportunity to be part of a very healthy working environment. During my internship I experienced many highs and very few lows. My best memories are of the time I spent with my colleagues and getting the opportunity to work with so many fellow women, which is very rare in our field of work. I am currently working as a government officer in Nepal Electricity Authority, as a Civil Engineer. My goal in ten years is to reach new professional heights and make a difference in every work I do* (Fig. 8).

Fig. 7 Ms. Sanju Thapa Shrestha

Fig. 8 Ms. Poonam Pandey

Fig. 9 Ms. Beebika Sharma

Ms. Beebika Sharma, a graduate intern in 2020. *"My internship at the bridge programme has been instrumental in catapulting my career to new heights, not only by providing hands-on experience in the engineering field but also by fostering a profound boost in my confidence. Throughout my time as a trainee, I distinctly recall young girls approaching me, expressing their aspirations to excel in their studies and pursue a career in engineering, inspired by the work we were doing. At present, I am based in Lumbini Province, fulfilling the role of Provincial Bridge Engineer for Motorable Roads Bridge Programme of the GoN, technically being assisted by the SDC. In a decade, my goal is to hold a Master's degree and utilize my expertise in bridge engineering to make a positive impact in the lives of others, ideally here in Nepal* (Fig. 9).

4.5 What Was the Learning?

The learning in the beginning was that when we take the best graduates, they will also have applied for other opportunities. Some interns might consider leaving the programme without completing it.

To avoid this disruption, we strictly applied the rule that—One must complete the programme and failure to do so would require repayment of the stipends received.

Second was the flexible timing for interns—trainees are given flexible time for their personal growth.

Most of the trainees continue to study for the civil services examinations of the GoN in parallel to pursuing the internship programme. Hence, it is also strictly monitored that interns follow the guidance provided by the mentor and supervisor and carry out all tasks and assignments given to her and get feedback regularly.

Assignments must be completed on time to achieve good performance results.

In the early stages, it was necessary to adjust the timing of intake, theoretical sessions, and field deputations to account for the halt in construction work during the monsoon season. The programmes therefore ensure that the interns learn the practical work such as site layouts, construction works and supervision works during the construction seasons while during the monsoon season, interns tend to be more occupied with office based work, such as reviewing designs, cost-estimates, studying the contract documents and so on. Finally, we also grew in knowledge that it was very important to provide interns with a safe working environment in which they were also protected from any potential physical and mental harassment (Fig. 10).

Fig. 10 **Top left**: Female interns in the field. Photo Credit: Thibault Gregoire. **Top right**: Intern graduates for the year 2023. Photo Credit: Vinita Thapa, SDC. **Bottom left:** Graduate interns with the Swiss Ambassador. Photo Credit: Jun Hada, SDC. **Bottom right:** A female intern carries out the survey of a bridge. Photo Credit: SDC

4.6 Moving Ahead

One of the SDC supported programmes presented here unfortunately came to an end in November 2023 and one will end in December 2024 respectively. However, the SDC has mandated the other SDC funded programmes to start internship programmes in other sectors as well. As a result, the other programmes, particularly, the programmes on technical and vocational skill development, have also started working together with the private sector industries and the academics for promoting on-the-job apprenticeship programmes to produce qualified technical human resources as demanded by the markets and industries.

4.7 Conclusions and Key Takeaways

This internship programme is very beneficial to all three involved parties – graduates, employers/industry, and academic institutions. Graduates end up better prepared for immediate employment and the confidence that they gain from the internship programme prepares them to further build their career goals.

Employers gain from the improved quality of capable and inclusive human resources with practical experience who are ready to get their hands and feet dirty on the ground.

The wider engineering industry also benefits ~~by~~ with a diversified workforce by employing competent female engineers, ~~with~~ who has the same or better efficiency and outputs then their male counterparts.

Academic institutions also gain from the goodwill of these programmes because organisations partner with them and hire ~~take~~ their fresh graduates. Considering their academic excellence in each year. This gives an energy boost to the institutions as well as students to excel year on year.

Therefore, this is a very successful example of "industry-academic" partnerships leading to better preparation of graduates for employment.

5 Industry Linked Delivery of the Postgraduate MSc in Engineering Leadership & Management at Swansea University—Presented by Dr Clare Wood, Dr Vasilios Samaras and Mr Ian Hobson

5.1 About the Programme

The MSc in Engineering Leadership and Management (ELM) programme at Swansea University represents a unique and dynamic collaboration between the Faculty of Science and Engineering and the School of Management.

This synergistic approach provides students with a holistic perspective, enabling them to bridge the gap between technical excellence and effective leadership within the engineering domain. ELM is not an isolated academic pursuit but rather a dynamic conduit to the world of industry. Through direct links with prominent industrial players, students gain access to real-world insights, industry best practices, and invaluable networking opportunities. These connections offer a first-hand look into the challenges and opportunities that engineers encounter in their careers.

The core mission of ELM is to nurture a deep understanding of management, leadership, strategy, and operations within the context of engineering so that it provides the students with a smooth transition from the academic campus to the professional world.

Beyond theoretical knowledge, the program places a strong emphasis on practical skill development. Students are guided in improving essential competencies such as effective communication, problem-solving, project management, and strategic thinking. These skills empower graduates to navigate complex engineering challenges and drive innovation within their organisations.

5.2 The Structure of ELM

The ELM programme adopts a condensed, high-intensity format, where a total of nine distinct taught modules are compressed into two- or four-weeks intensive teaching blocks, as displayed in Fig. 11.

The exception to this structure is the Strategy Module, which serves as the backbone to the programme.

This critical module is delivered every two weeks and within it, students engage in major group project work with tangible, real-life challenges presented by external organizations. This hands-on experience provides a dynamic learning environment that fosters practical skills and encourages students to apply their knowledge to actual business scenarios.

A notable advantage of the ELM's condensed schedule is the regular interaction it facilitates between students and the programme leadership team, including the

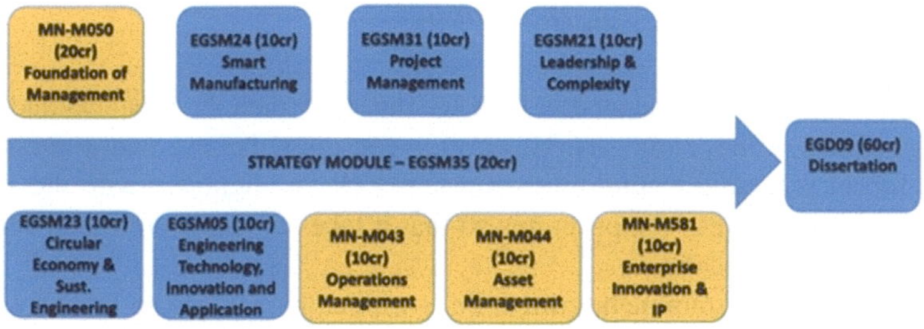

Fig. 11 Degree structure of ELM

Program Director and Academic Mentor. These frequent touchpoints create opportunities for meaningful discussions and continuous feedback from the students. This ongoing dialogue proves invaluable in identifying areas where the program can be refined and enhanced to better meet the evolving needs of the participants. Often, the feedback received from students leads to immediate improvements that can be implemented within the same academic year, ensuring that the ELM programme remains current, responsive, and aligned with the ever-changing landscape of executive leadership education.

5.3 *Entry Level Assessment*

Our educational institution prides itself on fostering a vibrant and inclusive community, comprising students from diverse geographical origins, backgrounds, and a wide spectrum of experiences. ELM reflects a dynamic blend of participants which includes fresh graduates' eager to embark on their academic journeys, professionals aspiring to enrich their knowledge, and ambitious entrepreneurs managing their own businesses.

Owing to this diverse composition of our classrooms and collaborative spaces, students can draw from a multitude of perspectives, tap into various skill sets, and gain insights from the unique journeys of their peers. This multifaceted environment cultivates creativity, adaptability, and a deeper understanding of the global landscape.

Recognising the uniqueness of our student body and also acknowledging the limitations of relying solely on traditional academic qualifications as a measure of a candidate's suitability, we employ a holistic approach to our admissions process. Our commitment to inclusivity and fairness is evident in our practice of conducting interviews with individuals who may not meet the traditional academic criteria. During interviews, we delve into the background and experience of applicants, considering their motivations, life experiences, and aspirations. We believe that this approach allows us to identify promising candidates who may have been overlooked by more rigid admission criteria and fosters an educational community where each student's unique contributions are celebrated and valued.

5.4 Teaching Innovation and Impact

The ELM programme is designed to leverage the strengthen the Business and Engineering schools of the University. However, we strongly believe that it is essential that as for an academic institution to constantly check alignment with the market requirements and refresh the programs so that it takes into account the changes in the professional landscape.

Therefore, the programme has direct links with employers via an Industry Advisory Board, as well as a great many other collaborative contributions from Industry. This provides a diverse range of inputs including projects, supervision, guest lectures and the like.

Companies involved in such collaborations in the recent years includes the international manufacturing company TATA and SEDA UK, a prominent packaging products company in the food industry sector.

Interactions such as this allow the ELM cohort to gain exposure to a range of organisational culture and environment. Moreover, the programme develops the students not only academically, but also as individuals. We systematically seek employers for feedback on our students whom they have hired and the feedback we receive, is that our students are well-rounded in their capabilities and ahead of their peers in meeting aspirations. This serves as a moral boost for the institution, faculty, and students alike to engage better and exceed expectations.

Each year is a unique journey, characterised by a diverse array of subjects and topics that are dynamically shaped to cater to the specific needs of our student cohort. One of the key principles that guides our approach is customization, tailoring the educational content to match the evolving requirements of our learners.

Despite the ever-changing landscape, some recurrent themes persist, illustrating the fundamental values we instill in our students. Effective team collaboration, for instance, remains a cornerstone of our curriculum, as we believe that this skill is not only essential for academic success but also for personal and professional growth.

Our students are accustomed to being intellectually challenged, and they not only welcome it but also expect it as part of their educational journey.

5.5 Student Experience

We take immense pride in our commitment for delivering a truly "tailored" educational experience to our students enrolled in the ELM programme. This dedication extends to every facet of our curriculum and the way we nurture our students' growth and development. Our foremost priority is to ensure that the academic content encompassed in our syllabus is meticulously designed to prepare our students for their chosen career paths. We understand the importance of keeping pace with the ever-evolving landscape of business and industry. To achieve this, we continuously review and adapt our course content to remain aligned with the latest industry trends and underpinned by up-to-date theory.

One of the distinguishing features of our program lies in its flexibility, allowing us to cater to the unique needs of each candidate. We firmly believe that possessing certain capabilities is only half the equation; the ability to effectively apply those capabilities is equally vital. To bridge this gap, we focus on developing our students' competence by engaging them in real-world activities that foster growth and development. This collaborative effort extends to our industry partners, who play a pivotal role in providing hands-on experiences. Each student embarks on a personalized 1-to-1 program, where they openly share their areas for self-development. Our dedicated mentors provide valuable input based on their observations. Areas of improvement often include assertiveness, presentation skills, navigating difficult conversations, active participation in lectures and group debates, and making a lasting impression. Tailoring our tutorials around these identified areas allows us to nurture both—their confidence and competence.

Feedback on the ELM programme from a Director Member of the ELM Industry Advisory Board: *"What I saw was special, I'm truly impressed with what you have created in such a short period of time—in my mind this type of programme is exactly what academia is missing! You should be so proud of what you're doing, and I have no doubt it will go from strength to strength as you build on feedback from students. I'm really looking forward to being part of the industry board and would love to continue to support the programme going forward."*

Student feedback on the industry-linked project in the MSc ELM Programme: ***"Being able to associate with TATA steel during the course (ELM) was a great asset in helping determine my future career.*** *Not only were we given various real organisational challenges to address, which cemented the knowledge learnt through ELM, but it also felt TATA's board of directors fully took on the recommendations that we presented. Working on the TATA steel assignment helped me improve myself both as an effective individual and as a valuable member of a team. Other benefits included being able to increase my confidence during presentations, conducting interviews and utilising the skills and knowledge from ELM in a practical way."*

5.6 What Have We Learned from the Programme?

Over the years it has become very clear that the content that we deliver must evolve year on year. We cannot constantly deliver the same content as it quickly becomes outdated. of date. The theory changes very little, but the application changes significantly, which is in-line with the general market demand. As teachers, we ~~also~~ too need to refresh our knowledge to be able to guide the students better. The beauty is that ~~and~~ we have as much to learn from our students as they do from us. The synergy is that we have a very good learning relationship that is open to discussions and honest debates.

We always build feedback into our programs and discuss how we can improve the content and delivery of the program. It is important to recognise that this program is successful because of the experience and knowledge of the lecturers and the ability to work with small cohorts, which enables a tailored approach to the program content.

We believe that collaboration with the market keeps ELM relevant year on year. Equally important is the collaboration with our students. They are the leaders of the future and if the market loses sight of the expectations of these future leaders, then they will fail.

6 Conclusions and Recommendations

In today's world, those involved with engineering education throughout the world are fully aware of the responsibilities to provide an enabling learning environment that delivers a first-class education. But what about our responsibility to the employing organisations?

Because if graduates' skills are not aligned to the requirements of the employers, then we are missing the point; a strong relationship between the educators and the employers therefore is essential. As we have presented in this Chapter, Industry-Academic partnerships are critical factor in enabling this.

- The model utilized by WomEng Eswatini has connected academic institutions and industry to empower women and girls in STEM. The beneficiaries have received skills and knowledge to prepare them for the workplace, with targeted interactions and support provided from as early as primary school. The model can be transferred to other locations and can be used for all genders.
- The Swansea University study on the impact of the placement year in undergraduate engineering shows that a gain in academic performance after the placement period is notable: + 5.7% for students returning from placement to year 3 of a bachelor's in engineering (3 year) degree, with students in the lower grade categories prior to the placement benefitting most from the year in industry. The year in industry in engineering is therefore demonstrably beneficial to students.
- Targeted female STEM internships in Nepal have proven highly beneficial to all three involved parties—graduates, employers/industry, and academic institutions.
- The exemplar presented of postgraduate Engineering Leadership Management training at Swansea University has demonstrated how close partnership in delivery can provide a transformational experience in developing theoretical knowledge and capability for future leaders in engineering organisations.

References

1. Practera (2023) Powering experiential learning. Available: https://practera.com/how-to-prepare-graduates-for-employment/
2. Khan I, Hou F, Le H, Ali S (2021) Do natural resources, urbanization, and value-adding manufacturing affect environmental quality? Evidence from the top ten manufacturing countries. Resour Policy, 72
3. Inceoglu I, Selenko E, McDowall A, Schlachter S, (How) Do work placements work? Scrutinizing the quantitative evidence for a theory-driven future research agenda. J Vocat Behav
4. Tanaka Y, Carlson K (2012) An international comparison of the effect of work-integrated learning on academic performance: A statistical evaluation of WIL in Japan and Hong Kong. Int J Work-Integr Learn 13:77
5. Duignan J (2002) Undergraduate work placement and academic performance: Failing by doing. Proceedings of the 2002 Annual International Conference of the Higher Education Research and Development Society of Australasia (HERDSA)
6. Duignan J (2003) Placement and adding value to the academic performance of undergraduates: reconfiguring the architecture-an empirical investigation '1'. J Vocat Educ Train
7. Rolland S, Jones J, Bunting G (2023) The impact of a year in industry on academic outcomes in higher education (engineering), Eur J Eng Educ, 48(4)
8. World Economic Forum, "Global Gender Gap Report 2023," 2023. Available: https://www.weforum.org/publications/global-gender-gap-report-2023/
9. World Bank, "Data—Population, female (% of total population)—Eswatini," 2023. Available: https://data.worldbank.org/indicator/SP.POP.TOTL.FE.ZS?locations=SZ
10. The engineering council, the accreditation of higher education programmes, UK Standard for professional engineering competence, 2014
11. The Engineering Council, UK-SPEC—The UK Standard for Professional Engineering Competence, The Engineering Council, 2013

Integration of Professional Skills (Communications, Teamwork, Etc.), Ethics, Empathy Lifelong Learning into the Student's Educational Experiences

Dawn Bonfield, Mohamed G. Hassan, and Juan M. Palmieri

Abstract Engineering, as a rapidly evolving field, thrives on innovation and technological advancement. Beyond the needed technical expertise, engineers require a diverse set of professional skills such as effective communication, teamwork, problem-solving, critical thinking, and management. There is also a demand for a complex set of abilities and attitudes such as ethics, integrity, adaptability, curiosity, compassion, and empathy. Engineering schools should strive to equip their students with these skills, abilities and attitudes thereby fostering the development of a 'lifelong learning mindset' that will prepare future professionals to rapidly adapt to actual and future, hence unknown, engineering (and personal) challenges. While addressing this responsibility, universities will necessarily face another challenge, change. This chapter is aimed to provide engineering institutions and educators effective strategies along with worldwide examples about engineering schools, programs and teaching initiatives that took this challenge. From different but complementary landscapes and viewpoints, this chapter analyses how to integrate professional skills into engineering careers and courses along with the needed compliance to local and international accreditation frameworks. Furthermore, several hints and takeaways are shared regarding how institutions and teachers can address this challenge and succeed.

Keywords Engineering education · Professional skills · Ethics · Empathy · Lifelong learning · Educational experiences · SDG

D. Bonfield
Aston University, Birmingham, UK
e-mail: d.bonfield@aston.ac.uk

M. G. Hassan
School of Chemistry and Chemical Engineering, The University of Southampton, Southampton, UK
e-mail: mghs1v19@soton.ac.uk

J. M. Palmieri (✉)
Universidad Tecnológica Nacional, Buenos Aires, Argentina
e-mail: jpalmieri@frba.utn.edu.ar

R. Kandakatla et al. (eds.), *Academic Leadership in Engineering Education*, Lecture Notes in Networks and Systems 1097, https://doi.org/10.1007/978-3-031-68282-7_11

1 Introduction

Everything is always changing around us. The speed and depth of change have become so evident in every field of our lives that we cannot look the other way anymore. Massive changes have occurred recently on the technology side from Artificial Intelligence, Big data, machine learning and robotics. But also, wider societal changes like climate change, biodiversity loss, the covid pandemic, and increasing levels of conflict and inequalities around the world. As individuals and as a society, we need to find the means to understand these changes, act effectively and therefore, change. International leaders and institutions are raising their voices and shaping present and future agendas, such as UN 17 Sustainable Development Goals (SDG) [1].

Whilst some these changes require adaptions to our teaching contents and practices, the pace of change is so fast that perhaps the most important skill as teachers is not trying to keep up with it, but teaching students how to navigate our world, how to pivot in the face of change, and acquire the attitudes and abilities required to continue learning throughout their career. Educational institutions and engineering schools should consciously integrate this attitude towards change into their learning programs as top-level principles, addressable through specific learning outcomes, and strive to equip students with the willingness, attitude, and skills to embrace change in their professional and personal "toolbox".

But in addition to the need to prepare future engineers for this changing world, engineering institutions should also recognise that the needed skills are changing as well. Whereas knowledge itself was king a decade back, we now find that attitudes, behaviours, and other non-technical but nevertheless critical skills are gaining in importance, alongside with the technical content. Therefore, engineering education should have as fundamental goal to empower stu-dents with not only the essential skills but also the willingness and mindset to readily adapt to change, both in their professional and personal lives. This adaptability should become a key component of their "toolbox". In this scenario, the role of academic leaders assumes unprecedented importance. Their attitudes and capacity for envisioning and planning attainable paths towards these goals are pivotal for guiding and encouraging teachers and students to attain these new skills and mindset.

The very first step towards this goal is challenging the idea of "no change" and providing students the learning environments and opportunities to develop flexible, curious, creative, and compassionate minds. As engineering teachers, we can envision that our students will face unknown situations as engineers, but also as human beings. So, we can honestly ask ourselves: (1) How can we help our students to develop these qualities and mindsets? (2) How can we integrate these experiences and opportunities into our engineering learning plans and courses, and (3) How can we (institutions, teachers, students, and society) be aware of our success in this quest?

As learning farmers, the most important seed that we can plant in our student's minds and hearts is love towards their learning. Lifelong learning is then one of our essential tools that will allow students to develop versatile and curious minds capable of learning through change. We learn from, with and for others, hence carefully integrating empathy, diversity and ethics along learning experiences could foster student's loving and compassionate minds.

As these enormous changes occur in our world what we have also seen recently is the huge impact that our engineering and technological decisions have on people. The covid pandemic showed us that different parts of society did not fare equally, and that many minoritized and marginalized groups and poorer communities were much more badly affected due, in part, health, housing and build environment inequalities. Recognition also of the need to ensure that the transition to renewable energy is a just transition, and the SDG mantra of 'leave no one behind' has increased our awareness of the fact that people need to be at the heart of our engineering choices. The ongoing dialogues regarding engineering ethics, inclusion, equality, diversity, sustainability, and the ethical utilization of AI serve as a few instances of the meaningful contributions of academic leaders in shaping this education focused on individuals within the engineering field. This has led to Industry 5.0, where the focus lies in ensuring human, environmental, and social aspects that are central to the digital revolution, and that stakeholders as well as shareholders are the priorities.

From this viewpoint, attributes, and skills such as teamwork, collaboration, entrepreneurship, communication, and empathy serve as valuable avenues for engineering students to cultivate the essential qualities required not only for their careers but also for their overall life experiences. As these skills are becoming mandatory in most of the international engineering accreditation frameworks, engineering institutions and educators must reflect on effective ways to integrate them into the student's educational experiences, learning plans and assessment methodologies. Fostering these skills from the commencement of the course of study and incorporating them into as many subjects as feasible could assist students in effectively cultivating these skills and preparing them for the ever-evolving professional landscape. Once more the role of academic leaders ought to be esteemed as visionaries and planners of these objectives, working in tandem with educators who craft opportunities for students to achieve them.

This chapter is aimed to provide engineering leaders, institutions, and educators effective strategies along with worldwide examples about engineering schools, programs and teaching initiatives that took this challenge.

From different but complementary landscapes and viewpoints, this chapter analyses how to integrate professional skills into engineering careers and courses along with the needed compliance to local and international accreditation frameworks. Furthermore, several hints are shared regarding how institutions and teachers can address this challenge and succeed.

2 Teamwork, Collaboration, and Empathy Learning Experiences.

2.1 A Collaborative Experience in Undergraduate Courses in Argentina

Argentine engineering schools are currently discussing the needed updates to their curricula and methodologies due to the recent approval of new engineering careers accreditation standards by the National Ministry of Education.

These standards compel schools to switch from content-based curricula to competency-based and outcome-based ones. In this transformative process, there is a necessity to shift away from the conventional paradigm that places exclusive emphasis on the contents students are expected to master. Instead, schools should adopt a novel approach that prioritizes the development of technical, professional, and personal skills throughout the students' learning journey. Addressing the development of professional skills, such as teamwork, written and oral communication and ethics could be challenging for faculty. However, these are among the most required skills for engineers in the industry and they are described as social competences that engineering students should develop in their course of study.

While this discussion is still taking place, since 2019 a group of undergraduate professors from two Argentine universities decided to work collaboratively to systematically integrate those professional skills into three courses from 1st year Industrial, 2nd year Electrical and 4th year Systems Engineering careers delivered at National Technological University-Buenos Aires School (UTN) and National Southern University (UNS). Upon starting our collaboration, while sharing our experiences we established the following drivers that would guide our innovation process to date:

- Teamwork and communications skills should be intensively addressed in our courses.
- As these skills should be practiced by students, we should integrate them in the laboratory and practical activities.
- We should use a coherent set of active learning and authentic assessment methodologies to effectively promote them in our courses.
- These methodologies should become a framework to enable its adaption to each course needs and its replication into other engineering educational experiences.
- Flipped learning and backwards design methodologies, as posed by [2], are mandatory to guarantee quality time during classroom and synchronous sessions.

While our collaboration moved forward, we analysed active learning and assessment methodologies that could foster those professional skills in our courses. With this goal in mind, and among others, we evaluated well-known research works related to engineering education innovation [2, 3], active learning [4, 5] and authentic assessment [6, 7] methodologies. As a result, a set of methodologies (summarized in

Table 1) were selected to be systematically integrated with the necessary adaptions according to each course learning outcomes, agenda, and practical activities.

Since 2019 these methodologies have been used aiming to address each course learning outcomes while fostering the forementioned professional skills. During this innovation process and from the professor's perspective, several adaptions have been introduced into the way this framework is used verifying its' adaptability to the needs of each course. On the other hand, Fig. 1 presents students' satisfaction level regarding the achievement of specific professional skills in the course Communication and Networks (2020, 2021 and 2022 editions). Extensive information regarding this innovation experience and results are available in [17] and [18].

As this innovation process continued, other professors from UTN and UNS became interested in our experiences and framework. Even more, other Argentine universities invited us to share our findings. At this point it was evident for us that integrating a competency-based approach (CBA) and fostering professional skills in engineering students require a new kind of teacher. So, then the focus was to upgrade the faculty so that they align with said approach.

Table 1 Framework of the selected active learning and assessment methodologies

Teamwork and collaboration	Oral and written communication [8]	Assessment
Felder and Silverman Learning styles Test [9] [10] [11] (for team building process)	Individual and collaborative minute paper (to reflect about personal y group learning)	Interactive applications like Mentimeter [12] and Socrative [13]) (for short Q&A sessions)
Team agreement (promoting commitment and responsibility)	Think-Pair-Share / Write-Pair-Share (for short active learning moments)	Written reports Rubric (considering group and individual participation)
Team roles (rotative through different practical activities)	Meeting and decision-making summaries (for sharing team decisions and responsibilities)	Oral presentation Rubric (considering technical knowledge and oral communication skills)
Team clinics (for solving team conflicts) and coaching (fostering empathy and respect)	Group written reports and essays using collaborative documents like Google Drive [14]	Collaborative applications like Slack [15] and Zulip [16] (for teamwork and group assessment)
Team assessment rubric (Peer and self-assessment)	Planned and spontaneous oral presentations (in classroom and during synchronous sessions)	Teacher's assessment Rubric (used by students to assess their learning process once per year)
Collaborative research projects and realistic problem-based activities addressing specific topics	Oral group debates and Peer review discussions	Students' learning process assessment Rubric (used by students twice per year)

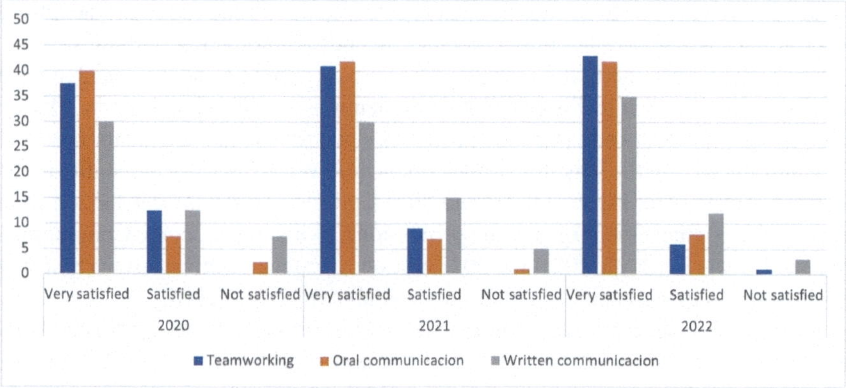

Fig. 1 Communication and Networks course (2020, 2021 and 202 edition) students' opinion regarding professional skills improvement

2.2 *Upskilling Faculty Competences in Latin America*

The competency-based approach is in constant discussion in Argentine and Latin American universities as well. Even though this approach is extensively adopted by Latin American institutions, a significant number of faculty is still unable to effectively foster professional skills and a competency-based approach in their courses.

The International Engineering Educator Certification Program (IEECP) is being offered in Latin America since 2018 aiming to upskill engineering teachers' competences and to give them access to IGIPs' International Engineering Educator certificate. Since then, more than 300 professors from Argentina, Chile, Colombia, Costa Rica, Mexico, Paraguay, Peru, and Puerto Rico completed the program which is based on a flexible set of modules. According to Latin American needs, InnovaHiEd and IGIP Argentine Section Training Centers adapted the IGIP curriculum as detailed in [19].

Among others, the modules named Effective Teamwork, Presentation and communication skills and Neuro-Linguistic Programming for Education aim to upskill teacher´s abilities to foster those professional competences in their courses while promoting a student-centered and empathetic teaching environment. Along the duration of the Program, each participant should pose, deploy, and assess an innovation project in a real educational scenario. This requirement to complete the IEECP and obtain IGIPs' certificate, compel teachers to design courses and learning activities which can effectively promote competencies and professional skills among their students with the necessary evidence. Even more, some of these innovations became reference cases that occasionally promoted new innovations in many Latin American universities expanding the impact of the Program within each school.

Extensive information about the IEECP initiative and results is available in [19]. Nevertheless, some evidence is worthy of mention:

- Educators positively appreciate the formative process in relation to its objectives, as well as the collaborative, empathetic, and understanding environment fostered. A highly valued aspect is experiential learning, facilitated through collaborative groups that undertake applied projects within their respective educational settings and are evaluated by their students.
- Institutions also express a positive assessment of the contributions made by certified educators, encouraging them to assist their colleagues in the process, while also demonstrating an interest in developing similar teacher updating processes.

3 Ethics, Equality, Diversity, and Inclusion Learning Experiences

As we have mentioned already in this chapter, the Engineering Higher Education sector has undergone a transformation over the past 20 years from content-based education to outcomes-based education, where 'Graduate Outcomes' (similarly called Learning Outcomes, or Graduate Attributes) describe what students are expected to know and be able to do by the time they graduate. These attributes then develop into a set of competencies which are required by engineers to become professionally registered.

Many accreditation frameworks exist around the world, some covering single countries, whilst others cover multiple countries enabling engineers to become internationally mobile, and practice in different countries where their attributes and competencies are seen as substantially equivalent.

One such global framework is administered by the International Engineering Alliance (IEA) [20], where some 30 countries have become signatories to the various Accords which relate to Professional Engineers (the Washington Accord), Engineering Technologists (the Sydney Accord), and Engineering Technicians (the Dublin Accord).

In 2021, the IEA established a working group in collaboration with the World Federation of Engineering Organisations (WFEO) as representatives of industry and the professional engineering institutions to create a new framework for engineering graduates, called the Graduate Attributes and Professional Competency (GAPC) Framework [21]. This framework reflects the changing face of engineering education, and the requirements for engineers to incorporate new technologies and engineering disciplines, new pedagogies, and key values including sustainable development, global responsibility, diversity, inclusion, and ethics to support engineers in building a sustainable and equitable world.

In the United Kingdom (UK), the current accreditation framework is called AHEP 4, the fourth version of the Accreditation of Higher Education Programmes [22], which was implemented towards the end of 2021. This framework brings in, for the first time, the requirement ***'to adopt an inclusive approach to engineering practice and recognise the responsibilities, benefits and importance of supporting equality,***

diversity and inclusion', and propose a sharper focus on the inclusion of sustainability and ethics.

This International and the UK case are both examples of how the requirement for engineers to possess these competences are increasing in importance, and these frameworks are key drivers in ensuring that they are systematically taught as part of engineering degrees.

The importance of these values in engineering education is not in doubt, but the question remains for us to determine how we can teach them, and how they are assessed. Several different new models of teaching engineering are gaining some attention in the UK, which seek to embed vital professional skills such as communication and teamwork, and at the same time create future engineers who are globally responsible, and capable of designing sustainably and ethically, are inclusive, and empathetic. These courses have the common factors of including significantly more project-based work than has previously been the case.

An example of this is an organisation called The Engineering and Design Institute London (TEDI-London) [23] which offers a global Design Engineering qualification which is predominantly project based. In this program students work in teams with industry involvement, and the engineering content is delivered as required by the project (as opposed to the more traditional longitudinal teaching method). Teaching comprises workshops, industry-led expert input, group discussions, design reviews, and is always aligned with making and testing. These teaching practices are supplemented by individual online learning regarding meaningful content. Assessment is a mixture of group project reports, individual reports, pitches, presentations , demonstrations, short videos, posters and websites.

The New Model Institute for Technology and Engineering (NMITE) [24] is another example of change largely based on the successful Olin College of Engineering model, where sustainability is woven throughout the degree courses and students are solving real problems while becoming career ready right from the start. Assessment is ongoing through a portfolio of projects, and workplace experiences, rather than conventional exams.

University College London (UCL) delivers an Integrated Engineering Programme (IEP) [25], a cross-faculty teaching framework amongst the undergraduate educational offering which uses a series of Project Based Learning activities to integrate real world problems within a more conventional modular degree (Project Based Learning, as opposed to Problem Based Learning, which allows students to complete a project rather than work on an open unbounded problem). This culminates in a final intensive two-week project entitled *How to Change the World* which focuses on socially complex real-world topics associated with the SDGs that are created in partnership with external experts, from policy, industry and the third sector.

And finally, Aston University is an example of a university which delivers, the Concepts and Conceives, Design, Implementation and Operation (CDIO) [26] initiatives. It is also delivered in USA, Europe, Canada, UK, Africa, Asia, and Australia, which integrates design-build-test projects throughout the degree. It facilitates additional skills such as teamwork and communication in a structured and supported way, featuring interwoven active and experiential learning opportunities.

These four UK examples show how engineering education is using innovative project-based education to deliver the skills required of our future engineers, whilst still complying with accreditation requirements.

3.1 Teaching Ethics

Ethics is a key tenet of the engineering sector. Engineers as well as the profession of engineering is considered one of the most trusted professions, but the teaching of ethics has struggled to find a place in the engineering education curriculum. It is often being tagged on to individual and group project work.

In a more recent initiative, emphasis placed on the teaching of ethics in the UK has been spearheaded by the Royal Academy of Engineering (RAEng) [27] and the Engineering Professors Council (EPC) who have developed a series of fictional case studies and teaching aids to help embed ethics teaching in a more structured manner.

In the UK, as in the rest of the world, where the engineering profession is regulated, all professionally registered engineers are expected to uphold the code of ethical conduct according to their respective registration body.

These codes bring together the values and principles that the profession upholds, including –

- To act professionally and competently according to one's expertise and within the law.
- Work to given standards.
- Be honest and have integrity.
- Respect life and nature
- Built environment, and
- Display leadership.

These principles and values should apply to all engineering students, who are training to become part of the engineering profession. The endeavor is not just to teach such principles as a part of the curriculum, but to instill them in other creative ways in our education system. The resources developed by the RAEng and the EPC aim to cover the principles contained in the Code of Ethics, by means of illustrative fictional examples, backed by teaching resources and further links, which explore some of the current day ethical dilemmas that engineers face, such as the ethics of facial recognition, power to food, use of chatbots and AI, solar for oil, smart metering, geo-engineering, and water wars. [28]

Finding other ways to incorporate these values-based attributes into our curriculum by 'nudges', as opposed to its complete redesign or dedicated teaching activity, are a practical way to embed these concepts throughout the student experience. For example, along with diversity, inclusion and sustainability, ethics references can be added to the concept, the content, the delivery, the assessment, the student feedback mechanism, and the course evaluation and review. This approach could create a whole system which reflects the values and behaviours that we are striving to embed.

A planning tool for doing this has been developed for CDIO programmes [29] where the emphasis of Learning Outcomes of programmes are subtly but distinctly changed to reflect the context of problems, and a process of constructive alignment is required to ensure that learning outcomes, teaching activity and assessment are all consistent and no discrepancy occurs (Table 2).

Corresponding learning and teaching activities could include giving space for discussion and sharing viewpoints. Assessments would identify how students have included reflections, critical thinking, and justification of ideas into their projects and overall activities (Table 3).

In some cases, it will be possible and straightforward to assess competences such as empathy and inclusion by simply assessing the evidence which shows whether designs have been validated with the user, or whether an Equality Impact Assessment has been completed, for example. But assessing whether the student engineers themselves have changed their mindsets to be more empathetic or ethical at the end of their qualification than they were at the beginning, will be more difficult. This is where 'behavioral markers' can be useful as means to assess values, beliefs and mindsets through the observable traits displayed by the students. Alternatively, more discursive, and subjective forms of assessment, such as reflective analyses, essays, or one-to-one interviews will be valuable and also a useful way of ensuring that students have not become too heavily reliant on the help of AI tools such as ChatGPT!

Table 2 Ways to embed context and values-based attributes to learning outcomes

Original learning outcome	Revised learning outcome
Use CAD, modelling and additive manufacturing in the product development process and embed control sensors, actuators and physical hardware into a complete system	Use CAD, modelling and additive manufacture in the product development process and embed control sensors, actuators and physical hardware to design a safe and complete system to address a societal need

Table 3 Constructive alignment of learning, teaching and assessment activities with revised learning outcomes

Learning & teaching activities, to include	Assessments, to include
Give space for discussion, allowing all perspectives to be heard, and lead activities that allow for a variety of viewpoints with no right answer	Identify the markers for inclusive behaviour during discussion sessions, observe values and behaviours of tolerance and empathy, and allow space in the assessment for students to reflect critical thinking and justification of their ideas

4 Real-World Challenges, Social Responsibility, Ethics, and Intercultural Learning Experiences

The landscape of engineering education and the engineering job market is undergoing an extreme transformation, driven by rapid technological advancements, and shifting global dynamics. Graduates must now possess a distinct set of competencies that extend beyond traditional analytical and critical thinking skills. This transformation is stressed by insights from books like "The World is Flat" by Thomas Friedman and "A Whole New Mind" by Daniel Pink. These authors argue that the evolving economic landscape demands engineers with emotional intelligence, creativity, and interdisciplinary thinking. As a response, engineering education is reshaping itself to equip graduates with the attributes necessary to thrive in this changing environment.

4.1 The University of Southampton's Chemical Engineering Department Case Study

The essence of chemical engineering resides in the art and science of transforming raw materials into valuable products. It is a process that entails the intricate transposition of optimal synthesis routes and the precise operation of processes tailored for their intended applications. Grounded in the synergy of science, mathematics, and economics, this core facet of chemical engineering serves as the backbone of industrial progress. However, in the contemporary landscape, where environmental concerns and technological advancements converge, the core principles must be augmented by a judicious blend of scientific exploration and engineering innovation. It will yield sustainable and groundbreaking solutions as further learn, to overarch the core.

This paradigm shift towards sustainable and innovative solutions is championed by the University of Southampton's (UoS) Chemical Engineering department, as depicted in Fig. 2.

This new approach is informed by a comprehensive set of objectives, among which is the unwavering dedication to serving society through the education and training of the engineers of the future. The endeavor is to be uniquely equipped to navigate the intricate tapestry of modern challenges and opportunities. This educational mission is underscored by a commitment to instilling in students a profound understanding of the fusion between scientific knowledge and engineering prowess, which is imperative in steering industrial development towards responsible and forward-looking pathways.

The Career Department embraces a novel perspective that transcends the traditional boundaries of the discipline. It recognizes that while the core principles remain constant, the complexities of the modern world necessitate a holistic approach that draws upon multiple facets of science and engineering. This multidisciplinary framework forms the bedrock upon which sustainable and innovative solutions are constructed. This comprehensive approach allows UoS chemical engineering

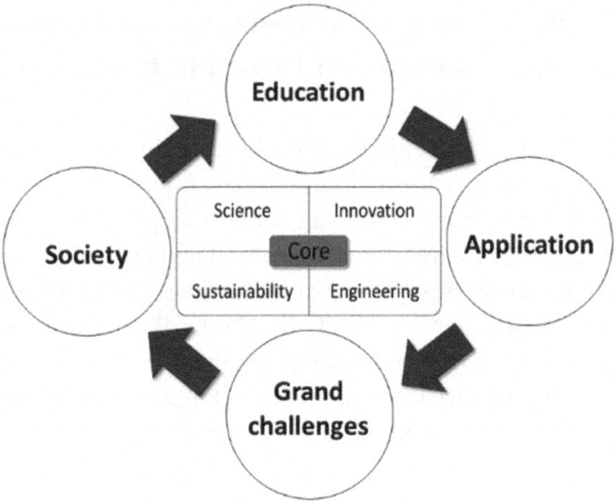

Fig. 2 New dimensions in chemical engineering education

students to bridge the gap between theoretical knowledge and pragmatic application, thus producing graduates who are primed to become agents of positive change within their industries and beyond.

By leveraging the diverse tools at the disposal of modern chemical engineers, such as advanced computational methods, cutting-edge analytical techniques, and a deep appreciation for the interplay between the spiritual elements such as ethics, equality, diversity, inclusion and the worldly elements of economics and technology, the department empowers its students to approach challenges from multiple angles.

This ability to tackle complex issues holistically is encapsulated in Fig. 2, which envisions a dynamic synergy between the core principles of chemical engineering and the diverse sciences and engineering domains that strengthen the pursuit of sustainable solutions.

This forward-thinking approach reflects the department's recognition that the trajectory of industrial development is inseparably linked with the principles of sustainability. As global obligations to reduce environmental impact and optimize resource utilization intensify, the role of chemical engineers becomes increasingly pivotal.

The UoS commitment to developing engineers of this caliber is clearly captured in its educational framework, as illustrated in Fig. 2. The intersecting layers symbolize the convergence of theoretical knowledge, practical skills, and an unwavering commitment to ethical and sustainable practices. Through a blend of structured learning, hands-on experiences, and exposure to real-world challenges, students are equipped with the tools needed to navigate the intricate landscape of modern engineering.

Core principles of chemical engineering lay the foundation for transforming raw materials into valuable products, a process that requires a harmonious blend of science, mathematics, and economics. However, the evolving landscape of global challenges and opportunities demands a broader perspective. Through this innovative education, the Department mold engineers who are not only adept at traditional practices but who possess the vision and skills needed to shape a better future for society. This novel thinking in education represents a leap forward in equipping the next generation of engineers with the tools to serve as catalysts for change and progress. The scope of the design can be categorized in Fig. 3.

The curriculum structure within the department is designed to nurture a comprehensive understanding of the field while facilitating both breadth and depth of knowledge acquisition. Over the course of four years, students undergo a progressive journey that spans from foundational concepts to specialized advanced topics, culminating in the potential for an integrated master's degree. Figures 4 and 5 give an overall backdown:

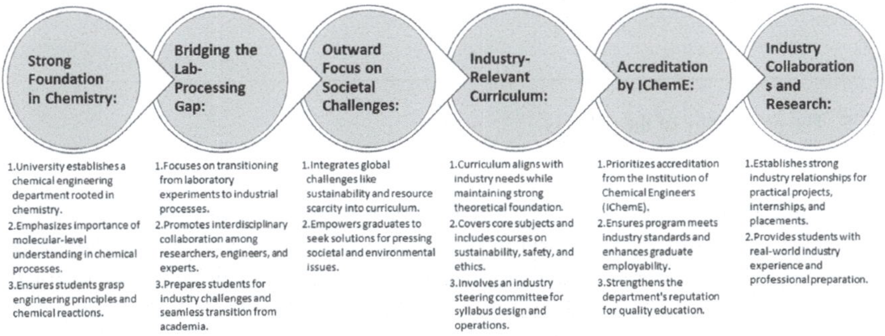

Fig. 3 The scope of the design

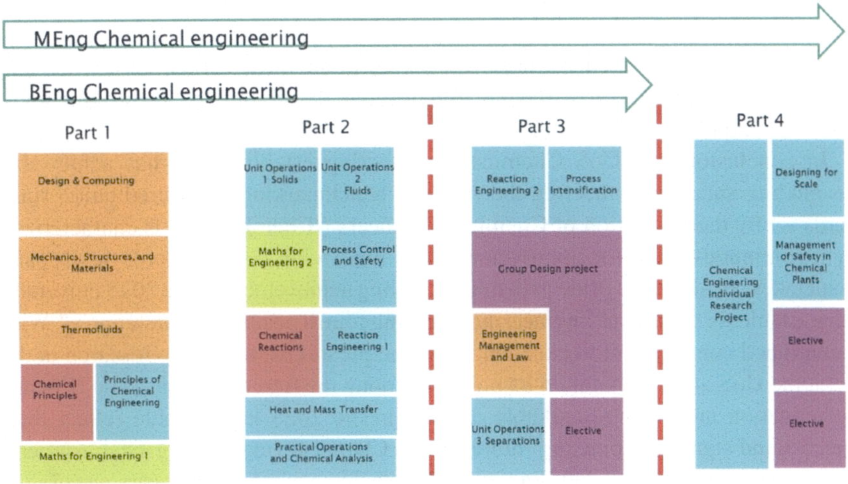

Fig. 4 Comprehensive overview of the syllabus structure for the 3 Year BEng and 4-year Meng course red lines represent opportunity for a placement in industry

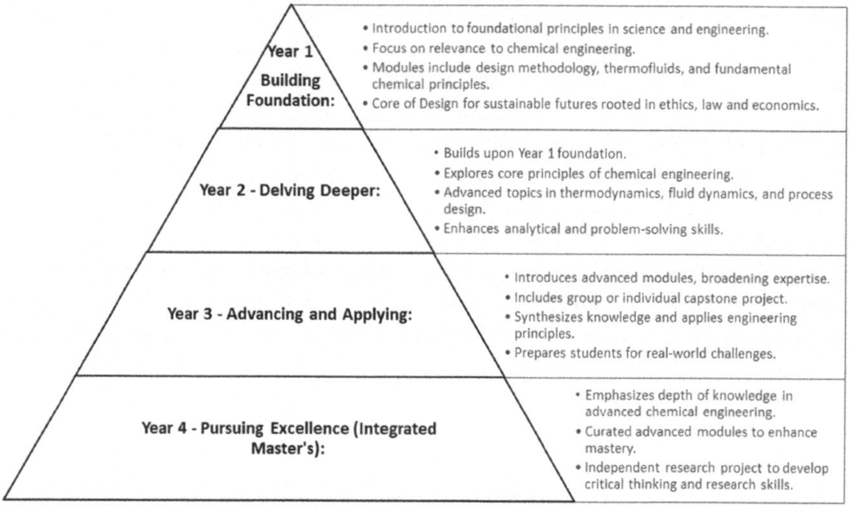

Fig. 5 The philosophy of the delivery

As students near the end of their academic journey, they are introduced to the practicalities of commercialization in the context of chemical research. They delve into the interface between science, engineering, and entrepreneurship, thus acquiring a holistic perspective on the application of their expertise. Moreover, students are given the freedom to select from a range of electives, enabling them to tailor their learning experience to their specific interests and aspirations.

Furthermore, the Department's emphasis on research serves as a catalyst for innovation and knowledge creation. Research initiatives within the department focus on a wide spectrum of areas, ranging from sustainable energy production to novel materials design. This research not only contributes to the advancement of science and technology but also exposes students to innovative developments, inspiring them to think beyond the confines of the classroom.

In conclusion, the UoS Chemical Engineering establishment has achieved a resounding success by meeting the rigorous benchmarking for accreditation standards set by the Institution of Chemical Engineers (IChemE). This is characterized by a continuous drive for improvement, and innovation. Furthermore, the Department takes considerable pride in their outstanding achievement in the 2023 university league table for Chemical Engineering, securing fourth position as a new entry.[1] With a continued commitment to excellence and the pursuit of innovative research in the field, the UoS aims to maintain its position among the top institutions. The IChemE accreditation and their remarkable ranking in the 2023 league table reflect their success and signal a promising future for the Chemical Engineering establishment, as they eagerly anticipate building upon these achievements in the years ahead.

[1] https://www.thecompleteuniversityguide.co.uk/league-tables/rankings/chemical-engineering.

5 Lifelong Learning and Professional Skills in Engineering Courses: Takeaways and Learnings for Teachers and Schools

Engineering, as a rapidly evolving field, thrives on innovation and technological advancement. Beyond the needed technical expertise, engineers require a diverse set of professional skills such as effective communication, teamwork, problem-solving, critical thinking, and management. But also, a complex set of abilities and attitudes as ethics, integrity, adaptability, curiosity, compassion, and empathy.

In this chapter we delve into the critical importance of the development of this needed mindset and professional skills within engineering programs from a variety of perspectives, educational scenarios, geographical locations, and systems. The experiences and insights shared highlight that there is no right or wrong way to integrate these skills, abilities, and attitudes in programs and courses. Engineering schools could start with specific courses, upskilling teachers who are willing to try something new, redesigning a career program, or comply to an accreditation framework. Schools should analyse their starting point and set realistic objectives or goals to upgrade the quality and value of their programs as well as the foundations for future innovations. As a result, universities will progressively develop a change-culture and increase their social responsibility by graduating quality engineering professionals.

Lifelong learning not only ensures the competitiveness but also enables engineers to pioneer cutting-edge projects rooted in science, societal policy, and economic frameworks.

To develop the 'lifelong learning mindset' among engineering students, leaders and educators play a pivotal role providing learning environments based on teamwork, collaboration, communication, realistic problem-solving situations, critical thinking, ethics, and creativity. Teacher's course-design should consciously aim to develop these skills and assess students in such a way that they, as well as the students, become aware of their progress, needed support, and prepare for a lifetime of self-improvement.

Curriculum design in Engineering programs should integrate these professional skills, along with technical contents, as key drivers to promote the required mindset through innovative courses, active learning, student-centered, and authentic assessment methodologies.

These programs should be flexible enough to adapt to changing scenarios and integrate new technological or social demands.

Courses and tracks should be outcome-based, and students should be aware of their expected learning achievements in the career, in each course or project to increase their commitment and active role.

Integrating industry partnerships and collaboration with social stakeholders into the program design is an important part of staying relevant through real-world exposure to professional engineering practices.

Engineering schools and teachers could feel that they should start to integrate these skills from 'scratch' or, on the other hand, obediently comply to every accreditation framework because 'it is the law'. From our point of view this is a thoughtful process as well as a creative one. This chapter explains recent and ongoing experiences from the UK and Latin America, along with the references to expand the information. As institutions, leaders, and teachers, we don´t need to 'reinvent the wheel'. Innovations in many fields of engineering education are published; we can reach out to those pioneers' institutions and teachers to learn more through collaboration and networking.

A final remark regarding accreditation frameworks is that we may consider them a constraint or a check list we must comply. We can also choose to look at them as allies that provide valuable information regarding society expectations about a certain degree.

At this juncture, we could articulate several takeaways to inspire and guide engineering schools, as well as their leaders and educators who aspire to have their students cultivate the professional competencies necessary to become valuable professionals and individuals in both the present and future societal context:

- Strive to equip your students with a robust and flexible 'lifelong learning mindset' to prepare them to rapidly understand and adapt to actual and future challenges.
- Establish feasible starting points with the aim of identifying routes of least resistance that could culminate in enduring outcomes.
- Provide numerous opportunities to foster the development and practical application of professional competencies. Whenever feasible, design both vertical and horizontal meaningful pathways throughout the academic journey.
- Enhance teachers' proficiency in incorporating this vision and addressing these needs within their instructional design and assessment methodologies. Foster collaborative efforts among educators to develop enduring learning experiences aligned with the objectives.
- Assess, communicate, and publish your results. Work collaboratively in your schools and with social stakeholders whenever possible.

Some years ago, Steve Jobs quoted "The people who are crazy enough to think they can change the world are the ones who do". Before him, Albert Einstein said that "Insanity is doing the same thing over and over and expecting different results". If engineering is called to find solutions to real-world problems, Engineering schools and teachers should consider expanding their boundaries, prompting new kind of questions, and responding to them creatively and responsibly while embracing change and innovation. If we want to teach these skills to our students, it is ethical to practice them as well - as Mahatma Gandhi once said, "Be the change you are trying to create".

References

1. United Nations, "The 17 Goals Homepage,". Available: https://sdgs.un.org/goals. Accessed 22 8 2023
2. Wiggins G, McTigghe J (2005) Understanding by Design—Expanded 2nd ed, ASCD
3. Felder R, Brent R (2016) Teaching and learning STEM: A Practical Guide. Jossey-Bass, Hoboken
4. Felder R, Brent R Active learning: an Introduction. Available: https://www.engr.ncsu.edu/wp-content/uploads/drive/1YB2KK3wLqP3EhXyYdKtE9-4mBJzc2rc2/Active%20Learning%20Tutorial.pdf. [Accessed 31 8 2023]
5. Felder R, Brent R (2015) Handouts with gaps. Available: https://www.engr.ncsu.edu/wp-content/uploads/drive/1l5p3UW6e7oQ_JBOpvMEPcpz8APhxj2Ph/2015-r_HandoutsWithGaps.pdf. [Accessed 30 8 2023]
6. Dawson P (2017) Assessment rubrics: towards clearer and more replicable design, research and practice. Assess Eval High Educ 42(3):347–360
7. Drezek K, McConnell E, Horan H, Zimmerman B, Rhodes T (2019) We have a Rubric for That. Assoc Am Colles Univ, Washington
8. Center for Teaching Innovation, Getting started with active learning techniques. [En línea]. Available: https://teaching.cornell.edu/getting-started-active-learning-techniques. [Último acceso: 29 8 2023]
9. Felder R, Silverman L (1988) Learning and Teaching Styles in Engineering Education. Engr. Education 78(7):674–681
10. Felder R, Soloman B Learning styles and strategies. Available: https://www.engr.ncsu.edu/wp-content/uploads/drive/1WPAfj3j5o5OuJMiHorJ-lv6fON1C8kCN/styles.pdf. Accessed 29 8 2023
11. Felder R, Soloman B Index of learning styles questionnaire homepage. [En línea]. Available: https://www.webtools.ncsu.edu/learningstyles/. [Último acceso: 29 8 2023]
12. Mentimeter Hompage. Available: https://www.mentimeter.com/es-ES. [Accessed 31 8 2023]
13. Socrative Homepage. Available: https://www.socrative.com. [Accessed 31 8 2023]
14. Google Drive Homepage. Available: https://www.google.com/intl/es-419_ar/drive/. [Accessed 31 8 2023]
15. Slack Homepage. Available: https://slack.com/intl/es-ar. [Accessed 31 8 2023]
16. Zulip Homepage. Available: https://zulip.com. [Accessed 31 8 2023]
17. Cukierman U, Graselli de Lima D, Palmieri JM (2020) An experience of educational innovation in engineering education: 10 years promoting the Approach by Competences in the undergrad. Available: https://www.laccei.org/LACCEI2020-VirtualEdition/full_papers/FP619.pdf
18. Cukierman U, Palmieri J (2014) Soft skills in engineering education: A practical experience in an undergraduate course. In: Proceedings of 2014 International Conference on Interactive Collaborative Learning, Dubai
19. Palmieri JM (2023) Competency-Based approach and learning plans in moodle. A Case of International Engineering Educator Certification Program (IEECP) in Latin America. In: Learning in the Age of Digital and Green Transition, Springer, pp 553–564
20. International Engineering Alliance Homepage. Available: https://www.ieagreements.org. [Accessed 14 10 2023]
21. IEA, "IEA Graduate attributes & professional competencies," 06 2021. Available: https://www.ieagreements.org/assets/Uploads/IEA-Graduate-Attributes-and-Professional-Competencies-2021.1-Sept-2021.pdf. [Accessed 14 10 2023]
22. Engineering Council, "Accreditation of Higher Education Programmes (AHEP) Homepage," 2020. Available: https://www.engc.org.uk/ahep. [Accessed 14 10 2023]
23. London The Engineering & Design Institute Strategic Plan Homepage. Available: https://tedi-london.ac.uk/about/strategic-plan/. [Accessed 14 10 2023]
24. New Model Institute for Technology and Engineering Homepage. Available: https://nmite.ac.uk. [Accessed 14 10 2023]

25. Integrated Engineering Programme Homepage. University College London, Available: https://www.ucl.ac.uk/centre-for-engineering-education/integrated-engineering-programme#:~:text=The%20Integrated%20Engineering%20Programme%2C%20introduced,and%20analysis%20of%20engineering%20systems. [Accessed 14 10 2023]
26. Conceive, Design, Implement & Operate Homepage. Aston University. Available: https://www.aston.ac.uk/eps/teaching/cdio. [Accessed 14 10 2023]
27. Royal Academy of Engineering Homepage. Available: https://raeng.org.uk. [Accessed 14 10 2023]
28. Royal Academy of Engineering, "Engineering Ethics Explorer Homepage,". Available: https://engineeringethics.epc.ac.uk. [Accessed 1014 2023]
29. Junaid S, Serreau, Y, Gwynne-Evans A, Granholm P, Fee K, Hitt S (2023)Using a constructive alignment tool to plan ethics teaching. Available: https://epc.ac.uk/toolkit/using-a-constructive-alignment-tool-to-plan-ethics-teaching/. [Accessed 14 10 2023]

Preparing Institutions to Become Knowledge Hubs for Research, Innovation, and Entrepreneurship

Fostering Research Culture, Collaborations, and Support Systems Leading to the Production of Quality Output

Edward Berger, L. Pratap Reddy, and Brian Self

Abstract This chapter describes the ways in which academic leaders can promote research cultures that lead to quality outcomes for stakeholders. The ideas are presented within the context of brief case studies about the research cultures of three institutions with different missions, geographic locations and cultural characteristics, and availability of/access to funding to support research projects. The chapter focuses on the ways the research culture manifests at each institution, given priorities about student outcomes, research impact, the stage of development of the research culture, and overall workload for researchers. The chapter concludes with reflections and recommendations for academic leaders on how they can enable the creation of a thriving research culture that generates quality outcomes.

Keywords Research culture · Research outcomes · Institutional case studies

1 Introduction

Higher education globally faces serious challenges related to perceived value, cost, and alignment of student outcomes with workforce needs, now and for the future. Engineering education in particular faces acute structural problems related to the historically-rooted nature of the curriculum, the sense of equity and inclusion in the environment and culture, the specific content that today's students need to learn (e.g., artificial intelligence (AI) tools and implications), and the content that engineering

E. Berger (✉)
Purdue University, West Lafayette Indiana, IN, USA
e-mail: bergere@purdue.edu

L. P. Reddy
Jawaharlal Nehru Technological University Hyderabad, Hyderabad, India

B. Self
California Polytechnic State University, San Luis Obispo California, CA, USA

R. Kandakatla et al. (eds.), *Academic Leadership in Engineering Education*, Lecture Notes in Networks and Systems 1097, https://doi.org/10.1007/978-3-031-68282-7_12

faculty are prepared to teach. Taken together, these challenges present both real problems and substantial opportunities for higher education professionals to re-engage with our core mission of education.

The research enterprise, too, faces many challenges in our globalized world, especially related to fostering a culture of open collaboration, relationship building, and co-equal partnership among researchers around the globe. Researchers need opportunities to connect with like-minded international colleagues, funding to support research collaboration, publication venues suitable for disseminating such research, and a clear understanding of the implications and further impact–beyond publications–of the collaboration. Research in engineering education exposes a particular tension in this space, wherein education is in many ways a highly-localized endeavor that relies on alignment among instructor and student perspectives and goals, the organizational and community cultures in which the education takes place, and the norms of the institution about relationships, collaboration, pedagogical approaches, assessments, and the like. Nonetheless, education research also elevates several transcendent processes and outcomes, many of which are related to pedagogical approaches. Active learning, problem-based learning, project-based learning, many forms of collaborative learning, and learning in high-quality blended environments all have been studied quite deeply in various contexts, and they all show benefits to students when deployed by skillful instructors.

The challenge, then, when considering international collaborations in engineering education research that ensure 'quality' outcomes, is to clearly define the outcomes and markers of 'quality' that are relevant to and shared by the stakeholders in the collaboration. Modern conceptions of engineering education research include: (i) a content, assessment, and pedagogy focus; (ii) an emphasis on identity and belonging; (iii) a deep understanding of equity and inclusion from various perspectives; (iv) the emergence of academic cultures as enablers or suppressors of student success; (v) the incentive structures and faculty professional development required to deliver a modern, relevant engineering education to students; (vi) the sources, amount, and accessibility of funding to support educational research. In each case, the outcomes vary from performance-based measures for individual students, to affective measures of those students, to institutional and organization cultures, and beyond. As such, both 'outcome' and 'quality' are fraught with difficulties in terms of both definition and measurement. This chapter elevates these challenges by addressing several specific dimensions of each, with an eye toward ensuring research quality as well as research-to-practice translation.

2 Relevance for Academic Leadership

This chapter is especially relevant now, as academic leaders across the globe wrestle with the vital challenges of providing high-quality education in relevant STEM subjects to ensure availability of a strong technical workforce. The economic vitality of a municipality, country, or region depends upon the existence of a strong, competent, and ambitious workforce to meet the needs of employers. Engineering education and STEM education more broadly also hold particular value, in many nations, for national security as the world becomes more complicated, contentious, and in many cases adversarial. The technical supremacy of the workforce in support of a national mission–whether focused on national defense, space exploration, counterintelligence, or other key areas–provides a strong incentive for higher education leaders to create and sustain the most effective education environment possible.

However, higher education institutions all have, in their own ways, legacy cultures, norms, policies, and practices that exist in tension with the demands of the modern world described here. Table 1 illustrates some of the features of institutions which influence their ability to create and sustain an engineering education research culture focused on quality outcomes. There are many more considerations than can be included here, and some of these considerations are more relevant for certain contexts than for others. We present these ideas to frame the significant and complicated challenges faced by academic leaders today when it comes to institutionalizing evidence-based educational practices based upon research and scholarship in engineering education.

Academic leaders play a central role in building a thriving, collaborative research culture that explores engineering education research topics, emphasizes research-to-practice activities, and elevates student outcomes. Specifically, leaders should help individual researchers shape the definition and metrics of 'quality' relevant to their institution and students, its goals and mission, the available resources, and its aspirations for the future. Building a set of policies, practices, and incentives that encourage this kind of engineering education research culture is non-trivial, and the remainder of this chapter explores several facets for academic leaders to consider.

3 Case Studies on Research Culture

This chapter's authors reflected on the research cultures at their own institutions, and in this section, we present brief case studies to explore and explain current realities at three institutions spanning diverse missions and contexts, situated within two countries. The institutions represented in this section span multiple missions and contexts, and they face quite different expectations, incentives, and constraints when it comes to research culture. While brief, these case studies highlight some of the ways that institutions can implement practices that support a thriving research culture by focusing on the needs, goals, and aspirations of the researchers themselves, and connecting outcomes to their institutional mission and stakeholders.

Table 1 Institutional features and implications for building engineering education research cultures focused on quality outcomes

Institutional feature	Implications
Time-based education (i.e., 'credit hours', 'study hours', etc.)	Time-based systems have inherent limits to the amount of content covered in a specific increment (e.g. a 3-credit course conducted over a single academic term) as well as the total amount of time to earn a degree. These time constraints force faculty to make difficult and often contentious decisions about content, priorities, and trade-offs, particularly where the arts and humanities are concerned. These time constraints also sometimes inadvertently reinforce existing practices within courses and curriculum, rather than encouraging educational innovation and research-to-practice updates to content, assessment, or pedagogy
Faculty role and job security	Incentive structures for individual faculty advancement, which in some cases include indefinite employment contracts ('tenure'), often do not align with educational innovation, modernization, curriculum reform, and so forth. Instead, such incentives revolve around research endeavors in the individual's area of expertise. These incentives for individual achievement often misalign with the goals of collaborative research cultures
Institutional mission	Higher education institutions exist on an activity spectrum defined by several axes, including teaching, research, outreach, commercial endeavors, and several others. Depending upon the current institutional mission and position on this spectrum, creating a thriving, collaborative research culture (or shifting the current research culture) may be challenging
Funding model	Institutional funding models come in essentially infinite variations depending upon the public or private nature of the institution, access to research funding from multiple sources, availability of philanthropic support, and so forth. In all cases, funding models–which operationalize as incentives for certain activities–do not always align with education innovation or a thriving, collaborative culture of engineering education research

3.1 Building Research Culture—a Time-Based Perspective

Experiences in Jawaharlal Nehru Technological University Hyderabad

Building a strong research culture is based on the amount of time that an individual spends on research activity. This time allocation may be different when attempting to maintain an *already-established* research culture. In the culture development phase, it is important to assess the present situation on allocation of time for research and build a model to increase the allocated time; Jawaharlal Nehru Technological University (JNTU) Hyderabad continues to build and refine its model and expectations for time spent on research. A survey was conducted among 1200 research scholars during 2013, who registered for the PhD degree in JNTU Hyderabad in various fields. The survey concentrated on the facilities that are available for every research scholar. Based on the study, it was identified that certain facilities like computational facilities, laboratory infrastructure, and access to publications and books had to be substantially improved. Leadership responded swiftly by making strategic investments in facilities and infrastructure to remedy this situation. The more difficult element of research culture identified in the survey was about time, tempo, and progress for research scholars.

Structured mentorship and milestones to ensure timely progress. JNTU leadership tackled the challenge of time allocation and research progress in a four step approach, focused on research trainees. *First*, senior scholars introduced a Research Methodology course (apart from other courses connected to the field of research) for every scholar as a mandatory subject (now this subject is offered even in postgraduate course). *Second,* leadership introduced a structured process for Research Review Meetings (RRMs) to track and provide feedback to emerging scholars about their progress. Every scholar shall go through two satisfactory RRMs at appropriate points in their work. In the first RRM, the scholar's progress is reviewed with regard to acquisition of core knowledge in their field and the depth of information that the scholar possesses based upon substantive review of prior work. The efforts of scholars are reviewed by a Doctoral Review Committee (DRC) of the respective department including the Research Supervisor. The second RRM reviews the scholar's progress in their proposed research topic. Evidence of scholars' attempt to share the knowledge through publications is considered with utmost care. In between RRMs at least 6 to 12 months gap is mandatory. These RRMs are open to all research trainees as a structured process for their studies and progression. *Third*, scholars are encouraged to publish their work in reputed platforms, thus ensuring appraisal of the quality by an outside expert. There is a minimum required number of publications for any scholar to progress through this third phase. *Fourth*, each research trainee leads a research colloquium open to the university community. The scholar presents the entire work in front of the DRC and a detailed review will be carried with regard to the depth of information in the respective area. If the result of the colloquium is unsatisfactory, a detailed report will be sent to the scholar and supervisor with suggestions to improve the work and present it again in a stipulated time period. After a satisfactory colloquium, the scholar is allowed to submit the thesis for a review and adjudication process defined by the University. There will be final oral presentation and defense after the satisfactory reports from three adjudicators. This structured process resulted in an increased time allocation for research trainees during which they received both mentorship and feedback. Milestones like these exist in many research traineeship programs, and this specific structure exists because of the data collection, analysis, and actions of leadership at the institution.

Explicit incentives for research outputs. Another attempt to build a thriving research culture was made in one of the institutions, which is a part of JNTU Hyderabad. Leadership established a cash incentive for every faculty who published a research paper in any journal. In two year period, more than 200 publications were reported by the faculty, and this was a rapid and positive improvement in university bibliometrics. However, the expectation for quality of the research publications was not well defined, and the proxy measurement for quality was that the paper had to appear in a journal (which means it went through some level of peer review). After three years, the incentives model was refined: only publications in a Scopus Indexed journal will receive the cash incentive. Introduction of this refinement sharply reduced the number of publications and therefore the cash payout by the institution. However, at the outset of building a research culture, the incentive program acted as an impetus to stimulate and focus research actions and behaviors. Despite the need for revision, the incentive model served an important purpose for this phase of the institution's creation of a research culture.

Research time constraints and challenges. In every institution, faculty are assigned teaching, administrative, and other duties. Sometimes social and community engagement adds workload expectations to the faculty. Within the context of this overall workload, the distribution of time for each activity will influence research culture creation and maintenance. Especially in Indian institutions, in the available time of 40 h per week, 16 h are allocated to teaching, 8 to 10 h are allocated to administrative work, and several more hours are allocated to community engagement. Research effort is often relegated to whatever time faculty have left after completing these other components of their workload. Leaders can therefore more clearly codify expectations for research not only in terms of productivity (e.g., number of papers published) but also in terms of time allocation. In this example, a 20% time allocation to research per week may be the best outcome possible, given the other elements of the institution's mission. Other universities might handle this differently, with a larger or smaller expectation for time allocation. Leaders must communicate clearly about the value of research within the overall operation of the institution, and ensure that faculty have a common understanding of an appropriate time commitment.

3.2 *Building Research Culture—a Mission-Based Perspective*

California Polytechnic State University

Educational research can be especially challenging for teaching-focused institutions, which often do not have doctoral programs–and thus a limited set of trainees to support a research mission. Many teaching-focused universities in the US and around the world are trying to build a more prominent research profile (because of both financial and reputation gains associated with larger research programs), yet may not have the resources, infrastructure, or facilities to achieve this goal. Early-career faculty in particular may be conflicted over initiating disciplinary research programs, delving into education-focused research, or committing deeply to their instructional practices. In this section, we will discuss how institutional features at a non-PhD granting public institution in the US affect the aspiration of building a robust engineering education research culture.

California Polytechnic State University, known as Cal Poly, is a comprehensive public institution that is well known for its "Learn By Doing" motto and its strong engineering and computer science programs. As with most US schools, Cal Poly has a time-based system based on credit counts. For each unit, a student is expected to spend one hour per week in class and to spend 2–3 h working on course material outside of class. Cal Poly operates on a 'quarter' system, offering 4 academic quarters per year with each running about 11 weeks. The teaching responsibility for faculty is 12 units per quarter, which could be three sections of a four-unit course or four sections of a three-unit course (counts are slightly different for laboratories and activity periods). Course sizes are typically 35 students, and faculty are required to offer at least four office hours per week for their students across the entirety of their course responsibility.

A focus on undergraduate engineering education constrains research time. As a result of this and the teaching focus at Cal Poly, faculty have very limited time to perform research. Typical teaching responsibility at research-focused institutions in the US might be 3–4 courses per year – the analogous responsibility at Cal Poly would be eight courses per year. Although undergraduate graders are typically used for grading homework and other low-stakes assignments, the lack of doctoral students means that faculty are responsible for grading midterms, final exams, projects, and research papers. This large teaching responsibility can make it difficult for faculty to implement evidence-based practices or to develop innovative instructional tools, further limiting their ability to perform educational research.

Fortunately, the institutional mission does focus on instruction, and the current administration in the college of engineering values educational research and teaching innovation. Instructors are a mix of tenure-track faculty and lecturers, who do not have a research expectation. Those seeking tenure must demonstrate excellence in teaching, service, and research, and the third pillar can be satisfied by performing educational research. Obviously, the research expectation is not nearly as high as at a research-focused institution, and promotion to Associate Professor might be achieved by publishing a peer-reviewed journal article once every other year in addition to some conference proceedings. Including undergraduate students in research is encouraged, and many faculty have students make presentations at regional conferences.

An emerging culture of interdisciplinary research collaboration. Although individual efforts in educational research are incentivized within the promotion system at Cal Poly, there is no system in place to encourage interdepartmental or cross-college collaborations. Cal Poly does not have an educational research center or department, and collaborations are entirely dependent on individuals making personal connections on their own. Some efforts have been made to establish interest groups, but without having some incentive for group leaders (such as release from a course), these efforts have not been sustainable.

Having educational psychologists or learning scientists on campus would greatly increase our ability to perform educational research. Some faculty in our *School of Education* perform research, but the program mostly focuses on credentialing of pre-service teachers to be placed in pre-college environments. This credentialing must align with state and local standards for pre-college teachers, and in the US these expectations are quite specific, detailed, and important, so the *School of Education*'s primary mission focuses on this training and credentialing work. Professors in the *psychology* department tend to focus on cognitive neuroscience, social psychology, clinical psychology, and developmental psychology, and only limited collaboration has occurred between the college of engineering and researchers in psychology. Similarly, researchers in the *Statistics* department experience serious time constraints due to their teaching responsibility, and they also have domain-based research agendas of their own. One interesting resource is their department's *Statistical Consulting Service*, whose mission is "to support and improve research infrastructure, enabling and accelerating quantitative research among students, faculty, and staff at Cal Poly." Although the faculty will not perform the actual analyses, some research teams in engineering that focus on educational research have had success hiring undergraduate statistics students to run numerous statistical tests using the advice of the free Consulting Service.

Research funding and availability for education-focused work. As with most US universities, funding models at Cal Poly focus on federal agencies such as the US National Science Foundation (NSF), with philanthropic donations towards educational research playing a much smaller role in the overall funding landscape. Obtaining competitive funding from a federal agency not only provides external validation for an individual to use in their case for promotion, but it also provides incentives such as reduced teaching responsibilities and potentially additional salary. Obtaining external funding can help faculty pay for undergraduate and master's degree student research assistants, fund travel to professional conferences where outside collaborations can be established, and hire outside consultants and research assistants. Cal Poly's focus on teaching makes it attractive to funding agencies, who recognize the institution's dedication to pedagogical innovations. Cal Poly and similar teaching-focused institutions also recognize that the US is fortunate to have federal funding agencies such as the NSF and the US Department of Education who provide monetary support for educational research activities; similar funding agencies are much less common in much of the rest of the world.

Collaboration across cultures. An interesting international collaboration that has had some funding attached has developed between Cal Poly and the Munich University of Applied Sciences. The German higher education system has two main types of universities, the Technical Universities (TU) that has high research activities and the Hochshules (or University of Applied Sciences) system that focuses on teaching and does not have a doctoral program. This is analogous to the California public system, which has the University of California research-focused universities (e.g., UC Berkeley, UC Santa Barbara) and the California State University (CSU) teaching-focused universities. Over the last twenty-five years, Cal Poly and MUAS have sponsored approximately 10 full-year faculty exchanges as well as numerous short-term visits and teaching assignments. Additionally, 5–10 students each year participate in student exchanges – typically just a single academic term.

This partnership has laid the foundation for potential education research collaborations in the future. Two faculty obtained funding from the Bavaria California Technology Center (BaCaTeC—http://www.bacatec.de/en/) to investigate how the two countries approached engineering design. The partnership was expanded through funding from the German Academic Exchange Service (https://www.daad.de/en/), which promoted additional exchange opportunities as well as high level discussion among academic leaders. Individual faculty certainly shared teaching ideas and practices, and the third author gave some workshops through the Bavarian Didaktik Zentrum and reported on his exchange experience with the MUAS author. To date, however, there has not been much collaborative education research performed, and this further emphasizes the challenges of cross-cultural collaboration—even when institutional missions align.

3.3 Building Research Culture—an Incentives-Based Perspective

Purdue University

Purdue is a large institution in every respect—enrollment at both the undergraduate and graduate degree levels, number of faculty and staff employees, physical size of campus, scope of mission and diversity of disciplines, and volume of research conducted. Purdue is a specific kind of institution in the US called a 'land grant' university (established by federal legislation in the mid/late 1800s), which means it was designated by the state as a beneficiary of several US federal actions designed to build capacity within the US to apply engineering and agricultural techniques to the growth of this (at that time) new nation. The land grant mission explicitly includes what we today call 'technology transfer', meaning that knowledge created at the university should be shared for the benefit of the state and the nation. For example, innovations in agricultural tools and techniques would be shared directly with farmers so that they could expand their efficiency and productivity. Because of this history, research and innovation are deeply rooted in the origins of land grant institutions in the US, and that research mission today has expanded to the include essentially all disciplines on Purdue's campus.

The institution expects research productivity for faculty. An explicit expectation for all faculty, and a central consideration for promotion, compensation, and career progression, is research productivity. This means: securing external financial support for their work, producing relevant scholarship and sharing it widely, and mentoring research trainees (undergraduate, masters, and/or doctoral students) who participate in the research. Specifically with respect to *engineering education research*, the work of the faculty and their research teams has both an external audience reached through publications and conference presentations, as well as an internal audience at the institution. The research-to-practice goals of engineering education research entail improving the teaching and learning practices writ large at the institution—a priority shared across institution types, locations, and missions. Engineering education research at a large, research-intensive institution like Purdue therefore plays a special role within the broader expectations for research productivity: the outcomes of the work are expected to make a fundamental contribution to the quality of teaching and learning at the institution.

The institution is designed for research productivity. Expectations for research productivity are high, and the university is designed to support research. Two key respects in which the institution is designed to support research relate to *research infrastructure* and *personnel time*. The research infrastructure element certainly refers to the physical and computational infrastructures required to actually perform the research, including laboratory space, central computing resources, and office space for researchers. But key research infrastructure also includes all the mechanisms by which external research funding is supported, such as: offices which support research proposal writing, processing, and submission; dedicated support for financial management of external grants; relationship management for non-federal external partners such as industries or philanthropic foundations; offices which ensure compliance with funder requirements and expectations (including US federal laws as relevant to federal awards); and, for large-scale research activities, project management support to ensure projects run smoothly. Moreover, the apparatus at the institution to manage graduate student recruitment, application, enrollment, and financial support (as available) significantly supports the research enterprise by delivering a seamless experience for research trainees at the masters and PhD level.

Resources, culture, and time. However, accessing these support resources, and understanding the expectations for research productivity, are necessary but not sufficient conditions to build a thriving research culture. One other key piece involves the implicit and explicit expectations about how people spend their time. In large research-intensive institutions in the US, the incentives to prioritize the research and innovation mission are very strong. Other workload responsibilities, like teaching assignments, are managed to provide ample time for faculty to conduct research and work with their research trainees. In institutions like this, faculty with research programs may be assigned 2–3 courses (6—9 units) per academic year, with the opportunity to decrease that assignment if research funds are used to pay a portion of their academic year salary. Taken together, in many US institutions with strong research missions and expectations, this system of infrastructure, workload policies and practices, and faculty incentives promote an overall culture the prioritizes research productivity and enables the success of that mission.

3.4 Institutional Research Cultures and Quality Outcomes

These brief case studies about the contexts and activities within three academic institutions provide a general sense of institutional orientation, priorities, challenges, and approaches. The connection of each institution to its goal producing quality outcomes is explored more deeply in this section. The 'quality outcomes' sought within institutions engaged in engineering education research span several themes:

- *Research trainees* should get strong mentored experiences that allow them to develop sharp research skillsets and contribute meaningfully to research projects.
- *Students enrolled at the institutions* should enjoy improved teaching and learning experiences because of the translation of research results into routine practice.

- *Faculty researchers* should derive intellectual and practical satisfaction from their research while also growing their research reputation.
- *Institutions supporting engineering education research* should realize practical benefits of improved student experiences while also reaping reputational benefits as institutions in which engineering education is being approached and delivered in a thoughtful, evidence-driven way.

Quality in the case study contexts. These outcomes manifest themselves in different ways across the three institutions considered here.

In the first case study (Jawaharlal Nehru Technological University), the institution itself recognized a need to reinvest in its processes and practices around guidance, mentorship, and expectations for research trainees. In that case, a central focus was a set of milestones and associated timelines that clearly defined the nominal growth trajectory expectations for students. For that moment in its history, the institution's most important path for shaping the culture in support of quality outcomes was to install a framework for trainee mentorship and growth. The description above conveys two components of the plan. *First*, the trainee experience is scaffolded with milestones that provide clear navigational landmarks to the students along their research trajectory. A suggested timeline for achieving these milestones is also present in the plan. *Second*, feedback to the trainee from multiple external authorities is central to the plan for mentorship. This feedback includes both on-campus avenues (the research review meetings and the doctoral review committees), but also truly external perspectives as embedded in the expectation for publishing in peer-reviewed venues. Taken together, these two features of the institution's plan align with quality metrics: they provide students with a clear set of expectations and milestones, and they embed expert feedback into the student developmental process.

In the second case study (Cal Poly), the institution integrates quality assurance practices into the educational research enterprise in two primary respects. *First*, from an internal perspective, the institution embraces and rewards high-quality teaching (according to the usual metrics, like course evaluations) as well as educational innovation. Within their routine process of delivering education to their students, they review the teaching and learning activities for various hallmarks of quality common in higher education settings. This enables evaluation of the quality of instruction, implementation of best practices, or advancement via innovation in a routine way. *Second*, the scholarship and grant related activities introduce an element of peer review from external experts into the research enterprise. As in the first case study, this external evaluation and connection to the broader research community provides another layer of quality assurance for the research and implementation.

In the third case study (Purdue), research quality measures rely substantially on various kinds of external peer review and feedback. For engineering education research, this peer review happens in two main ways. *First*, research grant support is such a central feature of the overall culture at the institution, success in securing competitive grants from US federal agencies is one powerful marker of quality for educational research. Securing funding from federal sources is a very competitive process in the US, and reviews of grant proposals are generally conducted by panels of experts who evaluate the work. *Second*, publication in peer-reviewed venues, as discussed earlier, conveys another mark of quality to research efforts. Promotion through the faculty ranks and external recognitions (i.e., professional reputation) are built upon both grants and publications, so in general these two perspectives on quality assurance align with institutional culture and practices around faculty activity.

Quality and the role of other incentives. We return briefly to one other element of the first case study, related to financial incentives for publications. In many higher education contexts around the world, institutions have experimented with financial incentives/rewards for faculty publications. Publication in the most prestigious journals (e.g., Science, Nature) can be very lucrative, while more 'routine' publication still results in a financial benefit. For many institutions, the value placed on bibliometrics is significant and affects their institution's ranking in various indices, or their reputations with the public. However, as described in the case study above, monetary incentives typically reward *publication* rather than identifiable *impact* (i.e., quality). The quality and impact evaluation uses peer review as a proxy, as seen in the stipulations about Scopus-indexed journals or any number of other criteria based upon rigor of peer review, impact factor, or prominence of journal in its field. This kind of incentive must be carefully calibrated to ensure that volume (in the bibliometric sense) and quality co-exist and are mutually supported by such incentive practices.

4 Technology as an Enabler of Research Cultures

Research cultures focused on quality outcomes are necessarily built upon the antecedent work conducted by researchers around the world and published in the open literature. The staggering volume of published research on virtually any topic presents serious challenges to the research culture that can at least partially be addressed through institutionalized use of modern tools for exploration of prior work. Modern AI-based tools promote powerful, customized search as well as provide summaries of the vast body of knowledge available. When used appropriately, such tools can form one element of a research culture that focuses on quality by ensuring appropriate consideration and attribution of prior work, while simultaneously accelerating a researcher's ability to understand prior work from their own and adjacent disciplines.

4.1 AI-Based Tools to Enhance Quality and Efficiency

AI-based tools provide researchers with new opportunities to explore the vast scope of prior research in an efficient way, and more importantly help them make sense of that prior work in ways that were previously too time consuming to achieve. AI-based tools continue to emerge, and they are too numerous to mention here, but their functions as powerful search aids, summarization tools, and synthesis engines provide researchers with new tools to support quality research outcomes. These tools play several specific roles that enhance research quality. *First*, they allow researchers to explore the vast, uncountable number of research publications in a given field with far greater efficiency than was previously possible. Researchers can effectively identify and sort through a massive number of research articles with unparalleled efficiency to identify the subset that are germane to their research topic. *Second*, AI tools can summarize each paper by providing researchers with details about topic, population size, method used, and results (e.g., statistical significance or effect size), again with staggering efficiency. The outcome is that researchers have a deeper sense of a specific article's research relevance with very little effort. *Third*, AI tools can search across a wider aperture than was previously possible, for instance by examining literature in a specific language or from a specific geographic location. This makes research more inclusive and connected, and it makes researchers more aware of what colleagues across the globe are contributing to the literature. *Fourth*, AI tools screen literature effectively enough that the time a researcher spends actually reading, scrutinizing, and understanding articles is better used because less relevant articles are filtered out by the AI.

Of course, all of this AI-based efficiency can lead to undesirable outcomes, including AI-authored articles, plagiarism, and other ethical breaches. Researchers must apply their best judgments and practices around research integrity, most specifically by keeping a human in the loop and approaching AI-generated output with critical thinking. Public (both peers and the public at large) trust in scholarly works is central to success to the research enterprise, and indiscriminate or careless use of AI can seriously undermine that trust. Moreover, research built upon a basis (or prior work) of incorrect, fabricated, or careless research compromises future developments in serious ways.

4.2 Implications for Research Culture

The ubiquity and ease-of-use of AI-based tools certain play a role in defining the practices and approaches embedded in the research culture, especially in the training of new researchers. The institutional culture around research, with a focus on quality outcomes as described above, needs to be resistant to the most powerful seductions of technology tools that invite individuals to cut corners, exploit the technology, and embrace efficiency over quality. Research culture must be bolstered by the same foundations of integrity that have governed the research enterprise for decades, calibrated for the new world of AI-enabled discovery. *First*, a strong foundation of governance within the institution about guidelines and policies for appropriate use of AI tools is absolutely compulsory. In the US, this begins with the principles of Responsible Conduct of Research (RCR), while in Europe this is governed by the European Code of Conduct for Research Integrity. China applies the Guidelines for Responsible Research Conduct. Many countries have strong governance in place for medical/clinical research, but standards for human subjects research in education and social sciences are less well developed. Nonetheless, a robust foundation and shared understanding of ethical principles and expectations for the use of AI in research is essential in today's research landscape.

Second, institutions must implement strong professional development and training for researchers, with periodic recertification, so that key ethical principles are understood, applied, and upheld within the research culture. *Third*, institutions should periodically audit research projects for their compliance with best research practices in general, and with AI usage in particular. Audits need not be intense or intrusive, but should focus on alignment of research team behaviors and actions with expectations about academic and research integrity. The cadence of such audits will vary by institution, but routine review of research practices every few years empowers researchers to conduct their work while ensuring institutional oversight and compliance with research standards. *Fourth*, institutions need to vet centrally-supported AI-based research tools to ensure they comply with reasonable standards for individual privacy, data security, reliability, and so forth. Institutions already do this for commodity computing tools like web browsers and email, and they should begin applying these standards to AI-based research tools to ensure the integrity of the research enterprise.

5 Reflections for Leadership

One important feature of academic institutions is that faculty have significant autonomy in decisions about how to spend their time and conduct their work. The faculty are central figures in shaping any dimension of academic culture because they are typically the highest-status individuals within the academic ecosystem. The challenge for leadership is to build policies, practices, and incentives that encourage high-status *individuals* with significant autonomy to engage in *collective* behaviors that build the desired research culture. The remainder of the section will describe some of the institutional levers available to academic leaders that can promote a culture of research with an emphasis on quality outcomes.

5.1 Challenges of Creating a Culture

This chapter so far has used the term 'culture' in a colloquial way; that is, a 'culture' is the set of beliefs, norms, traditions, viewpoints, and so forth that define the ways an organization operates. The culture of an organization, or even of a specific element of its operation, is difficult to define, and even harder to measure, because 'culture' is an all-encompassing concept. This chapter takes the view that 'culture' includes everything extant in the environment such as relationships, behaviors, policies, practices, values, and so forth (this is an anthropological perspective); culture is inclusive of organization performance (i.e., performance is not an *outcome* of culture; performance is part of the culture). As such, a 'research culture' includes all the essential ingredients of the research enterprise already discussed: people, infrastructure, funding, policies, practices, incentives, expectations, and outcomes—in this case, the goal is to promote 'quality outcomes'. Academic leaders therefore must work to shape the research culture by carefully investing in three broad categories of activities related to people, physical and intellectual infrastructure, and quality.

5.2 Leaders and People: Mentoring and Collaboration

The brief case studies presented earlier demonstrated the importance of research mentorship in multiple contexts, and of aligning institutional practices so that this mentorship happens in routine and high quality ways. Engaged and active mentoring is a well known feature of high-functioning organizations, and educational institutions hold the mentoring functions as central to their missions and success. The cases studies illustrate that ad hoc mentoring practices within the research enterprise are insufficient either to build a research culture or to ensure quality outcomes. Leaders must therefore make strategic investments in mentoring programs and processes in at least the following categories:

- *Research trainees* must have access to high-quality mentoring, a clear set of expectations and timeline, and an understanding of their expected deliverables. Moreover, trainees need active guidance on research methods, analysis techniques, how to draw appropriate inferences from research data, and so forth. The first case study from JNTU dramatically illustrates the potential positive outcomes associated with disciplined application of a mentoring framework to support nascent researchers.
- *Faculty researchers* should benefit from the guidance of their peers and mentors to develop high-quality research designs and to identify support resources and expertise as appropriate for a research project. Because engineering education research often takes place in formal education settings (classrooms, laboratories, etc.), leaders must also *encourage faculty to open their classrooms for research studies*; this achieves several important objectives. *First*, the researcher benefits from access to student populations and educational settings to conduct research studies. *Second*, the organization itself cultivates a culture of discovery and learning, and has significant opportunities for continuous improvement of its own educational practices as a result of the research. This translational component of research—the application of findings in one's own setting—is an essential element of building a research culture, and academic leaders play a central role in encouraging this kind of mentoring and collaboration.
- *Researchers at all career stages* would benefit from structured approaches to meeting potential collaborators from diverse disciplines, thereby promoting interdisciplinary research that integrates multiple perspectives, methods, and

approaches. Leaders should invest in creating opportunities for these connections to materialize, including approaches like institution-wide research symposia, poster sessions, informal networking, or other events that naturally bring researchers together in community. If resources allow, leaders can also establish seeds funding competitions for nascent collaborations, mentoring networks, or learning communities devoted to engineering education research.

5.3 Leaders and Infrastructure: Physical, Intellectual, and Financial Resources for Research

Leaders have great influence over infrastructure to support research via the physical spaces used to conduct research, the intellectual vibrancy of the environment, and the funding available to promote and enable research. The role of physical space for engineering education research is largely self-evident, relying on space for personnel to do their work, classroom or laboratory spaces in which measurements may be made, and potentially dedicated spaces to collect other kinds of data (ex.: interview or focus group rooms, or laboratory spaces for pilot studies on new technology such as virtual reality or wearable devices). The other elements of research infrastructure require more explanation in the context of engineering education research. *First*, the intellectual vibrancy of an institution specifically about teaching and learning is central to creating a culture of engineering education research. Teaching and learning, student outcomes, and the student experience must be routinely discussed, reviewed, deliberated, and so forth in order to promote a culture that values quality education outcomes. Leaders must facilitate these conversations and ensure that faculty and research trainees understand the importance of these outcomes to the institution's mission (as illustrated in the Cal Poly case study). *Second*, leaders should consider the financial elements of research support, in addition to funding considerations for the actual conduct of the research. Research support includes personnel dedicated to reviewing, approving, and monitoring ethical research conduct, support for financial elements of research management, and (for larger initiatives) professional project management personnel to help researchers plan, organize, and execute such projects. In addition, for researchers who apply for external funding—whether from governmental agencies, industry sources, or elsewhere—institutional support for proposal preparation in terms of research planning, budgeting, and compliance with sponsor forms or other requirements is an important priority for leadership. In each of these cases, leadership can support the research enterprise and the researchers themselves by hiring and retaining personnel who have specific skillsets (financial, project management, etc.) that the researchers themselves either do not already possess or do not have the time to acquire. The Purdue case study shows that this kind of infrastructure is particularly important for research activities that rely substantially on external (often governmental) funding.

5.4 Leaders and Quality: Impact

The most profound influence exerted by leaders on the research enterprise is their explicit commitment to high-quality outcomes, which can be operationalized in terms of bibliometrics, as research impact on key stakeholders, as technology transfer activities, or any number of other metrics. Notably, all these measures of quality rely upon external perceptions of research outcomes, whether through academic peer review, stakeholder satisfaction, or generation of intellectual property (via patents or licensing agreements). As always, academic leaders must align their priorities for research quality with the mission of the organization. For example, the Cal Poly case study shows that undergraduate students are key stakeholders in their engineering education research and innovation enterprise, and as such a quality outcome is one that makes a substantial positive impact on students and their experience. In high-volume research organizations, such as illustrated in the case study about Purdue, faculty career progression and promotion are explicitly tied to research impact, thereby codifying a quality metric into the employment expectations for faculty. As demonstrated in the JNTU case study, the tension between (financial) *incentives* for higher research volume (measured by number of papers published) and the expectations for publication quality and impact (characterized by the perceived quality of the publication venue) must be carefully managed by institutional leaders. The commitment to quality outcomes is also related to the intellectual infrastructure components of the previous section. When institutions and their leaders are committed to continuous improvement, especially in their academic programs, they create a culture in which discussions, deliberations, and improvements of the student academic experience routinely take place. In short, leaders can help their institutions become *learning organizations*, in which quality outcomes become a routine expectation within the culture.

6 Conclusions

The three brief case studies presented here illustrate multiple tactics and goals for building a culture of research that focuses on quality outcomes. Together, they show that no one approach applies to all institutions, and that academic leadership must intentionally align the approaches, incentives, goals, and quality assurance standards to the mission of the institution and its stakeholders. From the case studies, we extracted a more generalizable set of recommendations for leaders about three key categories of cultural influencers: people and mentorship, research infrastructure (including intellectual infrastructure), and a focus on quality and impact. It is imperative for academic leaders to make disciplined decisions about resource allocation to support thriving research cultures and ensure quality outcomes. The challenges, opportunities, and decision vary by institution, and these three ingredients are essential for the creation, growth, and ultimately the success of a thriving research culture.

Building a Pipeline to Encourage Innovation, Entrepreneurship & Incubation for Faculty and Students

Meghna Bandi, Hafeez Basha Ranipet, Sharath Chandra Aduri, and Fady Fadel

Abstract Institutions of higher education play a crucial role in fostering innovation, entrepreneurship, and incubation among both faculty and students. This abstract outline the construction of a comprehensive pipeline designed to cultivate these essential skills and mindsets within academic communities. The pipeline encompasses various stages, starting from awareness and education about innovation and entrepreneurship opportunities, to hands-on training and mentorship, and culminating in the provision of resources and support for the development and launch of innovative ventures. Key components of the pipeline include dedicated courses, workshops, networking events, access to funding and facilities, as well as partnerships with industry and government agencies. By implementing such a pipeline, institutions can create an ecosystem that empowers faculty and students to transform their ideas into impactful solutions, driving economic growth and societal progress.

Keywords Innovation · Incubation · Entrepreneurship · Startups

1 Introduction

Innovation is a fundamental concept characterized by the practical implementation of new ideas, methods, products, services, or solutions that bring about significant positive change. It involves the introduction of novel goods or services, improvement in existing offerings, and the creation of value for customers and stakeholders. Innovation is not limited to the development of new products but extends to processes, business models, and services within an organization.

M. Bandi
Aurora Higher Education and Research Academy, Hyderabad, Telangana, India

H. B. Ranipet (✉) · S. C. Aduri
Osmania Technology Business Incubator, Osmania University, Hyderabad, Telangana, India
e-mail: drhafeezbasha@gmail.com

F. Fadel
The American Business School of Paris, Groupe IGS Paris, Paris, France

R. Kandakatla et al. (eds.), *Academic Leadership in Engineering Education*, Lecture Notes in Networks and Systems 1097, https://doi.org/10.1007/978-3-031-68282-7_13

According to ISO TC 279 and various definitions, innovation is about creating new or changed entities that realize or redistribute value. It often involves the development of more effective products, processes, services, technologies, or business models that innovators introduce to markets and society. Innovation can take various forms such as technical innovation through engineering processes or disruptive innovation that fundamentally changes how value is created, delivered, or consumed.

Innovation plays a crucial role in helping companies remain competitive, adapt to changing environments, satisfy market needs, introduce new technologies, create new markets, and react to market disruptions. There are different types of innovation including incremental (small changes for efficiency), expansive (exploring new ideas for long-term growth), and disruptive (creating new products or business models that disrupt existing markets). The innovation process typically involves stages like design (idea generation), testing (validation with users), and scaling (implementation on a larger scale).

Successful innovation requires organizations to foster an innovative mindset, implement small incremental changes for continuous improvement, protect their innovations through intellectual property strategies, and capitalize on opportunities for growth and development. Innovation is not just about products but also about evolving business models, enhancing operations, and creating value across all aspects of a business.

Entrepreneurship: Entrepreneurship is the process of creating or extracting economic value through the creation or investment in businesses, often involving risks and rewards. It is characterized by the ability to perceive and create new economic opportunities, introduce ideas into the market, and make decisions on resource allocation.

Entrepreneurship is not limited to starting a business but also includes identifying opportunities, taking risks, and managing a business venture. Entrepreneurs are often seen as innovators and sources of new ideas, goods, services, and business procedures. They can be individuals or teams and are responsible for the launch and growth of an enterprise. Entrepreneurship can take various forms, including innovative, imitative, Fabian, and drone entrepreneurship, based on risk and type of business. It has several benefits, such as being one's own boss, making decisions, choosing whom to do business with, and having the ability to build equity. It also offers the possibility of significant financial rewards, involvement in the total operation of the business, and the prestige of being in charge. Additionally, entrepreneurship contributes to the local economy and, in some cases, contribute to the society as a whole through innovations.

The study of entrepreneurship involves understanding current concepts, evaluating entrepreneurs, their ventures, and the venturing environment. It can be approached from various perspectives, including academic research, publications, direct observation, and interaction with practicing entrepreneurs.

The field of entrepreneurship education has grown significantly over the past two decades, with an increasing number of colleges and universities offering courses related to entrepreneurship.

1.1 The India Scenario: Situational Awareness

Engineering Education in India has seen dramatic changes over decades of iterations in transforming from 'Traditional Book to Paper' to 'Outcome Based' Education and now trending towards 'Experiential Learning' into future.

This evolution of teaching and learning mechanism for Engineering Education has been derived from the changes in the environment in which it was being delivered. These environmental factors include social factors, political factors, technological factors, economical factors, legal factors and environmental factors. It is observed that, earlier students used to opt for Civil, Mechanical and then Electrical and Electronics Engineering with other engineering domains as fallback choices in admission process. Recently, shift of priorities was observed, with majority of them opting Computer Science & Engineering. Currently, they are opting in Emerging Technologies like Artificial Intelligence, Data Science, Internet of Things, Robotic Process Automation, 3D Printing/Additive Manufacturing, Unmanned Aerial Vehicles, Augmented Reality/Virtual Reality, etc. On the verge of transforming from Outcome Based Education into Experiential Learning, the emphasis is laid on Innovation and Entrepreneurship as the drivers of the new economic engine in India. Meaning, Academia would be witnessing more dramatic changes in the present times than occurred chronologically for decades. It is therefore recommended to periodically review the environmental factors that are contributing to the evolutionary changes in the Engineering Education to enhance the readiness and keep up with the nation's demand. To critically evaluate these changes, it is elaborated what each stakeholder is contributing in building a pipeline to encourage innovation, entrepreneurship and incubation for faculty and students.

In India, the startup ecosystem has been rapidly growing, with various government initiatives and private organizations promoting entrepreneurship. Several renowned institutions, including IIMs, IITs, State and Private Universities and Colleges, have established their own incubation and accelerator programs. Additionally, the Indian government's "Startup India" initiative has been driving the growth of startups by offering incentives, funding, and regulatory support. For students, faculties, and early- stage entrepreneurs in India, engaging with accelerators and incubators can provide a structured path to transform innovative ideas into successful businesses. These programs can offer a supportive environment, access to industry experts, and the resources needed to overcome challenges and achieve growth. It's important to research and choose the program that aligns best with the specific goals and needs of the startup.

The primary action is to be drawn towards defining the system and identifying the relevant stakeholders contributing or affecting the dynamic relationships with the system to be developed. The broader framework is already in place. It takes individual and organizational preparedness and executive alignment to effectively drive Innovation and Entrepreneurship forward.

Indian government's 'Startup India' initiative has provided support for startups across the country. It offers benefits such as tax exemptions and funding opportunities.

Many Indian states have launched their own startup and innovation policies that include funding support, infrastructure, and mentorship. The startup ecosystem in Telangana, particularly in Hyderabad, has been experiencing significant growth and development over the past few years. The state government's proactive initiatives, coupled with the presence of top educational institutions, technology parks, and industry collaborations, have contributed to the emergence of a vibrant and thriving startup ecosystem. These initiatives collectively create an environment that supports students, faculty, and early-stage entrepreneurs in Telangana.

1.2 The Local Scenario: Progress in Hyderabad, India

Government of Telangana has taken proactive steps to support startups through policies like the Startup Telangana, which offers incentives, funding, and regulatory support. This has attracted both local and national attention, encouraging entrepreneurs to set up their ventures in the state.

Hyderabad is home to numerous business incubators like Government operated Technology Hub (T-Hub), Women Entrepreneurship Hub (WE Hub), Research and Innovation Circle of Hyderabad (RICH), and various sector- specific incubators.

These platforms offer startups mentoring, funding, workspace, and access to a network of investors and industry experts. Hyderabad boasts several prestigious educational institutions like International Institute of Information Technology (IIIT Hyderabad), the Indian School of Business (ISB Hyderabad), Osmania University, Jawaharlal Nehru Technological University (JNTU Hyderabad), etc. These institutions contribute to the ecosystem by fostering innovation, research, and entrepreneurship among students and faculty. The city has seen a significant increase in funding from venture capital firms, angel investors, and corporate investors.

The city is home to various technology parks and innovation hubs like Genome Valley, Hyderabad Information Technology & Engineering Consultancy (Hitec City), and more. These hubs provide startups with infrastructure, networking opportunities, and access to a skilled workforce. The startup ecosystem in Telangana spans across diverse sectors, including technology, healthcare, agriculture, fintech, edtech, and more. This diversity reflects the state's commitment to fostering innovation across various industries. Various startup events, hackathons, conferences, and workshops take place in the city regularly. These events provide opportunities for startups to showcase their innovations, network with industry leaders, and learn from experts. Large corporations in Hyderabad actively engage with startups through corporate accelerator programs, innovation challenges, and collaborations. This provides startups with access to resources, mentorship, and potential customers.

Hyderabad's strategic location and well-connected infrastructure provide startups with access to national and international markets. This geographical advantage facilitates business expansion and growth. The city's growing reputation as an emerging startup hub has led to global recognition. Hyderabad's startups have gained visibility on both national and international platforms, attracting attention from investors and partners. Hyderabad offers a relatively lower cost of living compared to other major cities in India, making it an attractive destination for startups and entrepreneurs. Overall, the startup ecosystem in Telangana, particularly in Hyderabad, is characterized by a conducive environment, government support, industry collaboration, access to funding, and a diverse range of startups across various sectors. The government regularly organizes hackathons, competitions, and innovation challenges to encourage students, faculty, and entrepreneurs to come up with innovative solutions to real-world problems.

1.3 Global Perspective

The integration of innovation and entrepreneurship skills into engineering education is a key component of the global view on innovation and entrepreneurship in engineering institutions.

In order to meet urgent environmental concerns and make a name for themselves in the quickly changing engineering scene, engineers must adopt the strategy of being innovative and have an entrepreneurial mindset. Engineers can develop an entrepreneurial attitude that will enable them to become T-shaped professionals who can successfully conduct academic research, innovate in industry, or launch their own companies by fusing their engineering knowledge with business acumen.

Academic institutions are progressively promoting entrepreneurship among their staff and students by means of policy modifications. The global perspective on innovation and entrepreneurship in professional colleges emphasizes the importance of integrating entrepreneurial skills into education to prepare industry leaders for the dynamic landscape of innovation. Here are key insights from the search results:

Global Education Entrepreneurship and Innovation Certificate Program at Penn GSE: The Global Education Entrepreneurship and Innovation Certificate Program at Penn GSE is a unique and world-first online program designed for education industry leaders, school leadership teams, entrepreneurs, researchers, investors, policymakers, or anyone interested in addressing critical issues in education through innovation.

The program is led by renowned Penn faculty and industry experts and blends expertise, research, and practice from the Schools of Education, Business, Design, and Engineering. It provides critical knowledge, skills, and tools needed to seed, accelerate, and scale entrepreneurial practice in education.

The immersive week-long boot camp comprises customized real-time virtual sessions, paired with online asynchronous modules. It uses Penn's lean start-up curriculum, design thinking, and project-based learning. The program is ideal for any professional looking to address critical issues in a dynamic education landscape and create lasting change in education through innovation.

The program covers topics such as mapping the global innovation landscape in education, inspiring an organizational culture of innovation, activating entrepreneurial leadership and mindsets, creating an innovative program, product, service, or company, learning science tools and methods, and unleashing the power of journey stories.

Upon successful completion of all the modules, coursework, and projects, applicants will receive a certificate of completion from the University of Pennsylvania, independent of affiliation to any university/college/organization unless specifically mentioned. The program is designed for professionals in the education sector interested in creating leading-edge solutions in education. The program will educate professionals on education investment and how to find a product-market fit.

The program is delivered through a combination of real-time virtual sessions, online asynchronous modules, and an immersive week-long boot camp. The program is designed to provide professionals with the skills and knowledge needed to seed, accelerate, and scale entrepreneurial practice in education, addressing critical issues in a dynamic education landscape, and creating lasting change in education through innovation.

NYU SPS: The New York University School of Professional Studies (NYU SPS) is committed to fostering creativity in global contexts, global citizenship, cultural sensitivity, and leadership abilities. One of the most fascinating and active initiatives to hit the higher education scene is the Centre for Global Affairs (CGA), which provides demanding graduate and continuing education courses together with public programs that educate residents of the world to feel comfortable in any foreign setting.

A Master of Science in Global Affairs (MSGA) degree, which offers a comprehensive foundation in international relations, international law, international political economy, and analytical abilities for global affairs, is available to students via the CGA. The program enhances students' understanding of global challenges and international politics. It also helps student develop their professional resumes by offering eight-degree concentrations in cutting-edge areas within the field of international relations. Additionally, students can opt to focus on the United Nations' global risk, or data analytics, which will deepen their understanding of international relations and equip them with marketable skills.

The CGA offers professional enhancement courses, diploma programs, and public events that investigate current international policy trends and engage in spirited debate of important problems. These offerings are in addition to the MSGA degree. The goal of the program is to develop a community of global citizens who are able to transform the world. It is led by academics and professionals with firsthand knowledge of international affairs. Members of CGA is devoted and committed to the challenging global agendas and is passionate and enthusiastic about the initiatives.

Everyone interested in global affairs can find a place to participate and learn at the CGA thanks to its cutting-edge MS in Global Affairs, diploma programs, career growth courses, and thought-provoking public events. Students will gain from the global professional networks that NYU SPS has built and will learn transferable knowledge and skills.

In conclusion, the NYU SPS Centre for Global Affairs provides a variety of courses and activities aimed at fostering an understanding of other cultures, global citizenship, leadership abilities, and inventiveness in cross-border settings. A wide foundation in global affairs is provided by the MSGA degree, which offers specializations and concentrations that increase students' understanding of international policy and global challenges.

Institute of Innovation & Entrepreneurship at Hofstra University: Through its Institute of Innovation & Entrepreneurship, which cultivates a culture of creativity and business aptitude among students, Hofstra University encourages innovation and entrepreneurship. For students who want to pursue entrepreneurship as a career, the institute provides a variety of resources and initiatives, such as academic courses, hands-on learning opportunities, and mentorship. Students can major or minor in entrepreneurship at the Frank G. Zarb School of Business, which gives them a solid foundation in business principles and entrepreneurial abilities. The institution collaborates closely with the teachers and administration to provide students with opportunities for experiential learning, such as helping a group of student entrepreneurs start a new business. Capital One provides funding for these enterprises, with bankers acting as mentors for the duration of the academic year.

Symbiosis Entrepreneurship and Incubation centre (Pune, India): Within the Symbiosis family, the government-recognized and financed Symbiosis Centre for Entrepreneurship & Innovation (SCEI) as a start-up Technology Business Incubator (TBI) that supports people and/or organizations with business or product ideas from conception to commercialization. It is a platform that makes it possible for spinoffs and start-ups to be formed around business products or ideas and propels them to success by offering suitable services and mentorship at every stage of the commercialization process.

SCEI offers a range of programs and initiatives to support entrepreneurship and innovation, including pre-incubation and incubation programs, a certificate program in entrepreneurship, and various entrepreneurship activities within the Symbiosis International (Deemed University) community.

SCEI's pre-incubation program is designed to help early-stage entrepreneurs develop their ideas and prepare for incubation, while the incubation program provides a more structured and supportive environment for start-ups to grow and scale their businesses. The certificate program in entrepreneurship is a formal academic program that provides students with a comprehensive understanding of entrepreneurship and innovation, including the skills and knowledge needed to start and grow successful businesses. In addition to SCEI, Symbiosis University of Applied Sciences, Indore (SUAS) also has an incubation center that supports entrepreneurship and innovation. The SUAS Incubation Center (SUAS-IC) provides a range of services and resources to support the creation of new ventures, including consulting services, research projects, case studies, and funding networks. The SUAS-IC also provides physical and virtual incubation support, including workspace, machinery, infrastructure, and online start-up mentorship. The SUAS-IC is committed to creating a world-class ecosystem for start-ups and fostering a hub of start-up and entrepreneurship activities.

Society for Innovation and Entrepreneurship (SINE) at IIT Bombay: SINE at IIT Bombay is a technology-based incubator that aims to provide 'start-to-scale support' for business ventures.

Established about two decades ago, SINE has been instrumental in elevating Mumbai's startup ecosystem by supporting over 220 startups, including one unicorn and an IPO. The incubator focuses on segments like defence, healthcare, and other technology areas, with plans to create its own fund. SINE has been a pioneer in the academia incubation ecosystem, with notable successes like Gupshup entering the unicorn club and drone manufacturer ideaForge becoming a listed entity. The centre is planning to develop an aggregator platform called SINEdge to facilitate connections between startups, corporates, investors, vendors, and mentors, aiming to create a robust ecosystem for entrepreneurship. Other academic institutions looking to replicate SINE's success need to establish a similar autonomous structure with functional, operational, and financial autonomy to support entrepreneurship effectively.

Some of the services offered by SINE to startups include:

Incubation Support: SINE provides up to 3 years of incubation support to startups from various technology and science areas, offering both physical and virtual incubation options.

Grants and Monetary Support: SINE offers grants and other monetary support to startups, helping them navigate the challenges of product development and early deployment stages. This financial assistance can be crucial in averting crises like cash flow mismatches and supporting startups during their early phases.

Infrastructure and Facilities: SINE provides essential support and facilities to startups, including infrastructure, lab equipment, and other necessary resources to further their research and development efforts.

Mentorship: SINE reinforces startups through mentorship, providing guidance, advice, and support from experienced professionals to help startups navigate challenges and make informed decisions.

Networking Opportunities: SINE facilitates connections between startups, corporates, investors, vendors, and mentors through its aggregator platform, SINEdge. This platform aims to create an ecosystem for startups, enabling them to pitch their ideas, partner with corporates, and accelerate their growth.

The above cases highlight the need for building an Entrepreneurial Ecosystem so that young professional graduates choose entrepreneurship as their career option. The following are the key takeaways from the above examples:

1. Entrepreneurship, Engineering, Innovation, and Libraries: Universities play a vital role in cultivating innovators and entrepreneurs by offering entrepreneurship components in senior design courses. Collaboration between institutions and professional societies enhances student awareness of resources supporting entrepreneurship.
2. Developing Entrepreneurial Mindset in Engineering Students: Encouraging collaboration between engineering students and those from other disciplines, establishing industry partnerships, mentorship programs, and creating an entrepreneurship ecosystem within institutions are key strategies to fuel innovation among engineering students.

3. Integrating an entrepreneurial mindset into professional colleges equips students with the necessary skills to navigate real-world challenges, create innovative solutions, and succeed in a competitive global economy.
4. Technology incubation centres and mentoring plays a vital role in shaping the young entrepreneurs in the business world.
5. Mentoring by industry experts, alumni entrepreneurs meet/talks could help in building that entrepreneurial environment needed in the campus.

2 Research and Development Environment of India

India is aggressively working towards establishing itself as a leader in industrialization and technological development. Significant developments in the nuclear energy sector are likely, as India looks to expand its nuclear capacity. Moreover, nano technology is expected to transform India's pharmaceutical industry. The agriculture sector is also likely to undergo a major revamp with the government investing heavily for a technology-driven Green Revolution. Government of India, through the Science, Technology, and Innovation (STI) Policy-2013, among other things, aspires to position India among the world's top five scientific powers.

India initiated a landmark policy called Science, Technology, and Innovation Policy 2020 with a core vision of being decentralized, evidence informed, bottom-up, experts- driven, and inclusive. The policy aims to be bring-in the concept of 'dynamic policy' with a robust policy governance mechanism incorporating features such as periodic review, policy evaluation, feedback, and adaptation, and a timely exit strategy for various policy instruments. India's National Artificial Intelligence Strategy prepared by NITI-Aayog outlined a way forward to harness the potential of Artificial Intelligence (AI) in different fields. Accenture offers a framework for assessing the economic effect of AI for selected G20 countries in its latest AI research studies and forecast that AI will raise India's annual growth rate by 1.3% points by 2035. India now ranks 46 among 50 countries in the Global Innovation Index (GII). This is an improvement from the 48th position in 2020. The Government is extensively promoting research parks technology business incubators (TBIs) and (RPs), which would promote the innovative ideas till they become commercial ventures.

Intellectual property has always been a contentious issue for inventors, researchers, authors, and scientists around the world. The definition of intellectual property according to the World Intellectual Property Organization (WIPO) elaborates upon the terminology as "creations of the mind, such as inventions; The impact of trips agreement on foreign investment literary and artistic works; designs; and symbols, names and images used in commerce." The definition is deliberately worded in this way to be as inclusive as possible because a disagreement on intellectual property can arise on a variety of subjects and innovations. India stringently adheres to the TRIPS convention in Paris and the developments thereafter in Doha that was held recently.

India ranks third among the most attractive investment destinations for technology transactions in the world.

Modern India has had a strong focus on science and technology, realizing that it is a key element for economic growth. India is among the topmost countries in the world in the field of scientific research and is positioned as one of the top five nations in the field for space exploration. The country has regularly undertaken space missions, including missions to the moon and the famed Polar Satellite Launch Vehicle (PSLV). India is likely to take a leading role in launching satellites for the SAARC nations, generating revenue by offering its space facilities for use to other countries.

3 Role of Faculty and Students at Academic Institutions

a. **Faculty members are the first Mentors**

Faculty members can provide valuable mentorship, guidance, and insights to students and entrepreneurs. Their academic expertise and industry experience can help startups navigate challenges, refine their ideas, and make informed decisions. Faculty with expertise in relevant fields can offer technical, scientific, or business insights that contribute to the development of the startup's product or service. Their knowledge can help startups address technical complexities and ensure product-market fit. Faculty often has extensive networks that include industry professionals, potential investors, and collaborators. They can facilitate introductions and connections that can open doors for startups to access resources, partnerships, and funding. Faculty can provide critical feedback on business plans, prototypes and strategies. Their constructive criticism helps startups identify areas for improvement and refine their approaches.

In cases where startups are based on innovative technologies, faculty members can collaborate with startups on research projects. This collaboration can lead to novel solutions, product enhancements, and a strong research-industry link. Startups associated with academic institutions can access university resources such as labs, libraries, research facilities, and technical equipment. This can significantly reduce costs and accelerate product development. Some academic institutions run incubator or accelerator programs where faculty members serve as mentors or advisors. Their involvement in these programs enriches the startup's experience by providing academic and industry perspectives. Faculty members who have successfully combined academia with entrepreneurship can serve as role models for aspiring student entrepreneurs. Their success stories inspire students to pursue their entrepreneurial dreams.

Faculty may collaborate with startups on research projects, contributing to the startup's innovative edge. This collaboration benefits both parties by advancing knowledge and solving practical problems. It's important to note that not all startups have direct faculty involvement, and the level of engagement can vary. Successful startups often emerge from a combination of factors, including a strong idea, entrepreneurial drive, market demand, team dynamics, and external support from mentors, investors, and advisors. Faculty can play a crucial role in nurturing and guiding startups, but success ultimately depends on the collective efforts of the startup team and the broader ecosystem.

b. **Students are the End Beneficiaries**

Students gain practical experience in building and running a startup. They learn about various aspects of entrepreneurship, from idea validation and product development to marketing and fundraising. Accelerators and incubators provide access to experienced mentors who can guide students in developing their business ideas. These mentors offer insights, feedback, and industry knowledge that can be invaluable for students. Students get to connect with fellow aspiring entrepreneurs, industry professionals, mentors, and investors. These connections can lead to collaborations, job opportunities, and partnerships in the future. Many programs offer workshops, seminars, and training sessions on entrepreneurship-related topics.

Today, pitch preparation, market research, company development, and other topics are all available to students. Perhaps, certain programs for accelerators and incubators provide seed money or introduce participants to possible investors. With this money, students can develop their concepts into workable prototypes and enterprises. Students can validate their ideas and make sure they are addressing actual market requirements by interacting with mentors, peers, and industry professionals. Participating in accelerator or incubator programs helps students develop a variety of abilities. These abilities may include adaptation, problem-solving, communication, and team management. Students get the opportunity to apply what they have learned in the classroom to real-world circumstances by working for a company. Their ability to grow both personally and professionally may be greatly enhanced by this experiential learning.

Participating in a reputable accelerator or incubator program can enhance a student's resume and credibility. It demonstrates their initiative, entrepreneurial mindset, and ability to work in a dynamic environment. Through exposure to the startup eco-system, students may identify new interests, career paths, or business opportunities they hadn't considered before. For students wishing to launch their start-ups, accelerators and incubators frequently offer access to resources like office space, technological tools, legal support, and more. These resources can be extremely beneficial for their development both personally and professionally. Being part of an accelerator or incubator program creates a supportive community of like-minded individuals who share a passion for entrepreneurship. This network can provide ongoing support and encouragement.

Startups often face challenges and setbacks. Students learn to navigate failure, adapt to changing circumstances, and develop resilience a crucial skill in both entrepreneurship and life. For students interested in pursuing entrepreneurship as a career, accelerator and incubator programs can serve as a bridge between academia and the startup world, providing a structured path to launch their ventures. Some of the examples that can be found commonly are mentioned to help the students understand the market and its needs:

(1) Druva was founded in 2008 by Jaspreet Singh and Milind Borate, both alumni of the Indian School of Business (ISB).
(2) Bhavish Aggarwal and Ankit Bhati founded Ola in 2010. Bhavish Aggarwal is an alumnus of IIT Bombay. Ola initially started as a small-scale venture without for- mal incubation. Bhavish used personal savings to launch the platform and gained early recognition through word of mouth.
(3) Deepinder Goyal and Pankaj Chandah, both alumni of IIT Delhi, co-founded Zomato in 2008. While not formally incubated, Zomato's founders benefited from their educational background and network. They started with minimal investment and focused on building a user base and restaurant database.

In Telangana:

(1) Pigeon Labs: A startup incubated at T-Hub, Pigeon Labs focuses on AI-based solutions for improving customer engagement and experiences.
(2) Loop Reality: Loop Reality, a startup incubated at IIIT-H Foundation, specializes in providing virtual reality-based training solutions for skill development.
(3) Gourmet Garden: A women-led startup incubated at WE Hub, Gourmet Garden offers eco-friendly kitchen gardening kits to promote urban farming.
(4) Blue Semiconductors: Incubated at IIIT-H Foundation, Blue Semiconductors develops IoT-based solutions for water and energy management.

From the above examples we may conclude that the role of faculty and the incubation centers in professional colleges helps in promoting the Entrepreneurial ecosystem, which in turn gives scope for many student entrepreneurs.

It can therefore be concluded that the engineering colleges have a great role to play in nurturing students' venturing into future entrepreneurship. These can be pre-incubated in the campuses and then we have a new generation of Entrepreneurs passing out every year.

c. **Faculty and Students can work together to build a startup**

- Collaborating on a startup venture can provide students with real-world experience and faculty with practical application of their knowledge. It's essential to create a supportive and respectful environment where everyone's contributions are valued, and the focus is on achieving the common goal of building a successful startup.
- Ideation and Conceptualization: Brainstorm startup ideas that leverage the expertise of both the faculty and students. Identify a problem to solve or a market gap to address. Consider the unique skills and strengths of each team member.
- Team Formation: Build a diverse team that includes students with various skills (technical, design, marketing) and a faculty member with relevant expertise. Ensure clear roles and responsibilities for each team member.
- Market Research: Conduct thorough market research to understand the target audience, competition, and market trends. Gather insights to validate the idea's feasibility and potential demand.
- Business Planning: Develop a comprehensive business plan that outlines the value proposition, revenue model, customer segments, and go-to-market strategy.

Leverage the faculty member's industry knowledge to create a solid business strategy.
- Product Development: Utilize the technical skills of students to develop the product or service. Faculty members can provide guidance on technical aspects and ensure alignment with industry standards.
- Mentorship and Guidance: Faculty members offer mentorship, providing insights into market dynamics, industry trends, and regulatory considerations. Students bring fresh perspectives and innovative ideas to the table.
- Prototype and Testing: Build a prototype of the product or service to test its functionality and gather user feedback. Iterate based on feedback to improve the prototype's features and usability.
- Funding and Resources: Explore funding opportunities such as grants, competitions, angel investors, or incubator programs. Leverage the faculty member's network to connect with potential investors.
- Legal & Regulatory Compliance: Faculty members can guide the team through legal & regulatory requirements, including intellectual property protection & business registration.
- Marketing and Branding: Develop a marketing strategy to create awareness and attract customers. Students can use their creativity to design branding materials and promotional campaigns.
- Launch and Scaling: Launch the startup to the target market and gather user feedback. Faculty members can provide strategic guidance on scaling the business and expanding its reach.
- Continuous Learning: Both faculty and students should engage in continuous learning to stay updated on industry trends, technologies, and business strategies.
- Networking: Faculty members and students can tap into their networks to connect with potential customers, partners, and collaborators.
- Celebrate Success and Learn from Challenges: Celebrate milestones and successes along the journey. Use challenges as opportunities for learning and growth.
- Flexibility and Open Communication: Maintain open communication channels between faculty and students to discuss ideas, concerns, and progress. Be flexible and willing to adapt to changes based on market feedback.

4 Case Studies of Professional Colleges Across the Globe

Professional colleges that successfully encourage innovation and entrepreneurship have been included in the following case studies:

1. **Yale Centre for Engineering Innovation and Design (CEID):** This institution is dedicated to advancing innovative concepts and goods that influence society. It functions as an innovation hub for the campus community, providing resources, space, events, design and problem-solving classes to promote awareness of the "Maker Movement."

2. **ICESI University, National University of Colombia, and Pontifical Xavierian University (Javeriana):** These Colombian universities have worked hard to promote entrepreneurship teaching, knowledge sharing, and cooperation with outside partners. Initiatives to promote entrepreneurship education include entrepreneurship classes, academic programs in innovation and entrepreneurship, incubators, accelerators, and collaboration with businesses, government agencies, and accelerators.
3. **University of New Haven:** The University of New Haven has focused on promoting innovation and entrepreneurship within a college setting. Through initiatives like the Summer Undergraduate Research Program (SURF), students like Jonathan Spiegel have been involved in developing a better innovative and entrepreneurial ecosystem at the university. The university emphasizes hands-on learning experiences, collaborative settings, and engagement with innovation and entrepreneurship activities
4. **Shandong University of Science and Technology:** This university in China has implemented an Innovation and Entrepreneurship Training Program for undergraduate students. The program aims to enhance students' innovative abilities through project completion, foster entrepreneurial capabilities based on innovation, and improve overall innovation and entrepreneurship skills. It provides a platform for cultivating innovative thinking among students through project-based learning experiences.

These case studies highlight the diverse approaches taken by professional colleges worldwide to promote innovation and entrepreneurship among students, fostering a culture of creativity, problem-solving, and real-world application of knowledge.

5 Conclusion

- Every professional college must try and build the needed facilities within the campus for creating an awareness and building the spirit of entrepreneurship.
- Incubation centre needs to be established within the campus.
- The network of alumni entrepreneurs must be built. Successful entrepreneurs interacting with the present students is a big boost of morale to prospective entrepreneurs.
- MOUs cooperations with the industry and financial institutions must be built.
- Appropriate trainings and events should be regularly conducted in the college premises.
- Advanced leadership programmes for the students with an innovative and creative mind set must be provided.
- Institutions need to re-evaluate their programmes, industry alliances, events, etc. to allocate the resources wisely to meet the demand from stakeholders.

It can therefore be concluded that if the needed infrastructure and resources are provided in the professional colleges, more campus entrepreneurs will emerge. There needs to be a proper collaboration with the financial institutions, industry experts, venture capital funds, trained faculty, incubators and the student start-ups.

Increase Regional and Local Relevance of Engineering Institutions Through Community Engagement Aimed to Support Their Socioeconomic Development

William Oakes, Lelanie Smith, Rohit Kandakatla, and Wan Chin TAN

I alone cannot change the world, but I can cast a stone across the waters to create many ripples.
—Mother Teresa

Abstract Meeting the challenges of today requires different kind of professionals who are strong in their disciplinary knowledge, equipped with a broad set of professional skills and has the awareness and ability to apply these across socio-technical and cross-cultural contexts. Community engagement is a learning environment to equip graduates with these broad skills and to transform out institutions of higher education. This chapter provides four successful examples of community engagement in diverse contexts from the United States, South Africa, India, and Singapore. These diverse examples offer insights into how community engagement can be integrated into other contexts. Common lessons that have been learned from the experiences are summarized and shared.

W. Oakes (✉)
Purdue University, 701 W. Stadium Ave, West Lafayette, IN 47907, USA
e-mail: oakes@purdue.edu

L. Smith
University of Pretoria, R2–2 Engineering Building 3, Hatfield, Pretoria 0002, South Africa
e-mail: lelanie.smith@up.ac.za

R. Kandakatla
KG Reddy College of Engineering and Technology, Hyderabad, India
e-mail: rohit.kandakatla@kgr.ac.in

W. C. TAN
Ngee Ann Polytechnic, 535 Clementi Road, Singapore 599489, Singapore
e-mail: TAN_Wan_Chin@np.edu.sg

R. Kandakatla et al. (eds.), *Academic Leadership in Engineering Education*, Lecture Notes in Networks and Systems 1097, https://doi.org/10.1007/978-3-031-68282-7_14

Keywords Engagement · Community partnerships · Service-learning · Community-engaged learning

1 Introduction

Higher education institutions (HEIs) stand at a pivotal juncture in their role as agents of societal transformation. New approaches are needed in our societies to meet the United Nations' Sustainable Development Goals as noted in the 2023 report [1]. HEI's can play an important role in educating the future leaders as well as engaging faculty and students today. In nations, such as South Africa, grappling with a high unemployment rate of 29.1% [2] and marked by low educational attainment among young adults, the imperative for HEIs to address pressing community challenges becomes pronounced.

As the demands on HEIs to be part of change intensify, it is incumbent to adjust their strategic goals and curricular strategies. The mission lies in producing graduates imbued with a sense of social responsibility and equipped to contribute meaningfully to the multifaceted fabric of the global society [3].

Engineering institutions possess enormous resources that can be harnessed to address pressing community challenges. Integrating community engagement into their curricula engages students in real-world experiences while making tangible contributions to local or global development. From designing low-cost sanitation solutions to creating renewable energy sources, engineering institutions can leverage their academic prowess to transform theoretical knowledge into practical solutions that uplift communities.

Central to the effectiveness of community engagement is a collaborative partnership between HEIs and their communities [4]. Jointly delineating project objectives, interaction frameworks, monitoring mechanisms, and assessment ensures alignment with community needs. Furthermore, inculcating an ethos of reflection during and after projects bolsters students' transformative learning journey [5].

By incorporating service-learning and community engagement, HEIs can foster graduates who possess the skills and knowledge to uplift society while addressing the challenges facing our local and global communities [6, 7]. This is noted in the South African Education White Paper 3: A Programme for the transformation of Higher Education [8], which emphasized community service as an integral part of higher education but also underscored the role of HEIs in demonstrating social responsibility for the common good (p. 11). Research shows that these experiences equip graduates for future success in a wide range of careers [9].

2 Four Models for Engineering Community Engagement

Community engagement has seen an increase in acceptance as evidenced by the creation of Engineers Without Borders (EWB) organizations in 28 countries [10]. The American Society for Engineering Education (ASEE) has a dedicated division to Community Engagement [11]. This section includes four successful examples of community engagement within the United States, South Africa, India, and Singapore.

2.1 EPICS Program, USA

In the early and middle 1990's in the United States, many of the conversations about preparing engineering students for professional practice centered around moving away from the traditional classroom and integrating more experiences into the curriculum. While many approaches involved industry-sponsored design experiences, an innovative team from Purdue University, led by Professors Leah Jamieson, Ed Coyle and Hank Dietz created a different approach called Engineering Projects in Community Service (EPICS). In EPICS students developed designs for local community, nonprofit and governmental organizations. The idea was to create long-term partnerships within the community and engage students with the partners over multiple semesters or even years [12]. It began with 40 students from 3rd and 4th year from Electrical and Computer Engineering and grew into a multidisciplinary program with about 700 undergraduate students per semester from more than 30 majors ranging from their first to final year.

EPICS has two overarching goals. The first is to prepare students for their future careers and lives beyond the university. The EPICS environment integrates professional practices as much as possible and offers students the experiences that translates into their future careers [9]. Their preparation spans beyond their professional careers and prepares students to be engaged citizens in today's global society.

The second goal is to provide value to the community partners in the form of products that are designed and developed with the partners. The value to the partners also includes the long-term commitments of the program to multi-year partnerships. EPICS initially commits to a five-year engagement and many times it extends well beyond.

These two goals are held as equals and are interconnected. Because the students are working on real projects, they are placed into a setting that simulates professional practice. Because the students are engaged in professional practices, the quality of their engagement is enhanced as well as the value received by the partners.

EPICS has been recognized within the university for its value in education and engagement, aligning with the university's land grant mission. It is fully institutionalized as an academic program within the College of Engineering with dedicated staff, classrooms and laboratories. In 2022–23, the program engaged over 1300 students from more than 30 majors on more than 100 projects with 57 community partners.

Teams are vertically integrated with a mix of first-year to final year students. Each department determines how EPICS counts in their respective curriculum and it can be used for a variety of credits including a technical elective in all engineering disciplines and as a substitute for other courses including the capstone or final year projects in some majors. Data shows that the quality of the program has been maintained as it has scaled [13].

2.1.1 Sample Projects and Partnerships

Sample projects are provided as examples of the breadth of projects and partnerships that work with the EPICS Program [14].

Local Partners: The partnership with a local affiliate of Habitat for Humanity started in 1996 and is still active. The partnership began leveraging technology that digitized house plans and integrated energy costs. The partners co-developed a model sustainable home design that changed the building practices for the affiliate and impacted construction managers from across the state. Current projects are focusing on needs to make the volunteer construction process simpler and more efficient.

Regional Partners—Some partners are outside of the local area and students cannot frequently visit. One example is the partnership with an Indigenous Tribal College, the Oglala Lakota College (OLC) in South Dakota. EPICS and OLC have worked together to address food sovereignty on the Pine Ridge Reservation including a large greenhouse on the tribal campus. During the pandemic, projects expanded to include databases to help the tribal leaders manage health data. Current projects involve the repurposing of a large amounts of plexiglass. Plexiglasses were used on Purdue's campus during the pandemic and are being repurposed into small greenhouses and passive solar heating systems.

Global Partners–One partnership was started by an Anthropology colleague who worked with an indigenous community in the Amazon region of Brazil and co-teaches the EPICS section. The work that was being done was to digitally capture aspects of the tribal life and culture so as to help preserve that knowledge for future generations. The videos storage and archiving required power and the opportunity to engage EPICS teams. The colleague travels each year to the tribe with intermediate video meetings.

Global Teams—A model that blends local and global partnerships is the team working with a school for the blind and visually impaired near Purdue University. Purdue students visit the school during the semester and the design work is split between teams at Purdue and teams from an Indian university that started their own EPICS Program.

2.2 *South Africa-University of Pretoria, Joint Community Projects*

In 2005 the Faculty of Engineering, Built Environment and Information Technology at the University of Pretoria introduced a first of its kind in South Africa, Community-based Project Module, for all its undergraduate students. The module adopts a service learning, open-ended project-oriented approach and operates as an eight-credit module (equivalent to 80 h). The students are required to participate in 40 h of guided and collaborative community-based work, connected to written reflective assignments that support their integration of the experience [15].

The module, called Joint Community Projects (JCP), typically has two types of projects: (1) design and implement or (2) educational.

The design and implement projects rely on long term community partner relationships where the second year students form part of a bigger representation of the University. They act as the "hands" to design and build products ranging from apps to work with medical data to jungle gyms for rural schools. The educational projects require students to become involved in teaching mathematics, science, computer literacy, coding, and robotics. They are also involved in the mentorship of these young learners from the community so that they are being able to access Higher Education. This collaboration ensures a support community for the transition into the space of HE.

There are significant challenges when considering facilitation of active learning pedagogies like service and project-based learning to such a large cohort with a single coordinator. Students require coaching and real-time feedback loops to develop effective competencies. Most students, in their early years of schooling are conditioned to be passive learners. So, they require careful and constant framing to support them to develop agency, confidence and engaged participants in their learning activities.

The community partners, who are often from a different cultural and socio-economic background offer an additional challenge to students. While the partners are typically outcome-driven, there is a need to develop a relationship of trust and understanding with the students for the work to be brought to fruition. It takes time, effort, and coaching. But the whole approach is mostly not prioritized. In this back-drop, managing a project and handling the expectations of both–a large cohort of 30–50 community partners, and of about 350 students, require intentional design and management and an active participation from the leadership team.

There are 2 key features that JCP relies on to operate at scale.

2.2.1 Sustainable Partnerships and Centralized Projects

JCP provides a sustainable platform to initiate new or run existing projects through different partnerships. Centralizing projects and ensuring long term sustainability of these have been the primary focus over the last 2 years.

Centralising meant–

(1) Identifying and partnering with specific communities on a wider variety of projects rather than doing multiple small projects across multiple communities.
(2) Combining similar projects (as an example all the educational outreach programmes are run through a JCP-born NGO, Keep That Gold shining (KTG)), rather than multiple independent similar partners.

Partnering with KTG has led to a sustainable integrated approach where the same students that are mentored through high school into our programmes, return to their communities once they enter JCP, to work with the next generation of learners and after their time in JCP, become mentors for the next generation of JCP students.

Community partners are considered experts in their fields, contributing significantly by sharing knowledge and fostering reciprocal relationships. These partnerships evolve community partners into co-educators, challenging traditional representations and offering unique insights into their own stakes in service-learning projects. These transformative partnerships empower higher education institutions to better manage and evaluate service-learning and community engagement pedagogies.

Over the past three years, centralizing projects for sustainability and forging new collaborations have been key areas of focus, backed by a solid platform for ongoing initiatives [3].

Multiple new projects were developed in collaboration with industry and departments across Faculties. Partnership between Robot, Drone, Sensor School and Electrical, Electronic, Computer, Mechanical and Aeronautical Engineering programmes were established. Senior students develop the programme for high school learners. They teach the material to the 2nd year JCP teams who facilitate the programme to high school students from three township communities close to Pretoria through the support of KTG. These programmes also rely strongly on industry support for materials and kits to be developed and created. There are more opportunities similar to these, that specifically links to projects in family-medicine and occupational therapy.

2.2.2 Coaching Training in a Vertically Integrated Support Structure Framed as a "Company"

The vertically integrated coaching/mentorship team consists of 15 senior (typically 4th year) and 35 junior (typically 3rd year) students who have already completed JCP.

The training and development of the mentors takes place in collaboration with systemic transformation coaches and uses similar models for development as used for the JCP students. The junior mentors have two full days of interactive sessions developing skills in supporting and listening, giving, and receiving feedback, asking clarifying questions and caring responsibly.

A senior mentor, who has completed one year of training, has an extra set of skills developed in their second year (negotiation and responsible care) and are used to assist the training of the junior mentors. They are also partnered with junior mentors for the remainder of the year. These skills equip the mentors to be guides to the JCP students in their engagement with the community and their team members, as well as in their own self-discovery. Each senior mentor supports 3 junior mentors, and each junior mentor supports approximately 40 JCP students in 8–10 group projects.

The structure simulates a workplace, akin to a corporate setup, with the coordinator as the CEO, the community partners as clients, the senior and junior mentors as management levels and the JCP students as new employees.

All the learning exercises and casual engagements are framed within this context so that students can appropriately contextualize their learning. Functioning effectively through this framework scaffolds each student to seamlessly enter and be an effective and contributing employee to any organization.

The vertically integrated mentorship structure is fundamental to the success of the module at scale. Managing 50 students and designing engagements with them is pivotal to the success of this program. For this it is essential for the faculty to practice their own skills and support the students' ability to. engage with community and with effective development.

2.3 Unnat Bharat Abhiyan Program at KG Reddy College of Engineering and Technology, India

Unnat Bharat Abhiyan (UBA) is a national-level program launched by the Ministry of Education in India to encourage community engagement and foster social responsibility among higher education institutions (HEIs).

As part of the program, HEIs are recommended to initiate a dialogue with the communities in nearby villages and explore opportunities for collaboration. The UBA program aims to explore how innovation and research in the HEIs could be aligned to addressing the unmet needs and challenges of communities in the partner villages. The program is being implemented with a vision that HEIs, through active community engagement, can play an important role in the socio-economic development of rural India. HEIs themselves will tremendously benefit from the UBA program as it opens opportunities for meaningful research, innovation, and entrepreneurship, all directed to contribute to the development of society.

2.3.1 Unnat Bharat Abhiyan @ KG Reddy College of Engineering and Technology (KGRCET)

As part of the UBA program, KGRCET has partnered with five villages located close to the institution. Initial community engagement efforts were led by the faculty from the institution's Center for Innovation and Social Transformation (CIST) as they reached out to the Sarpanches (elected heads of each village) to discuss the possibilities of collaboration as part of the UBA program. Based on mutual consensus, KGRCET and the villages have formally agreed to partner through a formal association.

Gyan Shodh–3-day rural immersion experience to holistically assess Indian villages Before starting to identify problem areas in the village that needed technical interventions, the institution believed that it was important for faculty and students to get to know and build a holistic understanding of the partner village.

Faculty at KGRCET acknowledged their lack of any experience in engaging with rural villages and collaborated with Palle Srujana, a civil society organization in India with more than two decades of experience in scouting grassroots innovators in rural India. 'Gyan Shodh' is a village assessment model pioneered by Palle Srujana where a group of individuals (faculty and students in this case) visit a specific village and live with them for 3–4 days. During the stay faculty and students engage in informal discussions with different stakeholders in the village to learn about their lifestyles, culture, livelihoods, aspirations, and challenges. Gyan Shodh has been recognized for its asset-based approach to community engagement, as it enables the focus of faculty and students to expand beyond just look ing for problems and realize the naturally available regenerative resources in the villages. During the 3 days stay in the partner village, faculty and students were able to experience the lifestyles of the community members in their own environment. Living in the partner villages for over 3 days provided the faculty and students with opportunities to have deeper conversations with relevant stakeholders and empathize with their problems [16]. At the end of the 'Gyan Shodh', the team of faculty and students prepare a report based on their assessment of the village which includes their, aspirations, unmet needs and challenges, and reflections towards their socio-economic development. It is important to note that the faculty and students are expected to prepare the report in collaboration with a few important stake-holders from the village they visited.

Integration of Identified Problem Areas into KGRCET Engineering Curriculum

After the completion of Gyan Shodh, the village assessment report is shared with the village Gram Panchayat (elected governing body of the village), and further village visits are conducted to explore and mutually agree on problem areas that could be addressed with technical interventions. Faculty at KGRCET are expected to champion these problem areas based on the village they visited during Gyan Shodh. Multiple problem areas identified from different villages are introduced to engineering students as part of the 2-credit 'Social Innovation' course offered to students in their 2nd semester. During the Social Innovation course, students are expected to visit partner villages, interact with relevant stakeholders to scope the problem statement, and propose relevant solutions through 3D models. Respective faculty who agreed to champion the different problem areas scout and select the student teams with the best ideas as part of the student poster presentations conducted at the end of the semester. Students' teams with the best ideas are identified and supported in the subsequent semesters through open electives on "Product Design' and 'Product Development' where they develop the Proof of Concept and a working prototype to address the problem area in the village. The problem areas are introduced to the students at the freshmen level as the average project completion timeline is usually 2–3 years. This allows the student teams led by their faculty mentors to complete the design, development, and deployment of their solutions by the end of their undergraduate education.

Participatory Approach to Ensure Sustainable and Long-Term Engagement

The successful deployment and utilization of the technical solutions in the partner villages requires continuous engagement with local community members [17]. During the Gyan Shodh, faculty from the institution identify a few community members who can play the role of local champions and can support the faculty and students with valuable information such as indigenous knowledge and local wisdom from the villages. The local champions are also important in overseeing the sustainability and effective utilization of engineering solutions after their deployment in the villages. It is therefore important to continuously engage the local champions from the start of the project which includes the stages of problem identification, design, and development of solution. This includes inviting them for review meetings at the institution to get their feedback and input on the project. Faculty from KGRCET also regularly attend the 'Gram Sabha' (monthly meetings conducted by the leadership of each village), where the faculty and local champions can share the progress on various projects. Employing such participatory approaches and engaging with the community members significantly increases their buy-in and commitment towards the implementation of the project and also enables long-term engagement between the institution and partner villages.

2.4 *Singapore-Ngee Ann Polytechnic (NP)*

In 2016, Ngee Ann Polytechnic (NP) launched Service-Learning (S-L) as its signature pedagogy to advance its desired graduate outcomes, namely big-hearted students who are passionate learners and globally smart professionals. As a signature pedagogy, all students participate in at least one S-L activity or project, that is tied to a credit-bearing academic module or course to achieve three S-L pedagogical outcomes: academic learning, civic learning and personal growth [18].

S-L instructors work with partners (e.g. community residents, non-profit organizations, government agencies), regarded as co-educators, to ensure that projects meet identified community issues while achieving modules' learning objectives. Utilizing an Asset-Based Community Development (ABCD) approach that taps into community strengths, students and faculty organized projects covering diverse areas such as education and employability, eldercare, health, environment, and community development have been formulated. As a pedagogy used in academic modules, students are encouraged to play active roles in shaping their S-L projects. Emphasis is placed on the "hyphen" in Service-Learning to ensure continued connection between service and learning in the curriculum design. This connection is achieved through critical reflection. DEAL (Describe, Examine, and Articulate Learning) reflection model [19] is widely adopted in facilitated reflection sessions and written assignments.

S-L has been successfully incorporated in the School of Engineering through placement-based and project-based community activities relating to engineering technologies and human-centred design. Students are engaged with seniors, children and under-privileged families in educational programs relating to technologies like robotics, digital applications, and user experiences. Students can scaffold their S-L experiences in multiple years via activities like civic final-year or capstone projects, civic internships, and overseas community projects. Civic projects undertaken by engineering students in the past years included national projects addressing urbanization issues, education projects and enabling persons with disabilities with assistive technologies.

For example, first-year engineering students were engaged in introducing digital government mobile applications to seniors. Through the cross-culture and intergenerational activities with senior activity centres, the students learned about the push for digitalizing government services while addressing the concerns of citizens who might be left behind, e.g. the less tech-savvy seniors. Students apply cultural intelligence to craft lessons for the seniors to introduce and teach them digital government and electronic banking applications. They review the user-experience design of the applications and reflect on the roles in ensuring inclusive designs.

Students in the second year of Electronics and Computer Engineering applied micro-controllers, sensors and Internet of Things (IoT) devices to assist urban farmers in monitoring and adjusting environmental factors to support crop yield. The project taught students about utilizing technology to ensure sustainable food security for urbanized cities like Singapore. They were engaged in learning the microelectronics applications and were introduced to the tech-enabled green-economy jobs as well.

Finally, students in final year projects worked to improve the safety of pedestrians and cyclists in shared pathways and common areas through electronic warning devices. By applying the design thinking process, students understand the community issues, developed ideas, built mockups, and working prototypes to test their designs. The prototypes were put to test with encouraging results.

The engineering S-L courses and projects engage students in meaningful real-world projects and learning, while finding a purpose in applying engineering for social good. The community engagements also provide opportunities for students to work across different study disciplines and domains.

However, there were challenges faced in introducing the S-L pedagogies into the engineering curriculum at Ngee Ann Polytechnic. Before adopting the S-L pedagogy, the engineering curriculum did not prepare our students to engage with the community, understand socio-economic matters, and reflect on their experiences. Little was known about tapping on the community as co-educators, and how they can bring vast knowledge and experiences to our students.

With the assistance of the NP Office of Service-Learning, the S-L framework was introduced to the engineering faculty with support in finding projects and establishing community partnerships. A community of practice was established with strong support from the institution and other S-L practitioners in Asia and globally.

S-L has gained traction in our higher institutes of learning. While Singapore is a small country, there are opportunities around our island city for our students to be involved in regional Southeast Asian countries projects relating to education, post-pandemic recovery, climate changes, and sustainable goals development.

3 Key Takeaways from the Chapter for Future Academic Leaders

3.1 Drivers for Community Engagement

The four case studies highlight the drivers to develop or integrate community engagement as activities within a curriculum. Two primary drivers are highlighted–socio-economic development/reform and authentic development of graduate attributes [20]. The case study from South Africa and India both illustrate the drivers from a national perspective that aims to contribute to the country's socio-economic development whereas the US and Singapore examples speak to a development of graduate attributes to prepare them for the future world of work and leaders within their societies.

Partnering with communities opens up discussions and dialogue around real people with real needs and the facilitation of student engagement with these topics lead to experiential active learning opportunities. These kinds of activities inevitably have the potential to immerse students and allow them to develop attributes that are more difficult in a traditional classroom. It opens doors for collaboration with community and potential products, processes or projects that lead from those association, has the potential to benefit socio-economic development even when not addressed completely.

3.2 Challenges in Transformation to Enable Community Engagement

Higher Education institutions must carefully assess the costs and prerequisites associated with community engagement, including policies, faculty development, and student learning, to establish a practical and sustainable foundation. The most critical aspect to resolve is change management or transformation strategy. The effective implementation of community engagement in higher education institutions requires a careful evaluation of existing institutional structures. Without addressing these challenges, community engagement initiatives may fail to integrate, fail to become sustainable, or make a substantial impact.

Leadership plays a crucial role in the successful integration of community engagement into faculty development, drawing valuable insights from essential principles of transformation. Acknowledging that transformation is an ongoing, non-linear process characterized by the management of polarities is important. Rather than simply solving problems, leadership must embrace the inescapable state of tension transform ing it and not only derive meaning from it but facilitate and support the same for faculty.

3.3 Areas of Changes to Integrate Community Engagement

Strategic prioritizing socio-economic development: Many universities have traditionally focused on research and teaching, with community engagement viewed as an additional responsibility rather than an integral part of the mission.

We see in both South Africa and India where socio-economic development was the driver and that there is an institutional unit or office that coordinates the community engagement agenda across Faculties. It is critical for these units to develop language and narratives that can effectively convey our values across various communities, projects, and venues [21]. The success of such narratives relies on the interconnectedness of the stakeholders in a coordinated strategic and sustainable framework for community engagement.

deally, an ecosystem is conceptualized to consider community engagement initiatives in the context of the systems that govern them, the structures that define their operations, the design that underlines these structures, and the environments in which they operate [22–24].

A systems-based approach helps uncover the intricate interactions occurring across various levels of the system. The lack of integration severely limits the effectiveness and sustainability, especially in challenging economic times.

It can be helpful for advocates to situate the work of engagement within the context of the institution's priorities. The benefits of engagement are many that span student learning, student retention and recruitment to impact on communities and positive visibility of the institution. There are cases where the scholarship of engagement has become a pathway for promotion and recognition for faculty. In the U.S., some of the top research universities have worked to include this track as a path for enabling promotion to full professor [25].

Policy & processes to provide resources and practices; Adequate resource allocation is crucial for the successful integration of community engagement initiatives. In both cases–India with Gyan Shodh, and South Africa with the San Code of Research Ethics–aspects of mutual understanding, trust and transparency are critical for long term sustainable relationships. If an institution does not have an ethos of people-orientation the policy and processes can stifle any collaboration. At Purdue University, the EPICS Program began with the faculty looking for external funding through grants and collaborations [12]. Working within the institution to cost share grants and building institutional resources while documenting successes was an effective strategy for growth.

Faculty development as champions and leaders in this area; The success of community engagement initiatives is dependent on the effective development of faculty. Motivating and empowering faculty members stands as a pivotal yet frequently overlooked facet in the integration and conceptualization of community engagement.

Leaders should take a strategic approach to identify champions and change-makers within various departments and faculties, fostering a collaborative environment across different sections (students, faculty, community, administration, finance officers etc.).

Holistic and systemic transformation starts with the individual and if the institution wants to facilitate the shift, the burden is on the institution to create space and opportunities for staff to be guided through a personal transformation. The process of effecting change necessitates those individuals embark on introspection and self-transformation as a preliminary step before their influence extends to the broader ecosystem.

Within the academic community, there exist faculty and staff members who teach out of necessity rather than passion, potentially leading to skepticism and an unwillingness to grow. Conversely, passionate staff members can be overburdened and may become further exhausted by participating in such programs. Developing staff as facilitators of learning, rather than perpetuating existing classroom experiences, equips them and provides an opportunity to become a part of the strategic transformation process, ultimately fostering a sense of belonging and support for both students and staff.

Finding, building relationships and enabling community change makers; Gathering perspectives from all stakeholders (communities, faculty, students, industry) towards developing a systemic framework for community engagement is essential as the first step. The framework for community engagement should be underpinned by principles such as the interdependence of teaching, learning, research, and engagement. The framework should include valuing stakeholders, embracing different forms of knowledge, and mutual benefit through two-way knowledge flows. Key enablers for the framework include a shared understanding of community engagement, demonstrated commitment to engagement at all levels, ad equate resource allocation for capacity development, and co-designed initiatives for continuity. Successful implementation of such a framework involves embedding the community engagement philosophy, building capacity, stakeholder and context analysis, collaboration and co-design, implementation of plans, monitoring, evaluation, and learning for improvement and sustainability.

Commitments to relationships beyond the traditional academic unit is imperative. Communities seek partners that will engage with them along their journey. In many organizations, such as in Case 1 and with Engineers Without Borders (EWB) organizations, a five-year agreement is a start. The commitment is that it will include support of early projects and a commitment to continue to develop the relationships.

Protection against exploitation in collaboration is context-driven, and there is no universal "how-to" guide for different communities. Before setting up work with communities there needs to be a collaborative co-creation with community members to develop the relationship, which may include research protocols, to carefully ensure ethical practices and collaborations. Success ingredients include icons of leadership, supportive NGOs, legal support, and workshops that help participants gain new insights. Some key principles that could be considered are respecting the environment and cultural values, maintaining honesty and integrity, obtaining informed consent, ensuring mutual benefits, and forming relationships of trust [26].

Engineering institutions who are interested in community engagement must look and explore different avenues to find the right partners.

One of the most common approaches (as highlighted in case studies 1 and 2) is working with non-profit organizations who are already engaged with local communities and their representatives.

The non-profit in this case serves as an intermediate organization who is responsible for building and maintaining the relationship between the community and the institution. Engineering institutions must explore and identify non-profit organizations whose priorities align with the area of focus and strength of the institution. Another approach is to directly engage with nearby communities through their local government representatives, for example, case study 3, where the institution interacted with the local gram panchayat (administrative body) to initiate partnership with their village. In such cases, it is critical for the institution to find other individuals who are ready to champion and support the initiative on behalf of the village. It is advised to work with communities in close proximity to the institution so that there is the possibility of frequent interaction between the community and the institution. It is important to note that engineering institutions must partner with only communities that recognize the need for collaboration.

Reciprocity is considered as one of the cornerstones of service-learning and community engagement as it promotes mutual learning and benefits among all the stakeholders involved [27]. Engineering institutions must emphasize the need to build reciprocal relationships with their community partners through trust, care, and respect throughout the engagement. Careful attention must be paid to avoid exploitation of any community part nership through a sense of power and privilege among the faculty and students. Case study 3 presented the 3-days rural immersion experience which was organized to help students realize and appreciate the contributions of rural India to the country's economy. Partnerships once initiated must be monitored and nurtured through regular interactions to review the collaboration. Community champions must be invited to visit the institution and interact with the faculty and students about the community projects.

Preparing students for community engagement—authentic development of graduate attributes. Integrated student-centered learning is considered the state of the art in Engineering Education [28, 29] as a means to produce graduates who are responsive to sustainability goals, contributors to their communities, and who become leaders in their fields [30]. In the case studies, students are expected to collaborate with real community partners to address ill-defined problems while potentially involving negotiation with community members. Service-learning (S-L) is a cornerstone of all 4 case studies. It incorporates the use of appropriate technology to address community needs, promoting economic development, technology literacy, and infrastructure enhancement projects.

A wide range of positive attributes has been linked to the use of such active learning pedagogies ranging from teamwork, self-efficacy, autonomy, and decision-making to an improved conceptual understanding and synthesis of information [31–37] Reflection is integral [38], with guided sessions that help students connect their experiences to coursework as is shared by the 4 case studies. Working on real-world projects aligned with their studies, helps students develop practical skills and knowledge, while engaging with diverse peers that cultivates inclusiveness. S-L deepens the understanding of course content, promoting higher-order cognitive activities. It also instills civic and professional responsibility by connecting students with community partners and encouraging active roles in positive change efforts.

3.4 Opportunities Offered from Community Engagement

As a consequence of the challenges in 3.2 and the nature of the projects that form with community engagement there are opportunities that emerge.

3.4.1 Innovative Pedagogies

The four case studies outline 3 distinct educational frameworks in the delivery of community engagement: service learning, vertically integrated projects, and transdisciplinary projects.

Community engagement develops a range of effective attributes not typically possible in a traditional classroom setting. Most of these characteristics are not born from knowing, but from being. Much of the student's identity and attitudes are formed in childhood in relation to, among others, gender, fam ily background, and socio-economic realities. The socialization process of learning has an undeniable ability to reinforce or challenge this identity or attitude.

Community Engagement modules offer the opportunity for learning that challenges the student to critically evaluate and adjust their attitudes with:

(1) Ill-defined problems negotiated with non-technical community members.
(2) Immersion in a situation that mimics the real scenarios.

The diverse challenges in Community Engagement are the fertile soil for developing teamwork, leadership, and communication skills with no small amount of persistence and tolerance required. There is no limitation to the diversity in these teams ranging from different professional disciplines to the social diversities present in the student cohort (for example gender, race, and socio-economic background) to the diversity of the community they engage with (diversity in experience levels, age, international diversity, working styles, and modes of thinking beyond engineering). Students are immersed in a formative environment and learn different perceptions and mindsets.

For faculty, there is a wide scope for creativity of the content, project, and mode of delivery. EPICs and JCP for example, employ vertical integration for different reasons. Pai et al. [39] define vertical integration among students as: ***"provision of information aid and sharing learning roles through all learner stages, where students from higher levels are helping students from lower levels to understand engineering concepts and work collaboratively on projects"***. Within a vertically integrated project, students are active participants. Literature also suggests that vertically integrated education has a far greater ability to promote social support when compared with competitive and individualistic forms of education [16–18, 31, 39, 40]. EPICS uses vertical integration as a means to incorporate project hand-overs and allow students to be involved and mature in a project over multiple semesters [12], whereas JCP uses vertical integration as a mentorship structure to support the guidance of the design teams to manage their teamwork dynamics.

3.4.2 Socio Economic Impact and Transformation

All four cases show that partnering with community members positions universities at the nexus of real-world problem-solving, purpose-driven education, and entrepreneurial innovation. These enhances the academic experience and foster positive change in the community while preparing students for a dynamic and impactful future.

Real-World Solutions and Innovation: Collaborating with community partners provides universities with the unique advantage of addressing real needs within the local and sometimes global context. By engaging with community members who possess firsthand insights into the issues they face, universities can develop authentic, practical solutions. This not only fosters innovation but also ensures that the products and services created are relevant and effective. For instance, when working with a local neighborhood, a university could co-create sustainable urban development strategies, educational programs, or healthcare initiatives. The real-world impact of such endeavors transcends traditional academic boundaries, enhancing the reputation of the university as a catalyst for meaningful change.

Purpose-Driven Engagement: Partnerships with the community infuse purpose into the work of both students and faculty. Students who engage in community-based projects gain a deeper understanding of the real-world implications of their studies. This hands-on experience not only bolsters their academic learning but also instills a sense of social responsibility. Faculty members, on the other hand, find renewed purpose in their roles as educators and researchers when their work directly addresses community needs. This alignment between academia and the community not only enriches the learning experience but also contributes to the personal and professional growth of those involved.

Entrepreneurial Opportunities: As universities work with community partners, they often find opportunities for sustainable businesses. Students and faculty members can transition from problem-solving to entrepreneurship by identifying innovative solutions that can be further developed into viable products or services. It is important to remember that while entrepreneurial opportunities may emerge, it is not the main driver for engagement. The context of the engagement is to work with communities that have been left behind by the commercial market. Innovations may be identified that do not have a business case but can be disseminated. In these cases, the students learn the fundamental skills that can be applied in many settings and the community partners received value.

Innovative research: Engaging with communities opens up exciting opportunities for innovative research. Transdisciplinary research, a key facet of this collaboration, involves researchers from various disciplines to collectively tackle intricate issues and their interconnected components comprehensively, in mutually respected relation with community members. This approach fosters a dynamic knowledge community that transcends traditional disciplinary boundaries.

Transdisciplinary research is particularly valuable when addressing complex, multi-faceted problems that cannot be adequately resolved within a single discipline. It acknowledges the interplay of social, economic, and environmental factors, advocating for an integrated and interdisciplinary approach. All four case studies illustrate the extension of trans disciplinarity to education, where multiple disciplines and perspectives are integrated. As a result, transdisciplinary teaching and learning play a pivotal role in cultivating well-rounded and informed citizens prepared to address real-world complexities.

3.5 Pitfalls in Setting up Community Engagements

The emphasis should also be on the hidden risks and potential pitfalls of community engagement initiatives. The challenges mentioned in 3.2 have to be carefully considered and a strategic vision developed in collaboration with community, students and faculty, facilitated by systemic transformation coaches. Academics are not trained to facilitate transformation and we need to be honest and have the courage to be guided.

A pitfall linked to a lack of contextual perspective is linked to academic orientation that idealizes and reveres objective knowledge and the mindset of "we are helping or uplifting" communities. These are potential blind spots of individuals or groups that could lead to alienating communities, leading to a lack of relational trust and undermine the establishment of the intended community engagements. The emphasis should be on developing communities of trust (students, faculty, community etc.) to ensure sustainable relationships similar to the Gyan Shodh model mentioned in the India case study. Community members are in partnership with institutions. They are not research grounds or opportunities to be exploited which is often the case. As the partnerships develop, we all need to continue to listen and learn. As the relationships evolve, more is learned, and most may be expected as we learn to trust and grow with each other.

Transformation requires time. It requires consistent and clear narratives for institutional structures and culture to grow the capacity to resource and facilitate community engagement. The US and SA based case studies have the oldest and largest programmes in community engagements and have existed for 28 and 18 years respectively. Both of these initiatives started with small cohorts, showed success, and grew to their current sizes. The growth was not linear, and it was depended on capitalizing on the opportunities and alignment with the institutional culture. The lessons learned was to do the work well, document and tell the story. Aligning the story with institutional goals, which change with administrations, is extremely helpful. The impact of the EPICS programme has been significant, and the university has invested in the staff for the program and to help disseminate it to other institutions. The JCP module transformed over the last 3 years with an increase in resources and strategic direction provided from both the coordinators and from the new leadership in the institution.

The authors caution against adopting a tick-box approach to community engagement and emphasize on prioritizing the time, resources and support it requires. It may need to start small and grow and can be aligned with institutional values.

3.6 Key Points and Summary

Addressing the challenges within institutional structures and faculty development is essential for transforming higher education institutions into more community-engaged entities. Here are some critical steps to initiate this transformation:

- Mission and Vision Alignment: If it is possible, revisit the institution's mission and vision to ensure that they reflect a commitment to community engagement and social impact. Often this is already true, and the opportunities are to leverage these values to support the engagement efforts. In other cases, it may be necessary to connect community-engagement to the priorities of the current administration.
- Clear Policies and Processes: Establish clear policies and processes that promote and support community engagement. These could include guidelines for faculty and student involvement, resource allocation, and policies for partnerships.
- Engage Stakeholders: Involve all relevant stakeholders, including faculty, students, community partners, and administrative staff, in the planning and decision-making processes. Creating a shared vision is crucial. This may be done on a large scale or to create guidelines where these take place within individual partnerships.
- Faculty Support: Providing the necessary training, and resources to support the work and where possible, incentives to participate in community engagement effectively which will make the efforts much more effective. At minimum, removing disincentives for faculty who are interested in engagement.
- Assessment and Continuous Improvement: Implement mechanisms for ongoing assessment and evaluation of community engagement initiatives is important. Assessing student learning is important for academic accountability and the broader learning is useful to demonstrate impact. Evaluating the health of the partnerships is also important from an institutional level. Data from students and

the community can be integrated into scholarly products as well as to continuously improve programs and strategies.

Community engagement and service-learning have been accepted as legitimate approaches within engineering and engineering education. Data has shown a positive impact on students, faculty, and communities. Engagement has moved from novel to admired to now being critiqued. These days, the methods of engagement are being critiqued in regard to historical roles and power structures. Where teams of students would descend on a village to design and build a water system or a bridge, questions are now being raised on whether they are displacing local professors and laborers. The current state of the art is looking at decolonizing the methods of engagement and to move to an ever more participatory approaches with reciprocal partnerships. The field has moved from being excited about the work, to examining how we do the work the right way, and making sure we are doing the right work. This clearly raises the bar for engagement and can move engineering faculty out of their expertise and comfort zones necessitating interdisciplinary participation. This can be overwhelming, and some are choosing to disengage, rather than risk doing engagement wrong. The opportunities have never been greater for impact and contributions. As leaders, it is time to lean in recognizing that we may not do everything perfectly but have others on our journey who can help us, and together we can help make a small part of the world better each day.

References

1. UN DESA (2023) The sustainable development goals report 2023: Special Edition-July 2023. New York, USA, UN DESA. © UN DESA. https://unstats.un.org/sdgs/report/2023/
2. Oosthuizen A (2019) Youth unemployment: statistics and solutions. HSRC Rev 17(4):3–8
3. Jordaan M, Mennega N (2022) Community partners' experiences of higher education service-learning in a community engagement module. J Appl Res High Educ 14(1):394–408
4. Jacoby B (1996) Service-learning in higher education: concepts and practices, The Jossey-Bass higher and adult education series. ERIC
5. Eyler J (2002) Reflection: linking service and learning—linking students and communities. J Soc Issues 58(3):517–534
6. Osman R, Petersen N (2010) Students' engagement with engagement: the case of teacher education students in higher education in South Africa. Br J Educ Stud 58(4):407–419
7. Fenzel LM, Peyrot M (2005) Comparing college community participation and future service behaviors and attitudes. Mich J Community Serv Learn 12(1):23–31
8. Department of Education (DoE) (1997) Education white paper 3: a programme for higher education transformation. Government Printer, Pretoria
9. Huff JL, Zoltowski CB, Oakes WC (2015) Preparing engineers for the workplace through service learning: perceptions of EPICS alumni. J Eng Educ 105(1):43–69
10. https://www.ewb-international.org/. Accessed 26 November 2023
11. https://sites.asee.org/ced/. Accessed 26 November 2023
12. Coyle EJ, Jamieson LH, Oakes WC (2005) EPICS: engineering projects in community service. Int'l J Eng Educ 21(1):139–150
13. Oakes W, Pierce A, Abu-Mulaweh N (2019) Changes in student perceptions of course-based service-learning at large scale: EPICS at 23 years old. In: Proceedings of the 2019 ASEE annual conference. Tampa, FL

14. https://engineering.purdue.edu/EPICS/purdue/current-students/team-documents/teams. Acessed 26 November 2023
15. Jordaan M (2014) Community-based project module: a service-learning module for the faculty of engineering, built environment and information technology at the University of Pretoria. Int'l J Serv Learn Eng Special Ed 269–282
16. Kandakatla R, Penta S, B L, Sinha S, D C, K U S (2022) Role of empathy to identify unmet needs in Indian villages through service learning program—a case of the Unnat Bharat Abhiyan Program. In: 2022 IEEE IFEES world engineering education forum-global engineering deans council (WEEF-GEDC). pp 1–5
17. Radhakrishnan D, Kandakatla R, Delaine DA (2022) Participatory approach to engineering service learning programs—quality framework for the implementation of Unnat Bharat Abhiyan program in Indian engineering institution. In: IEEE global engineering education conference (EDUCON). pp 863−872
18. Bringle R, Hatcher J (2009) Innovative practices in service learning and curricular engagement. N Dir High Educ. https://doi.org/10.1002/he.356
19. Ash S, Clayton P (2009) Generating, deepening, and documenting learning: the power of critical reflection in applied learning. J Appl Learn High Educ 1:25–48. https://doi.org/10.57186/jalhe_2009_v1a2p25-48
20. Berger B (2009) Political theory, political science, and the end of civic engagement. Perspect Polit 2:335–350
21. Kandakatla R, Dustker SM, Bandi S, Oakes W (2023) Achieving Indian national board of accreditation engineering graduate attributes through project-based service-learning: conceptual analysis. Int J Serv Learn Eng, Humltarian Eng Soc Entrep 18(1). Article 1
22. Sturm S (2006) The architecture of inclusion: advancing workplace equity in higher education. Harv J Law Gend 29:247–334
23. Sturm S (2007) The architecture of inclusion: Interdisciplinary insights on pursuing institutional citizenship. Harv J Law Gend 30:409–424
24. Sturm S (2011) Reframing the equality agenda. Presented at Harvard LawSchool conference on Evolutions in Anti-Discrimination Law in Europe and North America, April 30, 2011.
25. Boyer EL (1996) The scholarship of engagement. Bull Am Acad Arts Sci 49(7):18–33. https://doi.org/10.2307/3824459
26. San Council of South Africa & SASI (2005) Implementation of ethics and protection in research. San Code of Research Ethics
27. Delaine DA, Redick S, Radhakrishnan D, Shermadou A, Smith MM, Kandakatla R, Wang L, Freitas C, Dalton CL, Dostilio LD, DeBoer J. (n.d.) A systematic literature review of reciprocity in engineering service-learning/community engagement. J Eng Educ. https://doi.org/10.1002/jee.20561
28. Graham RH (2018) The global state of the art in engineering education. MIT, Boston, MA
29. Kamp (2016) Engineering education in the rapidly changing world: rethinking the vision for higher engineering education. TUDelft
30. Mitchell JE, Nyamapfene A, Roach K, Tilley E (2021) Faculty wide curriculum reform: the integrated engineering programme. Eur J Eng Educ 46(1):48–66
31. Feldman ML (2006) Vertical integration of engineering education comprehensive report 2004–2006
32. Dunlap J (2005) Problem-based learning and self-efficacy: how a capstone course prepares students for a profession. Educ Tech Res Dev 53–1:65–83
33. Mills N (2009) A guide du Routard simulation: increasing self-efficacy in the standards through project-based learning. Foreign Lang Ann 42–4:607–639
34. Prince M (2004) Does active learning work? a review of the research. J Eng Educ 93–3:223–231
35. Gavin K (2011) Case study of a project-based learning course in civil engineering design. Eur J Eng Educ 36–6:547–558
36. Moliner L, Cabedo L, Royo M, Gámez-Pérez J, Lopez-Crespo P, Segarra M, Guraya T (2019) On the perceptions of students and professors in the implementation of an inter-university engineering PBL experience. Eur J Eng Educ 44–5:726–744

37. Savage RN, Chen KC, Vanasup L (2007) Integrating project-based learning throughout the undergraduate engineering curriculum. J STEM Educ 8(3&4):15–27
38. Dustker SM, Reddy BS, Kandakatla R, Joshi GH, Oakes WC (2021) Role of reflection in service learning-based engineering programs: a cross-cultural exploratory and comparative case study in India and the USA. In: 2021 ASEE virtual annual conference
39. Pai D, Kelkar A, Layton R, Schulz M, Dunn D, Owusu-Ofori S, Duraphe A (1998) Vertical integration of the undergraduate learning experience. In: Proceedings of the ASEE
40. Giralt F, Herrero J, Grau FX, Alabart JR, Medir M (2000) Two-way integration of engineering through a design project. J Engr Educ 89(2):219–229

Monetization of Intellectual Property Through Commercialization of Products and Services Developed Through Research and Innovation

Anand B. Kakade

Abstract This chapter outlines some of the best practices adopted at Rajarambau Institute of Technology (RIT), located in the Sangli District of Maharashtra state, India, for the monetization of intellectual property. Typically, intellectual property refers to legal rights obtained from the patent office to protect innovative ideas. However, in this chapter, intellectual property is considered in a broader sense, and not just limited to said legal rights. For any technical institute, a talented pool of students and faculty, serves as a source of intellectual property. Nevertheless, the institute must cultivate an ecosystem to foster applied research and innovation. This ecosystem includes a Fabrication Laboratory (Fab Lab), seed funding support, and a full-fledged Incubation center. It also includes a cache of highly motivated faculty and creative students. Transforming applied research into a commercial product is an exceptionally challenging task for inexperienced faculty. However, with systematic strategic planning, a selected few faculty member in the institute can be developed to undertake applied research for innovative product development or industrial consultancy projects. It is noteworthy that less than 5% of patents prove useful for industry, making the chances of earning from patents relatively low. Consequently, efforts must be made to convert the intellectual property available within the institute into successful startups. To achieve this, the incubation center must offer comprehensive support to ensure the success of startups. RIT is actively supporting promising startups by providing industrial sheds, since they may face challenges in investing in the construction of such facilities. This initial support is crucial for the success of startups.

Keywords Monetization · Intellectual property · Applied research · Startup · Consultancy · Incubation center · Innovation · Research ecosystem

---

A. B. Kakade (✉)
Rajarambapu Institute of Technology, Taluka Walwa, District Sangli, Rajaramnagar, Islampur 415 414, Maharashtra, India
e-mail: anandkakade75@gmail.com

R. Kandakatla et al. (eds.), *Academic Leadership in Engineering Education*, Lecture Notes in Networks and Systems 1097, https://doi.org/10.1007/978-3-031-68282-7_15

1 Introduction

> Monetizing intellectual property is not just about patents; it's about transforming inventive ideas into solutions that shape industries.

Innovation stands as the key driver, propelling the economic growth of any country. Intellectual Property Rights refer to the legal rights exclusively obtained from the patent office. Only 5% of globally filed patents successfully transition to commercialization or licensing. Higher Education Institutes (HEIs), traditionally focused on publishing their technological advancements, are now transitioning towards the patenting system. While many reputed institutes file patents, only a small percentage of them are either licensed or commercialized [1, 2].

A significant obstacle in Intellectual Property Rights (IPR) commercialization is the lack of collaboration between Higher Education Institutes and industry. Consequently, patents obtained by institutes may not align with industry needs [2, 3]. Patents, trademarks, copyrights, and trade secrets hold greater importance to corporate value than the manufacturing plants producing the products. The acquisition and protection of intellectual property are crucial for success and survival in the modern marketplace [3, 4].

Globally, startups are recognized as nation-builders due to their contributions to the economy through various channels, including employment, advanced technology, competitive products, technological innovations, and wealth creation. One of the ways of monetization of intellectual property is through startups. Startups require capital for business setup, expanding operations, and investing in research and development [5]. The challenges and strategies for promoting Intellectual Property Monetization in Developing Countries are outlined in [6].

In the domain of Higher Education Institutes (HEI), the intellectual capital residing within students and faculty becomes valuable asset when motivated towards high quality research. The foundation of quality research within any technical campus is built using the asset of self-motivated, highly qualified faculty members, along with students possessing an innovative mindset. Beyond the essential qualities of the individuals, the establishment of an adequate research ecosystem is important to facilitate and nurture research activities. This ecosystem surrounds various elements, including self-motivated faculties and students, state-of-the-art facilities like the Fab Lab, crucial seed funding support and the vital presence of an incubation center.

In the pursuit of quality research, the outcomes derived, possess the potential to go beyond the academics and be translated into tangible benefits. Monetizing these high-quality research outcomes serves as a dual-purpose mechanism, not only contributing to additional revenue generation but also fostering the brand building of the institute. The global landscape emphasizes the importance of cultivating such intellectual capital, with many countries actively promoting the development of startups within the premises of technical institutes. India, cognizant of this paradigm shift, lends substantial support to deserving technical institutes through government funding for the establishment and nurturing of incubation centers.

In this chapter, a case study at Rajarambapu Institute of Technology (RIT), located in Sangli District of Maharashtra, India, is discussed with emphasis on applied research activities and initiatives implemented to promote monetization of intellectual property. Notably, the conception of intellectual property is explained in its broadest sense, beyond the conventional limits of legal rights exclusively obtained from the patent office in the form of patents. This over all perspective surrounds around the spectrum of intellectual endeavors emerging from the research and innovation within the institute.

2 Applied Research Versus Theoretical Research in Technical Institutes

Technical institutes globally predominantly dedicate themselves to dispensing structured technical knowledge to undergraduate (UG) and postgraduate (PG) students through carefully designed curriculum. While the primary function of these institutes is educational, there exists a profound realization that the reservoir of talent residing within both students and faculty can be harnessed for substantial contributions to research. In essence, technical institutes serve as platform where the combination of academic diligence and innovative passion can yield remarkable outcomes. This collaboration is especially clear in the areas of applied research and theoretical research. These are two main types of research that have different purposes and goals.

The essential part of technical education worldwide often leans toward theoretical research, wherein the emphasis lies in the exploration and enhancement of fundamental principles and concepts. This aspect of research is typically manifested through scholarly paper publications, contributing to the theoretical framework of various technical fields. A considerable proportion of technical institutes globally moves towards this academic pursuit, aligning their research endeavors with the conventional path of advancing theoretical knowledge. However, the narrative takes a more dynamic turn when examining top-tier institutes that thoughtfully balance their focus on applied research.

Applied research, in contrast to its theoretical counterpart, is inherently geared toward addressing real-world challenges and finding tangible solutions. The distinct goal of applied research is the practical application of knowledge to overcome immediate issues or enhance existing processes, products, or services. Here, the essential part lies in the direct translation of research findings into real-world applications, development of innovative products, the acquisition of patents, and the formation of startups.

Esteemed technical institutes distinguish themselves by fostering a culture that not only values theoretical exploration but also places significant emphasis on applied research, thereby contributing to advancements in technology, industry, and entrepreneurship. In this situation, the gap between theory and application is bridged, and the research ecosystem benefits from a strong interaction between academic knowledge and practical innovation.

2.1 Applied Research

Applied research, a distinct facet of scholarly inquiry, primarily aims to solve real-world problems or shed light on specific problems. It stands as a beacon of purposeful investigation, concentrating its efforts on the immediate application of knowledge to confront tangible issues or enhance the efficiency of existing processes, products, or services. This form of research exhibits a direct focus on tangible problem-solving and is strategically geared toward offering practical solutions in industry, healthcare, technology, and other domains.

The ripple effect of applied research extends far beyond the academic realm, penetrating industries, healthcare practices and technological landscapes, thereby amplifying its real-world significance. The intended peak of applied research is not confined to scholarly publications alone; rather, its ultimate purpose lies in the implementation of its findings to produce solutions and enhancements in response to pressing challenges or to refine and optimize existing processes for greater efficiency and relevance.

2.2 Theoretical Research

The realm of theoretical research constitutes an intellectual pursuit where the important objective is our comprehension of fundamental principles and concepts that are inherent in a specific field of study. Unlike applied research, which is geared towards immediate practical applications, theoretical research is characterized by its inherent focus on the development or refinement of theories. It strives to contribute to the theoretical framework of a given discipline, without a pressing emphasis on immediate practicality.

The researchers immersed in theoretical research undertake the formidable task of proposing novel theories, models, or frameworks, aspiring to broaden the theoretical foundation of the subject matter under investigation. The distinguishing feature of theoretical research lies in its commitment to the abstraction and conceptualization of ideas, often not leading to the immediate requirement of practical implementation. While theoretical research doesn't have the direct and tangible outcomes characteristic of applied research, it plays a pivotal role in shaping the intellectual landscape of a discipline. The significance of theoretical research lies in its long-lasting impact on the foundational understanding and intellectual evolution of a given field.

3 Research and Innovation in Higher Academic Institutions

Research and innovation in academic institutions constitute an important aspect of the academic landscape, with a predominant focus on theoretical research leading to scholarly publications. The global landscape of technical institutes is characterized by a significant emphasis on theoretical pursuits, largely driven by the imperative to secure quality publications in reputed journals. The quest for such publications has become a necessity for institutes, which playing a crucial role in diverse ranking surveys that evaluate academic institutions.

While theoretical research stands as a foundation when it comes to institute ranking, it falls short in substantially contributing to revenue generation. But in applied research, which operates in the challenging ground of product development, emerges as a rewarding endeavor, fostering additional revenue streams and enhancing the institute's brand equity. The complexities inherent in crafting tangible solutions to real-world problems distinguish applied research, requiring researchers to gather considerable experience in product development.

The inherent contrast between theoretical and applied research manifests in the academic community, where there is a tendency to omit challenging aspects from research publications, and leaving them as to future prospects or future possibilities. In the realm of commercial product development, however, every aspect of functionality demands careful design. This challenging dimension extends the timeline required for technical institutes to fabricate reliable products, given that laboratory-tested prototypes may function optimally in controlled environments but face potential failure when deployed under the harsh conditions of industrial applications. Faculties and students embarking on applied research must cultivate the ability to comprehend the causes of product failure, engaging in rigorous testing and iterative processes to refine product designs and rectify issues.

Fostering a culture of innovative product development in technical institutes necessitates strategic planning and implementation. This strategic planning spans various aspects, including budget allocation for Research and Development, procurement of essential equipment and machinery for prototype and product development, provision of seed funding for innovative projects, motivation of faculties and students for product development and industrial problem-solving, and financial support for patent filing.

Inexperienced or fresh faculties can garner practical experience in applied research by initially focusing on developing simple products available in the market. Deploying these products in real-world applications, tests their reliability and serves as a valuable learning experience. However, it is acknowledged that products developed by faculty without prior experience are prone to failure in actual applications. Despite these initial struggles, the iterative process of testing and improvement is paramount. While experienced faculties may achieve product development success in fewer iterations, inexperienced faculties may need a greater number of iterations to develop reliable products. Faculties must systematically enhance their competencies by carefully analyzing the causes of product failure and iteratively improving failed products through a robust process of design refinement.

The primary goal of applied research, within the broader context of institutes fostering a culture of research and development, is two-fold: generating additional revenue and creating a significant impact for the holistic development of the institute. To attain desirable outcomes in applied research, institutes must cultivate a culture of research and development like industry practices. This strategic alignment ensures that research outcomes translate into tangible solutions, contributing to both the financial sustainability and academic prominence of the institute.

4 Research Ecosystem on the Campus

Establishing a robust research ecosystem on the campus is paramount to transforming innovative ideas conceived by faculties and students into tangible products. The initial phase involves the validation of ideas through the development of a proof of concept, commonly referred to as a working prototype. A crucial component supporting prototype development is the presence of an advanced fabrication laboratory, widely recognized as a Fab Lab, on the campus. The Fab Lab, equipped with fundamental tools and equipment essential for prototype development and to some extent, product manufacturing, serves as a foundational element. Key equipment and tools housed in the Fab Lab include 3D printers, 3D scanners, non-metallic laser cutting machines, TIG welding machines, MIG welding machines, drilling machines, grinding machines, various types of mechanical machining tools, high-powered computers, PCB fabrication facilities, a range of sensors and actuators, and microcontroller-based development boards.

In addition to the basic tools and equipment in the Fab Lab, the provision of specialized machines and facilities required for intricate product development further enhances the research ecosystem. Specialized equipment, such as sheet metal laser cutting machines, hydraulic CNC bending machines, powder coating facilities, Vertical Milling Machine (VMC), laser welding machines, and metal 3D printers, plays a pivotal role in advancing the capabilities of the Fab Lab. The combination of a sheet metal laser cutting machine, hydraulic CNC bending machine and powder coating setup proves particularly invaluable for prototype development and product manufacturing involving sheet metal work. Given that many commercial products and industrial solutions necessitate metallic enclosures and components, the efficient fabrication of high-quality metal enclosures and components within the Fab Lab significantly accelerates the development process. Moreover, providing services utilizing these facilities to external stakeholders, including nearby industries, presents an opportunity for additional revenue generation. This inflow of additional revenue not only fosters financial sustainability but also relieves the economic burden on the institute. The photographs given below depict some elements of ecosystem that is valuable for applied research (Fig. 1).

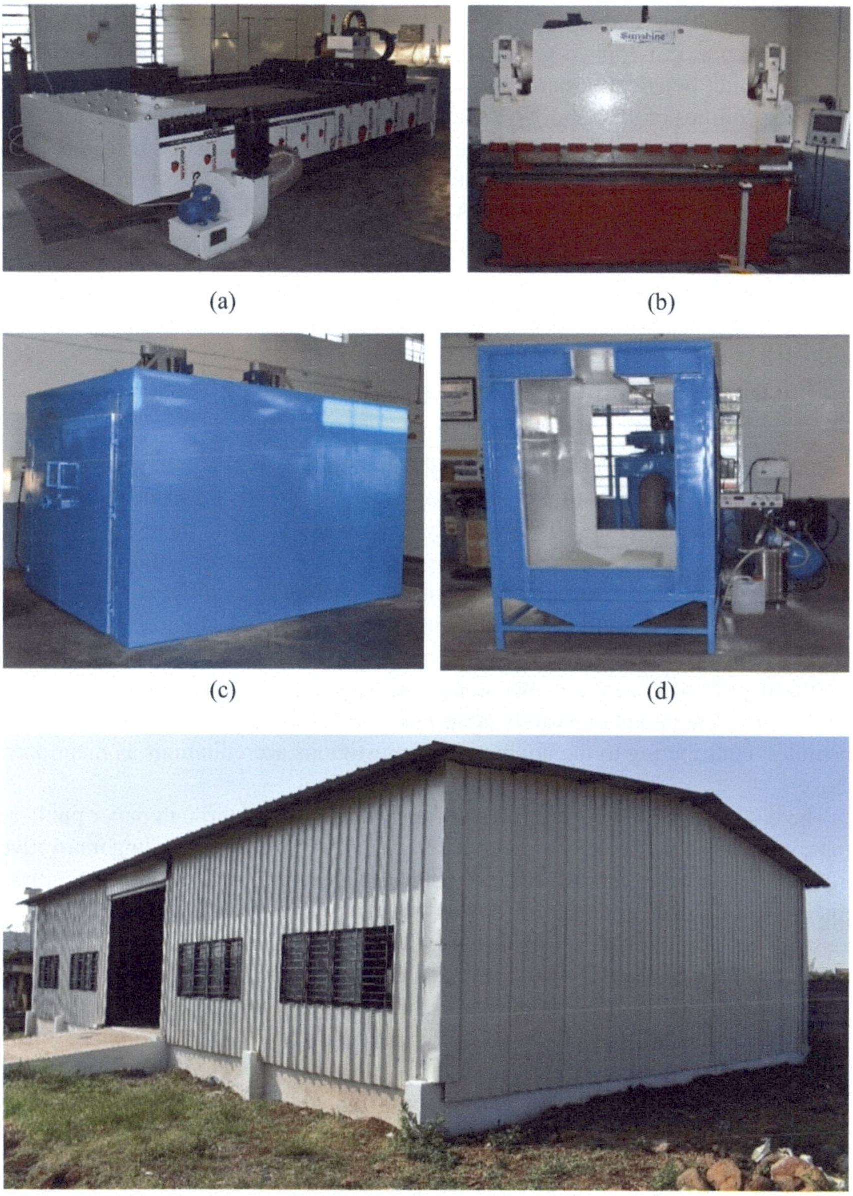

(a) (b) (c) (d) (e)

Fig. 1 Few elements of applied research **a** Sheet metal laser cutting machine **b** Hydraulic bending machine **c** Oven for powder coating **d** Powder spraying booth for powder coating. **e** Industrial Shed for startup

However, the mere availability of an advanced Fab Lab does not guarantee the desired outcomes of applied research. This big change depends on self-motivated faculties doing practical research as well as students who think creatively and the high-tech abilities of the Fab Lab. This trio can emerge as a game-changer for technical institutes, capable of yielding the desired outcomes in applied research, such as additional revenue generation, attainment of National Board of Accreditation (NBA) and National Assessment and Accreditation Council (NAAC) accreditations and securing favorable rankings in the National Institutional Ranking Framework (NIRF) survey.

5 Faculty Engagement in the Applied Research

Faculty engagement in applied research is an important aspect for the research landscape within technical institutes in India, where faculty recruitment traditionally emphasizes academic roles. The majority of faculties in many technical institutes allocate a significant portion of their time to academic and administrative responsibilities. Nonetheless, a select group of self-motivated faculties diligently invest their spare time in Research and Development (R&D) activities, often extending their research work beyond the institute's standard working hours and into weekends. Technical institutes engage in two comprehensive categories of research: theoretical research and applied research. Both faces of research are integral for technical institutes, contributing to the attainment of prestigious accreditations as mentioned above.

The outcomes of theoretical research manifest primarily through paper publications, whereas applied research yields a diverse array of results, including innovative products, industrial solutions, patents, additional revenue streams, and the cultivation of startups. True to its nature, applied research poses significant challenges and consequently, only a handful of faculties within an institute are actively engaged in this dynamic domain. Applied research can be pursued either through guiding students or through self-involvement. In the author's perspective, faculties should predominantly conduct applied research independently, reserving guidance-based research, for instances where the continuous availability of the same students for product development is a restricted. Students, though valuable contributors, often possess limited experience in the complex process of product development. Therefore, faculties must shoulder the load of conducting a substantial portion of their applied research autonomously, allowing them to cultivate technical competency in developing commercial products and addressing industrial challenges.

In the pursuit of fostering applied research, institutes can identify and support self-motivated faculties inclined toward commercial product development by reducing their administrative responsibilities and providing necessary resources, including in-house seed funding for product development and access to advanced facilities like Fab Labs. While it is crucial for faculties to be extensively engaged in applied research independently, involving students in product development is highly desirable as it contributes to their technical competency. The institute's strategic approach should involve identifying faculties with a leaning towards applied research, facilitating an environment favorable to their engagement and offering unwavering support, both financial and infrastructural. The tangible outcomes of applied research, such as innovative product development and effective problem-solving in an industrial domain, can be achieved if faculties are deeply immersed in these chasing. Acknowledging the learning curve, institutes must provide sustained support and encouragement to faculties embarking on applied research, recognizing that it typically takes a minimum of two years to develop the essential competency in product development. Through this strategic approach, technical institutes can position themselves as hubs of innovation and impactful contributors to the field of applied research.

6 Financial Support for Applied Research

Securing financial support for applied research is a critical component in fostering innovation within technical institutes. The required funds can be procured through internal fund allocation or external sources such as funding agencies and collaborations with industries. However, to achieve substantial and impactful outcomes in applied research, it is important to have a cadre of self-motivated and highly experienced faculty members dedicated to applied research on the institute's campus. In the context of Indian technical institutes, where the predominant emphasis is often on academic activity, a limited number of faculties exhibit a natural inclination towards applied research.

To catalyze a culture of applied research, institutes must proactively identify and support a select group of self-motivated faculties with an inherent interest in applied research. These identified faculties should be granted in-house seed funding to facilitate the initial stages of their journey into applied research. Particularly for new faculties lacking prior experience in applied research, a strategic approach involves tasking them with the development of simple products within their domain or exploring existing market products. This hands-on strategy aims to cultivate the necessary competencies essential for successful engagement in applied research.

Certain fundamental qualities such as logical thinking, basic technical knowledge and the ability to apply knowledge to problem-solving is essential for faculties engaged in applied research. These attributes are not developed overnight but require sustained efforts and consistency from faculties over an extended period. Recognizing the various nature of applied research, institutes can offer support to faculties during the initial phases, including attendance at basic training programs related to applied research. Financial assistance and on-duty leaves can be provided to facilitate faculty participation in such training programs, ensuring the acquisition of foundational knowledge in applied research.

It is crucial to acknowledge that while training programs contribute to faculty development, the skill set required for product development necessitates extensive practice, hands-on experience and self-involvement in applied research. Mastery of product development, being a complex skill, cannot be solely acquired through training programs but evolves through continuous engagement and active participation in applied research initiatives. Developing the technical competency of faculties in applied research, leading to the accomplishment of innovative product development, is a gradual process that typically spans two to five years. Institutes committed to promoting applied research must adopt various strategy enclosing financial support, targeted training, and a supportive ecosystem to nurture faculty development.

7 Challenges in the Monetization of Intellectual Property

The monetization of intellectual property, viewed in a broader sense beyond legal rights granted by patent offices, presents numerous of challenges. In this context, intellectual property encloses the innovative ideas and processes crafted by both students and faculties. When intellectual property takes the form of a granted patent, its monetization often depends on transforming the conceptualized idea into a tangible and commercial product. This transition from prototype to a market-ready product demands a substantial investment, a prospect that becomes feasible only if prominent industries express interest in the patented concept. The industry's inclination towards granted patents or innovative ideas largely depends on the market demand for the corresponding idea or product. Notably, the market feasibility of granted patents is low, with less than 2% deemed viable for earning potential.

Alternatively, intellectual property can be monetized through the commercialization of products or services via on-campus startups. Whether the intellectual property consists of prototypes, innovative ideas or granted patents, the process often involves proving concepts through working prototypes. However, the actual challenge lies in the conversion of these working prototypes into reliable and commercially successful products. Commercial product development is a complex undertaking, demanding extensive experience in product design and consideration of factors such as reliability, cost-effectiveness, aesthetics, certifications, and market requirements.

To overcome these challenges, on-campus support systems, particularly incubation centers, plays an important role. Startups incubated on the campus receive necessary assistance, including seed funding, industrial facilities for production, client connections and mentoring. Notably, the journey from prototype to commercially successful product is significantly eased when startups benefit from non-refundable seed funding and discounted rates on industrial facilities. These support mechanisms enhance the prospects of startup success, provided the developed product is not only reliable but also addresses a genuine market need. Basically, the monetization of intellectual property requires a strategic teamwork of industry interest, startup incubation and overcoming the complex challenges associated with converting innovative ideas into commercially viable products or services.

8 Monetization of Intellectual Property Through Patents, Consultancy, and Startups

In the preceding discussion, I underlined the essential need for a research ecosystem on the campus to facilitate applied research. Now, let's dig into how the monetization of intellectual property can be achieved through applied research, involving innovative product development, industrial consultancy, and the commercialization of products via startups. Institutes can strategically file patents for innovative ideas originating from the creative minds of faculties and students. These patents can subsequently be licensed or sold to industries, thereby generating additional revenue. However, the commercialization of patents or any innovative idea, poses challenges, as only less than 5%, may have the market feasibility required for substantial returns.

An alternative way for monetizing intellectual property involves offering industrial consultancy services. Self-motivated faculties with extensive experience in applied research can undertake consultancy projects for industries and address their technical challenges. In return, industries compensate the institute with consultancy fees. Establishing strategic plans for developing faculties with domain expertise is crucial for successful industrial consultancy outcomes. The process of cultivating faculties for such roles is not instantaneous. It requires two to five years to inculcate the necessary technical competency.

Another avenue for monetizing intellectual property is the creation of startups on campus. This involves the commercialization of innovative products developed within the academic environment. Startups, whether initiated by students or through collaborations between students and faculty, can be challenging but immensely rewarding. The success of startups lies in robust support systems, which can be provided through incubation centers. Support mechanisms include non-refundable seed funding, discounted co-working spaces, access to advanced machines, client connections and even mentorship. Institutes can earn additional revenue by offering services, such as access to fabrication laboratories to external startups. Additionally, the institute may take a stake or a percentage of the profits generated by the startups it supports.

Joint startups involving students and faculty offer a unique approach to fostering success. Graduating students often lack the experience necessary for commercial product development and running a business. Allowing faculty to engage in joint startups with students increases the likelihood of success. In such collaborations, the institute may negotiate for a stake or profit-sharing arrangement. Notably, at Rajarambapu Institute of Technology, where I currently serve as Dean Research & Development, we have two successful joint startups—TGP Bioplastic Pvt Ltd and Origin Tech. The former, initiated by Dr. M. V. Pisal and three B.Tech students, who secured substantial funding for developing biodegradable plastic. The joint startup showcased the impactful potential of collaborative ventures. The latter, founded by myself and Mr. Shubham Omase, who specializes in manufacturing electronics systems for industrial applications, demonstrating the success achievable through faculty-student collaborations.

Both startups have benefited from discounted industrial sheds and various support services provided by the institute. These real-world success stories underscore the transformative impact of joint startups on campus, enhancing the institute's revenue streams, achieving accreditations and contributing to the socio-economic fabric through innovation. The Biodegradable granules and Biodegradable plastic bags developed by TGP Bioplastic startup are depicted in Fig. 2.

Some of the commercialized electronic driver cards and systems developed by Origin Tech startup are shown in Figs. 3, 4 and 5. The showcased case studies clearly throw light on the potential for monetizing intellectual property that emerges from the creative students and faculties within the campus environment, specifically through the channel of startups.

Although this mode of monetization is undoubtedly challenging, but not impossible. The significant advantage is that, it liberates the institute from a dependency on external industries for the commercialization of innovative ideas. The success stories of these startups serve as a testament to the feasibility and effectiveness of this approach. In acknowledgment of the institute's vital role in fostering and supporting these successful startups, a mutual beneficial relationship can be established. Particularly, successful startups, as a gesture of appreciation, may choose

(a) (b)

Fig. 2 **a** Granules of biodegradable plastic **b** Bio-degradable plastic-bags manufactured from granules

Fig. 3 Infrared heater driver card

Fig. 4 Actuator driver card

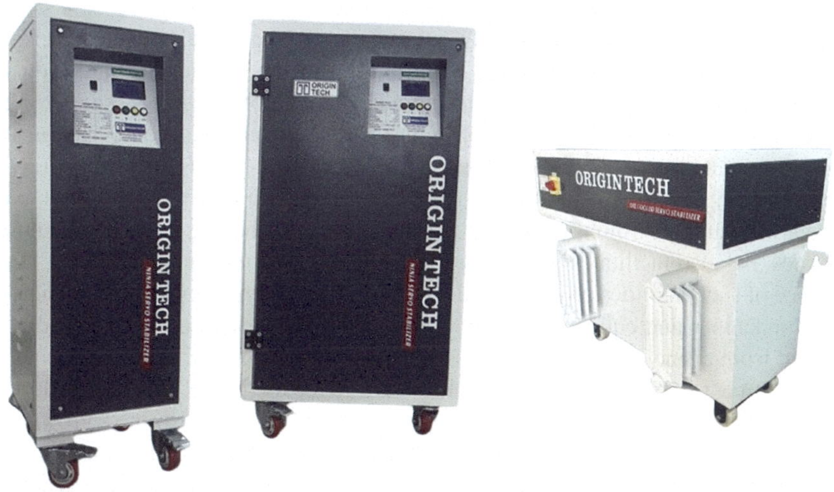

Fig. 5 Three phase servo stabilizers

to establish sponsored laboratories within the institute. These sponsored laboratories combined with the essence of innovation and research, can become integral contributors to the overall development and academic excellence of the institute. This collaborative dynamic not only benefits the institute's financial standing but also enriches the academic landscape by creating hubs of cutting-edge research and development, thereby fostering a culture of innovation and entrepreneurship within the educational ecosystem.

9 Benefits of Monetization of Intellectual Property to Institute

The monetization of intellectual property on campus yields a countless benefit for the institute with both tangible and intangible advantages. Among these benefits, substantial additional income generated, plays an important role in expanding facilities crucial for product development and for the R&D wings. This financial aspect is particularly significant in the evaluation criteria of major ranking surveys for technical institutes globally. Accreditation bodies such as the National Board of Accreditation (NBA) and the National Assessment and Accreditation Council (NAAC) in India, as well as the prestigious National Institutional Ranking Framework (NIRF) survey conducted by the Ministry of Human Resource Development (MHRD), consider the additional income earned through the exploitation of intellectual abilities as a key performance parameter.

The intangible benefits are equally noteworthy, as successful monetization of intellectual property attracts funding from both government agencies and industries. These funds are instrumental in upgrading laboratory facilities, supporting applied research initiatives, and establishing incubation centres. The development of a thriving startup ecosystem on the campus is a distinctive form of monetization, leveraging the intellectual skills of students and faculties. This ecosystem not only garners support from government agencies for strengthening incubation centre activities, but also facilitates the creation of Common Facility Centres (CFCs). The presence of such facilities provides students with unprecedented exposure to sophisticated machines and equipment commonly found in large industries. The funding received from agencies like the Maharashtra State Innovation Society (MSInS) and the Department of Science and Technology (DST) for the NETRARIT Incubation Centre at RIT underscores the tangible support available for the institute's growth.

These indirect benefits of monetizing intellectual property extend to the overall development of the institute, fostering an environment of innovation and research. Moreover, they contribute to the enhancement of the academic experience by offering students opportunities to engage with cutting-edge technologies and facilities, thereby nurturing a culture of entrepreneurship and applied learning on campus.

10 Key Takeaways

- This chapter delves into various strategies for intellectual property monetization, with a particular emphasis on cultivating an ecosystem for this purpose.
- It underscores the importance of Higher Education Institutes (HEIs) developing a skilled pool of faculty members capable of engaging in applied research, focusing on innovative product development, and offering industrial consultancy.
- Additionally, HEIs are encouraged to establish fully equipped incubation center with advanced manufacturing facility and FAB Lab to nurture startups.

- Initially, startup needs industrial sheds and electrical connection for manufacturing of their product which require substantial investment. By providing these facilities at subsidized rates, HEIs significantly increase the likelihood of startup success.

11 Conclusion

This chapter has outlined various strategies for monetizing intellectual property originating from the collaborative efforts of students and faculty members within the campus. The methods discussed herein are reflective of the initiatives implemented at RIT offering insights into the practical approaches employed to leverage intellectual property for financial gains and institutional development. The focus on attracting angel investors underscores the importance of external financial support in propelling startups to the next level of their evolution, contributing to the overall success of the campus's entrepreneurial ecosystem.

References

1. Bhardwaj M, Sandhu A, Ghumman N (2021) Patents commercialization profile of universities and higher education institutes in India. J Intellect Prop Rights 17:199–207
2. Nithyananda KV (2012) Alchemy and IPR–monetizing intellectual property rights. J Intellect Prop Rights 17:406–416
3. Rader RR, Voit T, Nagasawa K, Kim KJ, Wang MH (2019) Law, 'policy and monetization in intellectual property. Cambridge Scholars Publishing
4. Kulyagina EA, Kolozhvari1 YB, Koval SV An analysis of the forms of commercialization of intellectual property objects of higher education institutions: Russian and international experience. Adv Econ, Bus Manag Res 128:617−625
5. Panda B, Joy S (2021) Intellectual property rights-based debtFinancing to startups: need for a changing role of Indian banks. Vikalpa 4(3):143−152
6. Ghafele R, Gibert B Promoting intellectual property monetization in developing countries: a review of issues and strategies to support knowledge-driven growth. World Bank Policy Res Work Pap 6143

Empowerment of Faculty and Students for the Twenty-first Century

Academic Leadership for All: A Systems Thinking Approach to Pedagogical Development and Professional Learning for Engineering Educators

Xiangyun Du, Lelanie Smith, Khalid K. Naji, Sohum Sohoni, and Aida Guerra

Abstract This chapter highlights the importance of pedagogical development for engineering educators and its positive influence on their leadership journeys. It begins by problematizing the linear approach currently dominating the provision of training to engineering faculty members and proposing a system thinking approach to conceptualize engineering faculty's learning as a complex dynamic system of interrelated components, including personal professional growth and interactions with environmental factors such as institutional conditions and cultural values. Through three case studies of different approaches to pedagogical development and cross-case analysis, recommendations are made for future institutional activities aiming to support academic leadership for all. These include (1) organizing PD activities centering engineering educators' intrapersonal characteristics developed, based on their prior experiences and personal interests, and designs to engage their pedagogical knowledge to enhance their self-efficacy. (2) Providing opportunities for engineering educators to engage proactively and agentically in the decision-making and implementation processes regarding engineering professional development. (3) Creating space for engineering educators to use their expertise to influence others, from the individual classroom to the institution as a whole. Policies that recognize and reward teaching can be powerful in supporting engineering educators to sustainably engage. In general, engineering educators' professional learning and academic leadership can be developed through practical learning and professional agency enactment.

X. Du (✉) · A. Guerra
Aalborg University, 9000 Aalborg, Denmark
e-mail: xiangyun@plan.aau.dk

L. Smith
University of Pretoria, Pretoria 0002, South Africa

K. K. Naji
Qatar University, Doha, Qatar

S. Sohoni
Milwaukee School of Engineering, Milwaukee, USA

R. Kandakatla et al. (eds.), *Academic Leadership in Engineering Education*, Lecture Notes in Networks and Systems 1097, https://doi.org/10.1007/978-3-031-68282-7_16

Keywords Pedagogical development · Engineering educators · Professional learning as a complex dynamic system · System-thinking approach

1 Introduction

University teachers globally are on the frontlines of educational change as new demands on higher education presents both–challenges and opportunities–for learning [1, 2]. While some adapt more easily to such changes than others, perhaps due to their prior experience and skills or support from their institutional and social contexts, others struggle with the process of educational transformation [3, 4]. We contend that engineering education institutions will continue to encounter challenges due to growing complexity and mounting unpredictability caused by rapid societal changes. However, these challenges may also present opportunities to reimagine engineering education as a profession and enhance educator leadership in the field [5]. Such a view may provide space for educators to proactively and agentically engage in professional learning and practice, instead of passively receiving information through transition-focused training [6].

While recent decades have seen growing interest in the improvement in university teaching and learning in general, current pedagogical development initiatives focus on teaching activity techniques and new educational technologies, with an emphasis on information transmission. However, it provides with limited space for active participation, interaction, or negotiation of conditions and contextual factors such as curriculum and assessment regulations [6–9]. Current approaches, while providing training relevant to desired immediate changes in teaching practice, have been criticized for having little effect on educators' pedagogical beliefs and values [2, 8]. They also lack sustainable change outcomes [6–8].

Such critiques highlight concerns about the linear approaches to 'training' university educators, which is generally based on the assumption that they learn from short, context-dependent activities [6, 9, 10]. As an alternative conceptual perspective, the notion of teachers' professional learning as a dynamic system, offers a lens through which to explore the multiple interacting dimensions influencing teacher learning [10–13]. Engineering educators' professional learning can be conceptualized as individual processes emerging from fluid, situated, interactive, contextually bounded, and culturally dependent experiences [14–18]. Teacher learning for academic development comprises interrelated dimensions, including individual characteristics, such as motivation, intentions, cognitions, efficacy beliefs, and values in teaching, all of which may be largely influenced by prior experiences and teachers' relationships with students, colleagues, and institutional environments [2, 6, 9, 15–17].

The ongoing intense educational change, spurred by rapid societal change, provides a worthwhile context for examining how engineering educators can best be supported to proactively organize their professional learning and develop academic leadership. Specifically, work should be undertaken to better understand how engineering educators in different institutional and societal contexts initiate, act, and lead their professional learning towards educational change, and how they enact pedagogical leadership by influencing others.

In this investigation, 'professional learning' refers to activities and practices that university educators undertake to develop their teaching [19]. This definition highlights the agentic role of engineering educators in the process of educational change.

To explore engineering educators' leadership in professional learning, especially amidst emergent educational changes in higher education, this chapter provides conceptual insights on engineering educators' professional learning as a complex dynamic system [5, 10, 13, 20].

Three case studies offer illustrative examples of individual engineering educators' development of academic leadership, beginning with their pursuit of professional learning and continuing with their efforts to influence their peers and environments. These cases demonstrate the importance of engineering educators' professional agency and highlight the value of a ***systems thinking approach*** for institutions in supporting professional learning, collaboration, and engagement—all core to the academic leadership components [5, 12].

2 Narratives from Academic Leaders in Engineering Education

In the context of the complexity theory lens discussed above, highlighting the development of professional agency and academic leadership, this chapter takes a life course approach and reflective practice perspective [12, 14] to the structure of three narratives. The co-authors used the outline below as inspiration for reflective writing about their development of agency and leadership in professional learning.

Overall outline following the life history approach to narrative, as suggested to the co-authors for inspiration.

1. Your personal journey towards such a topic regarding engineering educators' professional learning for teaching advancement.
    a. Why did it become important in your own professional life?
    b. What did you do to improve your own teaching?
    c. How do you see your own pedagogical development engagement contributing to your career?
    d. What challenges have you encountered and how did you cope?
2. In which ways you have extended your own journey and learning to your surroundings? E.g.
    a. What did you do to influence your colleagues and institutions?
    b. What are the encountered challenges and developed coping strategies regarding your work of extending or constructing institutionalized activities for engineering educators' pedagogical development?

c. Prospectively, what do you consider as important for future works on engineering educators' pedagogical development, and beyond?

2.1 Case 1: Engaging in Pedagogical Development Through Self-Development

One of Dr. Sohum Sohoni's initial part-time jobs during his time in graduate school was as a mathematics tutor at the University of Cincinnati's University College, which catered predominantly to adult learners.

As a tutor, he helped anyone who showed up during his assigned hours in the tutoring room, and so the students ranged from pre-algebra to calculus. This was Dr. Sohoni's first exposure to college-level teaching. Although the job did not include teaching duties typical of a faculty member's work, such as planning lessons, creating assessments, or designing teaching aids for lectures, it gave him first-hand experience of the joy of helping fellow participants understand things with which they had previously struggled. After two decades in academia, Dr. Sohoni still finds that the joy of helping someone along their journey through life is what motivates him as a professor. It was this early tutoring experience, and later, helping fellow graduate students to master some of the more difficult concepts in computer architecture, as they prepared for exams or job interviews, that made Dr. Sohoni realize he wanted to be a professor.

After earning his Ph.D and starting his first faculty job as an assistant professor, Dr. Sohoni quickly realized that he had not taken full advantage of the teacher training opportunities available at the University of Cincinnati, such as the Preparing Future Faculty (PFF) program. Like many other newly minted Ph.Ds, he had gone from a research lab to the classroom, with limited training in teaching and almost no knowledge of pedagogical innovations or literature on effective teaching. He had picked up some good practices from his instructors and had made mental notes on what *not* to do from less positive experiences as a student. With the benefit of these experiences, a heartfelt desire to do the best for his students, a positive attitude, prioritization of his teaching duties, and a likeable personality, Dr. Sohoni got off to a good start, receiving positive student evaluations.

Despite the positive responses he received early on from students and colleagues, Dr. Sohoni felt inadequately prepared to excel in teaching. He was fortunate to have colleagues and a department chair, who shared a passion for quality education. The department had recently secured a substantial NSF Department-Level-Reform grant to improve teaching and learning practices. Dr. Sohoni was one of their first mentees and was quickly introduced to not only scholarly teaching but also the scholarship of teaching and learning. His work started with some rudimentary studies of the impact of groupwork and project-based learning, and he gave presentations at conferences such as the Annual Conference of the American Society for Engineering Education, Frontiers in Education, and the Association for Computing Machinery Special Interest Group on Computing Education (SIGSE). In 2011, Dr. Sohoni put together a team of his own, including faculty from English and Education, and received an engineering education grant from the NSF to design and evaluate a teaching tool [21].

Early in his career at Oklahoma State, Dr. Sohoni became aware of the National Effective Teaching Institute (NETI, https://www.neti-workshop.org/), a nationally recognized flagship program for effective teaching in engineering, run by Dr. Rich Felder and Dr. Rebecca Brent. He later reflected that participating in NETI was as transformative for his teaching as a 3-day program could be. While he continued to believe that the core of good teaching is the teacher's desire to serve students above all else, he also saw the enormous value in research-based tools and knowledge programs that NETI provides teachers to turn that desire into quality teaching practice. Just as engineering students need faculty to curate materials, create cohesive curricula, and assess learning, the faculty themselves need curated information, best practices, relevant examples, and role models to maximize their teaching capabilities. Participating in NETI was a milestone in Dr. Sohoni's journey as an engineering teacher, and it solidified his desire to become a role model for others and give back by providing such training.

During his formative professional years as Assistant Professor, Dr. Sohoni ran into Dr. Krishna Vedula at the 2011 ASEE Annual Conference in Vancouver, Canada. Dr. Vedula had founded an organization called the Indo-US/Universal Collaboration for Engineering Education (IUCEE https://iucee.org/). IUCEE had a significant influence on Dr. Sohoni and his career, thanks to its focus on faculty development and capacity building. Dr. Vedula's presentation clarified that IUCEE aimed to provide a much-needed service: training faculty to move away from the "sage on the stage" model of teaching (or worse, reading from a textbook) to one that was student-centered. Central to IUCEE was a faculty certification program, loosely based on NETI, which included a 3-day in-person launch event, a semester-long guided implementation phase, and a final reflective phase in which the participating faculty pitched their teaching philosophy statements. For many trainees, this was their first and only experience of in-service training. Over the past decade, this program has transformed many careers, with graduates spreading the knowledge they gained at hundreds of institutions across India. Dr. Sohoni served as an instructor for the in-person phase as well as the semester-long phase of the certification program for several years.

Excited at meeting many like-minded people from around the world in his professional circles, Dr. Sohoni found that he had failed to notice the rift in his department between those who prioritized teaching and learning and those who believed that a university professor's primary responsibility was research. Despite the legitimization of engineering education as a research discipline in its own right through the foundation of a directorate at the NSF and peer-reviewed journals supported by reputable societies such as ASEE and the Institute of Electrical and Electronics Engineers (IEEE), many in his department and at the institution did not think of engineering education research as "real" research at par with research in the engineering disciplines. Dr. Sohoni had not initially given much thought to a colleague's comment, made when he received one of the highest teaching honors at Oklahoma State, the Regents Distinguished Teaching Award as an Assistant Professor, that the award would be the "last nail in the coffin" in his case for tenure and promotion. He was denied tenure.

Dr. Sohoni tried to explain to potential employers that he was looking for a faculty position that valued his teaching expertise, and that he planned to change his research focus from computer architecture to engineering education research. Although it was a tough sell, he was able to garner a few jobs offers. He took one from the College of Technology and Innovation (CTI) at Arizona State University (ASU)'s satellite campus in Phoenix, Arizona. CTI seemed the perfect fit: The faculty cared about student success, there were rich conversations about teaching and learning at the faculty retreat, and Dr. Sohoni felt that he had miraculously found a place at a major public university where tenure-track faculty would not be under pressure to focus on research at the expense of classroom performance.

His first two years at ASU were positive. His colleagues valued discussions on teaching and learning, and there were many opportunities for in-service teacher training. In addition to in-house opportunities, he was able to participate in ASEE's virtual communities of practice (VCP, https://vcp.asee.org/), an initiative aimed at bringing together people, teaching in similar disciplines at different universities, to engage in conversations around best practices. This was a promising program and was well designed, but it dwindled over time and eventually dissipated. Dr. Sohoni felt that the participants and facilitators had tried to meet too many different professional obligations at once.

Turmoil arose in Dr. Sohoni's third year at ASU, as the College of Technology and Innovation was disbanded, and the new Polytechnic School was eventually formed. With the closing of CTI and the faculty's absorption into the research-intensive Ira A. Fulton Schools of Engineering, conversations on effective teaching were quickly replaced by conversations on research productivity in terms of Ph.D students graduated, publication and citation counts, and external research funding obtained. Research-focused faculty were hired to grow the budding engineering education research program at the Polytechnic School, and existing faculty, including Dr. Sohoni, were phased out. Nevertheless, Dr. Sohoni chose to focus on providing mentoring through IUCEE's certification program and continued to participate in conferences like ASEE's Pacific Southwest Section meeting, where one of his papers [22] received a best paper award. It was at events like these that grassroots-level changes in teaching and learning practices were taking place.

Towards the end of his time at ASU, Dr. Sohoni had the opportunity to co-teach a course with Dr. Shawn Jordan. Although he had previously taught lecture-based courses with substantial student interaction and semester-long projects that carried a majority of the course grade, this was his first direct experience of a completely flipped, project-based course without a single lecture. This was not an easy transition, and it provided a lot of insight for Dr. Sohoni to talk about flipped classrooms in in-service faculty training sessions in India.

In 2019, Dr. Sohoni joined the Milwaukee School of Engineering (MSOE), an institution with a long-standing reputation for its excellence in teaching and a focus on students. He currently serves as a Program Director and Professor of Software Engineering at MSOE, cherishing every opportunity to serve the institution and to act as a faculty mentor. MSOE is a predominantly undergraduate school with a clear mission centered around high-quality, relationship-rich education. Decision-making is rarely complicated by extraneous factors like research reputation and bringing in research funding. Faculty are encouraged to participate in ASEE's annual conference and are given funding to attend, whether they have papers accepted at the conference or not. There are regular opportunities to engage in discussions on teaching and learning within institutional, departmental, and program-specific online communities and in-person meetings. However, with a heavy teaching load and an expectation that students are given detailed feedback on their labs and homework, faculty are often pressed for time to engage in meaningful discussions. They are also much less inclined to engage in training–as trainees or trainers–outside of the institution, with some notable exceptions.

Thus, even at an institution, where excellence in teaching and learning is revered, ironically, there are still roadblocks to faculty engagement in in-service training.

Leadership skills here must come strongly into play and involved leadership skills that positively influences faculties and graduates must be displayed. Identifying problematic scenarios and situations of the involved parties, their challenges–which may be professional or otherwise, their shortcomings and mind-blocks, all has to be addressed and managed with an eye for progressive outcome, where all three parties that fall in the direct line of influence can be benefitted. The leadership team play a vital role in managing such roadblocks and advancing the core purpose of teaching.

More recently, Dr. Sohoni has transitioned to providing more customized training for faculty, aiming to provide examples directly relevant to the courses they teach. For example, he recently piloted a program with a group of faculties who were teaching introductory programming at Kalasalingam University in India, designing specific pedagogical interventions, and redesigning their assessments with a focus on improving student outcomes. These efforts are consistent with the DBER ambassadors concept that Dr. Sohoni has previously advanced [23].

Drawing together the narrative of his career, Dr. Sohoni focuses on four things

- First, it takes intrinsic motivation to seek in-service training and make time for it when the incentive system often works against it.
- Second, he was able to find mentors and collaborators everywhere he went, and they made invaluable contributions to his professional life and the good work he did.
- Third, institutions and policymakers should provide more opportunities and support for faculty who seek to improve their teaching.
- Finally, outside agencies such as IUCEE and REEN, as well as professional societies, should continue to provide programs to support faculty, as these will always be in demand.

2.2 Case 2: Engaging in Pedagogical Development Through Relations and Network Building

Dr. Lelanie Smith's personal journey as a teacher has several strands. Long drawn to education, she initially considered studying mathematics or psychology. She eventually settled for engineering because of her love for mathematics.

Throughout her undergraduate studies, Dr. Smith's free time was spent tutoring mathematics and coaching hockey. During her postgraduate studies, she completed a comprehensive two-year training program to become a yoga, meditation, and mindfulness teacher, and later completed a variety of life and team coaching trainings. In all of these, the focus was always on meeting the participant where they are and helping them connect with the content or facilitating a journey of self-discovery. Thus, in a sense, much of her training as a teacher occurred outside of higher education platforms, but in smaller groups where self-awareness and the creation of supportive environments were critical.

In her 15 plus years of teaching, the material Dr. Smith taught, included both technical content in disciplines such as fluid mechanics, and professional skills like team competency and civic responsibility. The first class she taught as a young academic had 900 students. Her classes over the next decade and a half, varied in size from 350 to 1650 students. Establishing personal relationships with such large classes is virtually impossible. Her supervision of 10 to 15 final-year design and research students, each year became a space, in which she could get to know each student, customize her approach depending on their motivations, learning styles, and needs, and support them in their journeys toward becoming professional engineers.

In 2014, Dr. Smith was approached by an international group (including members from Germany, the United States, and Finland) to have some of her final-year students join their collective development of a technological solution to rhinoceros poaching incorporating the use of drones. At first, this merely provided a convenient and contextualized problem for her students to work on, but over time, Dr. Smith began to see potential in allowing younger students to join the work. She set up a co-curricular vertically integrated structure that enabled students ranging from second-year undergraduates to postgraduates to work together on interrelated subsystems, in collaboration with local industry and international students.

Dr. Smith began noticing a difference in the students who were part of this project. Their approaches to the final-year capstone design and research projects exhibited greater agency. They showed more confidence and were more resourceful and creative in their approach to solving problems. By creating an environment in which students were free to experiment and fail within the safe space of a supportive learning environment, the project helped them collaborate and learn with their peers.

Dr. Smith began reflecting on her engagement with these students and the co-curricular structure of the project compared to the existing curriculum.

She was also nominated for a faculty teaching and learning award and was partnered with an educational psychologist who opened her eyes to pedagogical tools and frameworks she had been drawing on intuitively. Turning to literature on engineering education, she understood that the meaningful change she had achieved was creating a safe learning environment for students. She listened to them, supported their ideas–however wild they seemed–and allowed them to experiment even when she anticipated potential failure. Failure happened, as anticipated, but sometimes, she was also pleasantly surprised.

It became clear to Dr. Smith that developing a better understanding of what she had been doing instinctively gave her a chance to improve her practice and introduce new elements to support her students' and learning more intentionally, rather than merely intuitively. She began to see herself as a facilitator of learning, and to see her knowledge as less important than her ability to create a space for safe learning and experimentation through the use of guided and carefully scaffolded challenges (both technical and personal), and to provide support by listening and asking questions rather than instructing.

At this point in her journey, Dr. Smith began to question the generally accepted approach to curriculum development. The questions were based on her realization that engineering students rarely encounter such a learning environment, and that academic performances may not be the only–or even an accurate–indication of their future success as a professional engineer. How, she wondered, would they be able to develop these experiences at a large scale? It was during this time that Dr. Smith joined a Royal Academy of Engineering Masters class in Engineering Education in Cape Town, South Africa, where she met Prof. John Mitchell from University College London (UCL). He shared his experience in developing and rolling out the Integrated Engineering Programme (IEP). The programme ~~that~~ was designed to integrate real-world problem-based learning (PBL) at the curriculum level and, more importantly, at scale, with 1000 students in the first year. The engagements led Dr. Smith to clearly formulate ideas for transforming the Engineering School at the University of Pretoria. She shared her vision with the dean, and he supported her in visiting UCL to become familiar with the operational structure, resources, students, staff capabilities, and curriculum.

As Dr. Smith grew acquainted with UCL's approach, she became excited about the opportunities that adopting such an approach at the University of Pretoria would bring. On the other hand, she was also nervous about the potential impact of the contextual challenges she and her colleagues faced in South Africa. Chief among these were the excessive workloads faced by many academics and the fact that most had no experience in facilitating the kinds of learning engagements used in UCL's program. The UCL team made Dr. Smith aware of the staff development program at Aalborg University, and she subsequently spent several days in Aalborg speaking with Prof. Aida Guerra, Prof. Anette Kolmos, and others. She was struck by how diligently educational decisions were made at Aalborg University and by the professional development opportunities provided to faculty to support them in developing the fundamental skillset required of PBL facilitators.

This series of engagements completely redirected Dr. Smith's career as she left her role as a senior lecturer in the Department of Mechanical and Aeronautical

Engineering to lead the community engagement and curriculum transformation for the Engineering, built Environment, and Information Technology Faculty.

As Dr. Smith's primary development has been as an aeronautical engineer with very little formal grounding in engineering education research, she relied heavily on her community in engineering education–the South African Society of Engineering Education. The teams at UCL and Aalborg University became meaningful resources of support, encouragement, wisdom, and confidence. In 2020, when the Royal Academy of Engineering call for Transforming System Partnerships opened, Prof. John Mitchell suggested that he and Dr. Smith apply. The funding would offer the chance to make the ideas on integrated curriculum and teacher development she had formed during her engagement with UCL and Aalborg University a reality for engineering educators in South Africa.

The creation of staff development opportunities and an integrated curriculum of real-world problem-based learning for students became her goal over the next few years. She met with various well-known and respected engineering education researchers and practitioners in South Africa and formed a team to apply for the RAE grant, which they received in early 2021. The recognition that came from collaborations with highly respected institutions, such as the RAE grant and ongoing UCL partnership, gave the initiative a national profile, with the engineering deans of all 16 engineering schools in South Africa buying into the development of a contextual framework for integrated engineering and its eventual implementation at their institutions. The national team became a support system through which colleagues established a collective vision, in collaboration with contacts at UCL. The trust Dr. Smith felt from the members of the team boosted her confidence as a leader, despite her colleagues having more engineering education research experience. The team became a community of practice, built on strong personal relationships and a shared vision, something most team members lacked in their institutions. Dr. Smith also established a relationship with the Engineering Council of South Africa as the accreditation body for the programs the team developed, further strengthening her credibility as a leader and change-maker in engineering education.

Seeking to "practice what she preached," Dr. Smith took a position teaching one of the most challenging modules at her university, one characterized by complexities of scale, interdisciplinarity, community engagement, and affective development. She hoped to determine whether her ideas for authentic student development were practicable at scale. 1650 students across 18 degree programs were enrolled in the module, which required them to work collaboratively with community partners, resolving real-world problems. She was acutely aware that a new curriculum model would face concerns about its potential to successfully operate at scale with limited staff and resources. If she could singlehandedly implement the proposed approach in a module of 1650, with all the added challenges of inter- and trans-disciplinary working, she would have experiential knowledge to draw on when concerns were raised.

Dr. Smith recognized that teaching in a new program would require new teaching skills. She drew on her coaching background and community and started working closely with a systemic transformation coach, Mr Jaco Fourie from Curiosity

Campus, to co-create a process for ensuring staff resilience and capacity to navigate change. The challenge was creating an intentional engagement pathway for staff that encouraged them to reflect on their classroom practice and explore the tension between where they were and where they wished to be. Multiple pilot sessions were run to refine the process and ensure it is adequately and meaningfully contextualized. Finally, July 2023 saw the first national in-person workshop, starting with personal transformation and using the same tools in a series of workshops that followed to translate these skills to the classroom. The teachers' responses were positive, and 3 years of funding was granted from the University Capacity Development Programme (UCDP) to develop a personal and pedagogical development program for engineering educators across South Africa.

At present, the development and implementation of an integrated curriculum are being piloted at the program level at the University of Pretoria. Dr. Smith, Mr Fourie and Prof Karin Wolff, who has been at the forefront of multiple national development programmes and involved with a variety of different Engineering Curriculum projects, facilitate the process of curriculum redesign where both the curriculum and the faculty members' teaching skills are developed simultaneously.

With the engineering profession facing constant technological advancements that not only disrupt educational processes but also challenge engineers' resilience in response to the constant demands to adapt, faculty's personal development must be a fundamental focus. It is no longer possible to rely on content knowledge and ignore the pedagogical development of engineering academics. The more engineering students must grapple with constant change, the more important the development of their teachers' skills and sense of inner purpose will become.

Parker Palmer said, in his book "A Courage to Teach" which was published in 1997, ***"We teach who we are"***. Who we are and who we allow ourselves to become will be fundamental to the academic development of our field. Our ability to connect with others and our students, to take them on a transformative journey to becoming professionals, is not something technology can facilitate. So, our ability to facilitate learning and truly understand the level of vulnerability required from us as educators needs to be prioritized, supported, and protected by educational leaders.

2.3 Case 3: Engaging in Pedagogical Development Through Leadership Growth and Institutional Vision

Every educator understands the need for continuous improvement in teaching and learning. One reason this need exists is the constantly changing nature of contemporary society, which makes education a dynamic field. There is a constant demand for new educational research, new teaching technologies, and new pedagogies to create more relevant and inclusive educational experiences. Hence, educators must stay well informed through personal reflection, workshops, seminars, collaborative discussions, etc. Meanwhile, continuous improvement allows educators to refine and enhance their teaching skills, thereby improving students' engagement and learning outcomes. Professional development activities also give educators opportunities to collaborate with colleagues and enhance their job satisfaction and supporting their career advancement.

Dr. Khalid Naji's personal journey in engineering education began with a desire to address increasing student demands for new teaching pedagogies. One impact of the current wave of technological advancement, termed the 4th Industrial Revolution (4IR), has been heightened awareness of the difficulty conventional teacher-centered lecturing has in keeping students' attention and concentration. Twenty-first-century students are accustomed to fast-paced information delivery and easily lose motivation if a lecture is slow and unengaging. This dynamic was a wake-up call for Dr. Naji to undertake a professional journey to improve his scholarship in teaching and learning to benefit both himself and his students.

Dr. Naji believes that educators should continually–

(a) Reflect on their teaching practices.
(b) Cultivate a growth mindset by being open to new teaching strategies.
(c) Stay informed and updated through professional development activities or reading professional journals related to education.
(d) Establish clear expectations and routines in the teaching and learning environment to encourage all parties' responsibility and accountability; and
(e) Build strong relationships with students to foster a sense of community within the classroom and encourage collaboration and mutual respect among students.

With these values in mind, Dr. Naji worked to improve his teaching in several ways. After attending PBL workshops, he decided to implement PBL in his courses, despite having limited preparation time. He converted one of his technical elective courses to a PBL model with the help of subject matter experts in engineering education. He conducted weekly comprehensive assessments in this course to ensure the new model was working effectively and that any gaps or challenges in implementation were addressed. An example of such a gap is the rigidity of summative assessment policies at Qatar University, which required Dr. Naji to fine-tune his assessment approach to accommodate the requirements of both PBL and institutional policies. He also had to update resources for students, such as textbooks, technical videos, and teaching material related to the course. One of the most effective steps in the PBL implementation process was an initiative to bring industry mentors to the university to deliver specific technical material in PBL sessions. To manage students' expectations, Dr. Naji and his colleagues provided a weekly course-related PBL package for additional guidance.

These efforts helped Dr. Naji better understand his students and support them to enhance their technical capabilities. They also increased his academic confidence and supported his professional growth. Thanks to this experience of direct involvement in educational change, he is now better equipped to adapt to future changes in curriculum, assessment methods, and educational policies. He also found that his personal effectiveness as a teacher increased, making him more adept at engaging students, fostering a positive learning environment, and improving student performance. Through these experiences, Dr. Naji came to see proactive pedagogical development as an essential component of career enhancement for educators that not only contributes to personal and professional growth but also enhances the quality of education provided to students.

All educators in their academic careers face challenges that vary, depending on their professional context. Generally, these challenges are–

(a) Limited resources, e.g., budget, outdated material, and inadequate technology.
(b) Managing diverse learning styles, abilities, and backgrounds in the classroom.
(c) Large class sizes that do not support individualized feedback/attention.
(d) Policies that stress limited assessment types.
(e) Lack of professional development opportunities.
(f) Rigid and outdated policies and procedures.
(g) A lack of effective administrative support; and
(h) Student motivation.

Amidst such challenges, Dr. Naji feels that academic culture needs to encourage both pedagogical development and institutional improvement. Growing universities such as Qatar University should encourage research and development in engineering education by considering these areas of work favorably in academic promotion policies. Accreditation procedures should be revised with research-based requirements that do not create unnecessary rigidity in teaching and assessment practices.

Large-scale change requires a shared vision for the future. To that end, Dr. Naji has worked to help his colleagues see how changes in teaching and learning environments can contribute to the broader goals of the college and institution. On a personal level, he influenced many of his colleagues by supplying guides and lectures on pedagogical research and development. He also shared a PBL framework developed through pedagogical research activity at the College of Engineering. Dr. Naji highlights that starting small with new educational initiatives and building momentum over time allows educators to adjust their approaches gradually and build their confidence through a supportive culture. To help motivate educators, the College of Engineering at Qatar University introduced a Teaching Excellence Award, which includes seed funding for educational research activities. Dr. Naji also notes the importance of leading by example. He made changes in his own teaching practices and found that others were encouraged by his successes to follow suit. He hopes to foster a community of practice in which educators can share their teaching and learning experiences.

Academic administrators always face challenges when it comes to pedagogical development. One of the biggest obstacles is resistance to change from educators, often based on fear of failure or a lack of knowledge. Other challenges include limited time and resources, with educators often already overburdened with teaching loads, administrative tasks, research activities, and professional service to local communities. These challenges can vary depending on educational context, institutional policies, and individual factors. Overcoming them requires that academic administrators set realistic goals and put in the effort to organize or support workshop participation for educators. Informal sessions should also be provided for educators to share experiences and knowledge to create a supportive environment within the academic unit.

Effective pedagogical development requires a supportive environment that encourages creativity, student participation, and effective learning results. The first factor in creating such an environment is institutional support, particularly positive leadership, and policies. Institutions should create a culture that values and promotes pedagogical development and provides sufficient resources and time for these efforts. There must be clear and transparent communication about the expectations from pedagogy development activities, supported by collaboration and networking opportunities through in-house professional development programs. Institutional leaders should support the implementation of evidence-based practices to ensure effective pedagogical development. Finally, incentives and rewards should be put in place to recognize educators for innovation in the scholarship of teaching and learning.

3 Cross-Case Analysis and Collective Reflection

The narratives describe these three participants' journeys of growth as engineering educators. Despite their different backgrounds in terms of engineering disciplines, education systems, and institutional and national culture contexts, they shared several similarities, which we focus on in the following cross-case analysis. Following the complex dynamic systems perspective to understand professional learning, we structure our analysis and reflection by focusing on three interactive dimensions: intrapersonal characteristics, actions and relations, and interactions with environments.

3.1 Intrapersonal Characteristics

All three contributors, despite their educational background as professional engineers, express interest in teaching, care for students, and are motivated to improve teaching and learning in general. They recognize the dynamic nature of education and the need to adapt to new challenges and demands, along with the role of societal and technological change in driving the need for changes to teaching approaches. They link educators' pedagogical development to improvements in student motivation and engagement. The realization that traditional teaching methods may not align with the needs of contemporary students is a common motivator for change. Educators must continuously evolve their teaching methods to meet the changing and complex needs of students.

3.2 *Taking Actions for Self-Improvement and Change*

Building on their prior experiences at each stage of their careers, all three contributors cultivated their professional agencies and established clear goals for their individual teaching advancement, recognizing that a student-centered approach is crucial for effective teaching. They also highlighted that understanding students' needs and motivations and providing a positive and engaging learning environment contribute to better outcomes. In particular, Dr. Naji and Dr. Smith recognized the value of incorporating real-world PBL into engineering curricula. Engaging students with authentic problems in a safe and supportive learning environment enhances their creativity, resourcefulness, and ability to apply theoretical knowledge to practical scenarios. Creating that environment requires fostering positive relationships, encouraging collaboration, and providing resources that enhance the learning experience. The educators advocated for proactive engagement in professional learning activities, such as workshops, seminars, reading literature, mentorship or peer learning, and collaborative discussions, to stay informed and enhance teaching skills. They stressed the importance of educators taking initiative in their professional development rather than passively receiving information-focused training. All three highlighted relationships with students, with personalized engagement with students seen as crucial for effective teaching and learning.

As the contributors developed their experience, competence, efficacy, and professional agency through professional learning, they engaged in interactions with their environments and worked to influence others. The positive student outcomes they achieved, such as improved engagement and motivation, encouraged colleagues to join their journeys of change-making through teaching advancement. They also negotiated with their environments, finding ways to cope with resistance from students, colleagues, and managers, and with evaluation policies that failed to recognize and reward engineering educators for engagement in pedagogical development. Spanning a diverse set of countries and institutions, these case studies highlight common challenges within traditional education systems, including rigid policies, outdated assessment methods, and limited flexibility, hindering effective pedagogical development. All three agreed that the classroom educator should take a leadership role and influence structures from the bottom up, though this was difficult to achieve in practice. When Dr. Naji took on a leadership role as a dean, he was able to empower engineering educators within his institution. Collaboration with international partners and recognition from reputable organizations contribute to the success and impact of educational initiatives. Exchanging ideas and practices on a global scale also enhances pedagogical development quality.

Recognizing the interconnectedness of personal and pedagogical development is crucial. The educators in these studies underwent personal growth alongside their pedagogical development, highlighting the importance of aligning personal values with teaching philosophies and self-improvement with institutional change.

4 Recommendation–a Systems Thinking Approach to Capacity Building in Academic Leadership for All in Engineering Education

Based on the above discussion, we propose a system thinking approach to capacity building in academic leadership for all in engineering education. Recent literature has recognized professional learning as a complex and dynamic system, advocating a systems thinking approach to planning professional development [2–5, 15–17] and identifying characteristics of successful development opportunities–(summarizing from Taylor, [20]):-sustained and intensive, featuring varied activities, both formal and informal, learner-focused and curriculum/subject-relevant; collaborative and relationships-focused; practical, with opportunities for active learning; oriented towards experimentation, inquiry and research; enhanced by coaching, mentoring, co-teaching, and peer observation; promoted through networking and wider professional learning communities; supportive and encouraging of dialogue, thinking, and change; nurtured by recognition and rewarding policies, supportive middle and senior leadership, trust, coherence, and cohesion; and supported by external expertise.

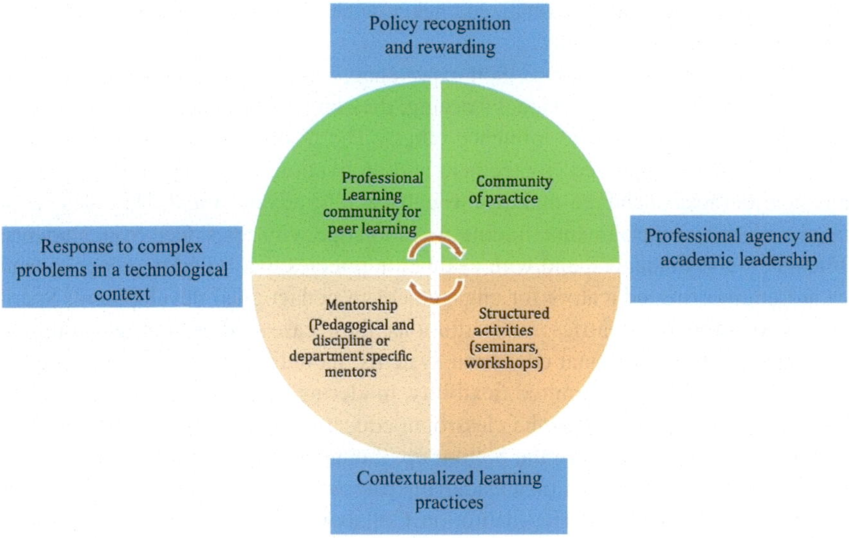

Based on these findings and the common features of our contributors' narratives, we propose a system thinking approach to developing activities to support academic leadership for all engineering educators.

In practice, we recommend:

First, the organization of PD activities should center engineering educators' intrapersonal characteristics, with material based on their prior experiences and personal interests. It should incorporate activities that motivate them to engage in pedagogical knowledge and skills development to enhance their self-efficacy as teachers.

Second, engineering educators should engage in proactive and agentic decision-making and meaningful approaches to implement change. Teachers should be empowered to self-organize, establish goals and plans, and engage in self-monitoring, process evaluation, reflection, and revision.

They should be supported to make sense of their learning by interacting with students, peers, and others. Supportive activities should include not only long-term structured initiatives like seminars and workshops, but also the establishment of communities of practice for peer learning and mentorship.

Last but not the least, when engineering educators gain relevant experience, they should have space to influence others from the classroom to whole programs and institutions.

Policies that recognize and reward teaching can be powerful in supporting engineering educators to engage sustainably in professional development leadership. In general, engineering educators' professional learning and academic leadership are developed through practice and the enactment of professional agency.

References

1. Vähäsantanen K (2015) Professional agency in the stream of change: understanding educational change and teachers' professional identities. Teach Teacher Educ 47:1–12
2. Chaaban Y, Al-Thani H, Du X (2023) A systems-thinking approach to evaluating a university professional development programme. Prof Dev Educ 1–20. https://doi.org/10.1080/19415257.2023.2193199
3. Chaaban Y, Qadhi S, Al-Thani H, Floyd A, Du XY (2023) Supports and constraints to middle leadership development in higher education: a q-methodology study. Educ Manag Administ & Lead 1–22. https://doi.org/10.1177/17411432231174092
4. Chalmers D, Gardiner D (2015) An evaluation framework for identifying the effectiveness and impact of academic teacher development programmes. Stud Educ Eval 46:81–91. https://doi.org/10.1016/j.stueduc.2015.02.002
5. Du XY, Guerra A, Chen JB, Lindsay E, Nørgaard B (2023) Exploring academic middle leaders' viewpoints on supporting change in higher education institutions using Q methodology–a case in Poland. Educ Manag Adm & Lead 1–26. https://doi.org/10.1177/17411432231206247
6. Du XY, Lundberg A (2021) Examining emic viewpoints on pedagogical development program's long-term effects. Stud Educ Eval 71:101088. https://doi.org/10.1016/j.stueduc.2021.101088
7. Du XY, Kolmos A, Ahmed MAH, Spliid C, Lyngdorf N, Ruan YJ (2020) Impact of a PBL-based professional learning program in Denmark on the development of the beliefs and practices of Chinese STEM university teachers. Int J Eng Educ 36(3):940–954
8. Du XY, Spliid C, Kolmos A, Lyngdorf N, Ruan YJ (2020) Chinese engineering instructors' development of critical reflection for transformative learning in a PBL based professional learning program in Denmark. Int J Eng Educ 36(40):1356–1371
9. Du XY, Naji KE, Ebead U, Ma JP (2021) Engineering instructors' professional agency development and identity renegotiation through engaging in pedagogical change toward PBL. Eur J Eng Educ 1–23. https://doi.org/10.1080/03043797.2020.1832444
10. Strom KJ, Viesca KM (2020) Towards a complex framework of teacher learning-practice. Prof Dev Educ 47(2–3):209–224. https://doi.org/10.1080/19415257.2020.1827449
11. Du XY, Chaaban Y, Al-Thani H, Lundberg A (2023) University teachers' professional learning towards becoming pedagogical change agents–a Q methodology study. Int J Acad Dev 1–18. https://doi.org/10.1080/1360144X.2023.2265891

12. Chaaban Y, Al-Thani H, Du X (2023) University teachers' professional agency for learning and leading sustainable change. Prof Dev Educ 1–17. https://doi.org/10.1080/19415257.2023.2229338
13. Garner JK, Kaplan A (2021) A complex dynamic systems approach to the design and evaluation of teacher professional development. Prof Dev Educ 47(2–3):289–314. https://doi.org/10.1080/19415257.2021.1879231
14. Lyngdorf NE, Ruan YJ, Chen JB, Du XY, Kolmos A (2022) University teachers' simplexity in professional learning and role identity change in an intercultural (study-abroad) study programme. Int J Eng Educ
15. Guerra A, Nielsen H, Chen JB, Du XY, Kørnøv L (2024) Teacher professional agency for education for sustainable development–a case of an Erasmus program collaborating with Eastern Asia using Q methodology. Int J Sustain Higher Educ 156–176. https://www.emerald.com/insight/1467-6370.htm
16. Du XY, Guerra A, Nørgaard B, Chaaban Y, Lundberg A, Lyngdorf NE (2022) University teachers' change readiness to implement education for sustainable development through participation in a PBL-based PD program. Sustain 14:12079 (2022). https://doi.org/10.3390/su141912079
17. Nørgaard B, Chen JB, Smink C, Guerra A, Du XY (2023) University educators' professional learning in a PBL oriented program in Africa–Q study. Eur J Eng Educ 1–21. https://doi.org/10.1080/03043797.2023.2250738
18. Chen J, Du X, Jiang D, de Carvalho Guerra AOP, Nørgaard B (2023) A systematic review of staff pedagogical development in engineering education. Europ J Eng Educ 49(2):299–329. https://doi.org/10.1080/03043797.2023.2290032
19. Saroyan A, Trigwell K (2015) Higher education teachers' professional learning: process and outcome. Stud Educ Eval 46:92–101. https://doi.org/10.1016/j.stueduc.2015.03.008
20. Taylor P (2023) The complexity of teacher professional growth–unravelling threads of purpose, opportunity and response. Professional Dev Edu 49(1):16–29
21. Mulia WD, Fritz DJ, Sohoni SA, Kearney K, Mwavita M (2013) PLP: a community driven open source platform for computer engineering education. Int J Eng Educ 29(1):215–229
22. Sohoni S, Mar C, Craig SD (2016) Comparing cooperative learning in online and in-person versions of a microprocessors course. In: Proceedings of the American society for engineering education Pacific Southwest section conference
23. Sohoni S, Craig SD, Vedula K (2017) A blueprint for an ecosystem for supporting high quality education for engineering. J Eng Educ Transform 30(4):58–66

Building a Campus Ecosystem to Foster Holistic Learning and Development

Poornima Panduranga Kundapur and Kip Kussman

Abstract In the evolving landscape of higher education across the globe, holistic learning emerges as a crucial paradigm for shaping transformative learning environment for engineering students. The challenge for institutions is to respond dynamically to technological advancements, embracing interdisciplinary learning and addressing the need for inclusivity, diversity, and those of the employment industries. With a focus on holistic development, the vision for engineers extends beyond technology to encompass academic leadership and global citizenship skills. The way forward is adapting the 21st-century skills framework encompassing learning and innovation, digital literacy, and life and career skills. In India, engineering colleges play a pivotal role in shaping the nation's workforce. Institutes are working on maintaining quality and are aligning with industry needs to provide opportunities for innovation and reform. Top-tier engineering colleges aspire not only to academic excellence but also to cultivate well-rounded individuals. Faculty members, as educators and architects, contribute to holistic educational experiences, emphasizing the interconnected growth of students and faculty. Engineering colleges are adapting holistic learning approaches to nurture individuals intellectually, emotionally, socially, and physically, creating tailored experiences for holistic student and faculty development. The chapter includes two case studies, one from Manipal, India and the other from Milwaukee, USA.

Keywords Holistic development · Experiential learning · Community building · Student leadership · Success coach

P. P. Kundapur (✉)
Manipal Institute of Technology, Manipal, Karnataka 576104, India
e-mail: poornima.girish@manipal.edu

K. Kussman
Milwaukee School of Engineering, Wisconsin 53202, USA

R. Kandakatla et al. (eds.), *Academic Leadership in Engineering Education*, Lecture Notes in Networks and Systems 1097, https://doi.org/10.1007/978-3-031-68282-7_17

1 Introduction

1.1 Engineering Education

The global engineering education scenario is characterized by a dynamic response to technological advancements, a shift toward interdisciplinary learning, and efforts to address challenges related to inclusivity, diversity, and the evolving needs of industry. Innovations in teaching methodologies and a focus on holistic development are shaping the next generation of engineers to meet the complexities of a rapidly changing world.

The vision, therefore, for engineers emerging out of a campus that promotes holistic learning and development, is to transcend the development and application of technology and enter the realm of academic leadership with necessary skills of a well-rounded global citizen in a sustainable and equitable global society [1, 2].

Peter Drucker, management consultant and author said,–

> The only skill that will be important in the 21st century is the skill of learning new skills. Everything else will become obsolete over time.

1.2 Twenty First Century Skills

India continues to be a global hub for technology and innovation, the role of engineering colleges, both government-funded and private, remains pivotal in shaping the future of the nation's technical workforce. These evolving global conditions require future engineers to prepare and equip themselves with twenty-first century skills. These skills are divided into three categorized in Bernie Trilling and Charles Fadel's book "21st Century Skills: Learning for Life in Our Times". Trilling and Fadel, 2012 mention these skills to include:

1. **Learning and innovation skills**: Critical thinking and problem solving, Communications and collaboration, and Creativity and innovation.
2. **Digital literacy skills**: Information literacy, Media literacy, and Information and communication technologies (ICT) literacy, and
3. **Life and Career skills**: Flexibility and adaptability, Initiative and self-direction, Social and cross-cultural interaction, Productivity and accountability, and Leadership and responsibility (Fig. 1).

Along with these twenty-first century skills, engineering education, not only in India but across the globe, has its share of other challenges such as maintaining quality, aligning with industry needs, and addressing issues of affordability and accessibility.

Zooming into a closer world view, these challenges also present opportunities for innovation, reform, and the continued evolution of engineering education in the country. There is a heightened sense of obligation in top tier engineering colleges to

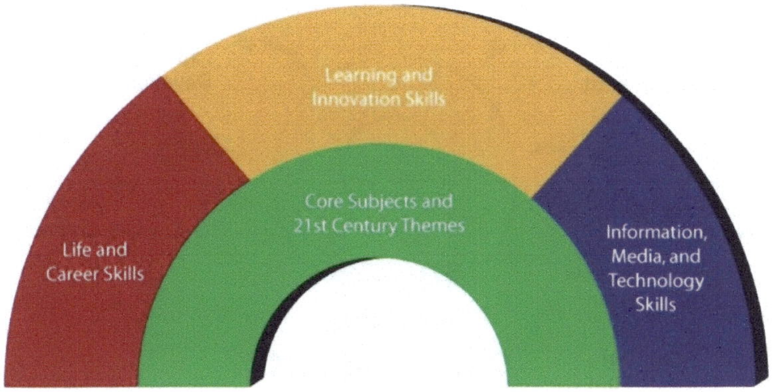

Fig. 1 Twenty-first century skills framework (*Source* https://www.oecd.org/site/educeri21st/40756908.pdf)

offer a diverse range of specializations and innovative teaching methods with a shared aspiration of fostering comprehensive *all-round growth* and learning experiences along with academic achievements of their engineering graduates.

Given the expansive canvas of engineering colleges globally, the pursuit of academic brilliance is intricately interwoven with a larger vision, one that extends beyond the confines of textbooks and classrooms. It is a vision that aspires to cultivate not just proficient engineers, but well-rounded individuals equipped with a spectrum of skills, resilience, and a deep sense of awareness of working for a community. Here, *faculty members are not only educators but architects* of a transformative campus ecosystem. A faculty's professional growth and development increases the three dimensions of a teacher–facilitation of learning, education, and service to all stakeholders [3]. Many empirical studies emphasize the pivotal role faculty play in shaping a holistic educational experience [3, 4].

Engineering colleges are adapting holistic learning and development to encompass a broad and interconnected approach to education, aiming to nurture individuals intellectually, emotionally, socially, and physically. These aspects are interdependent and contribute collectively to the overall growth and well-being of both students and faculty through collaboration and active engagement. These institutes are working towards a seamless and tailormade experience of holistically connecting students to the support mechanisms needed to reach their scholastic aspirations and professional career goals (Fig. 2).

To enrich the content and provide practical insights into the implementation of holistic learning and development at engineering colleges, this chapter includes two compelling case studies from different locations. The first one is situated in Manipal, India, and the other from Wisconsin, USA. These case studies not only provide valuable insights into the implementation of holistic learning and development but also offer diverse cultural and institutional contexts, enriching the discourse on engineering education.

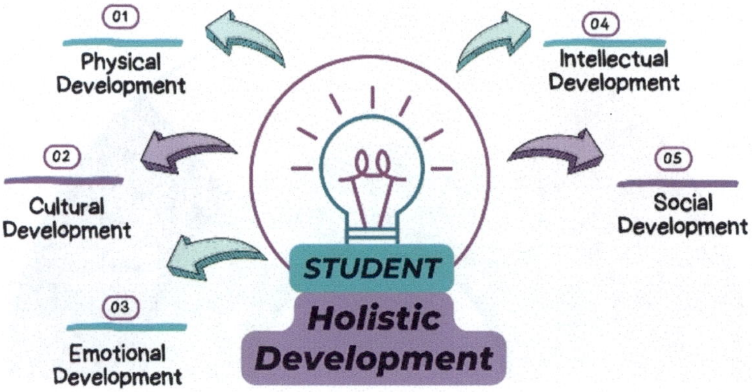

Fig. 2 Author's representation of holistic development of a student

2 Case Study 1: Nurturing Holistic Excellence at Manipal Institute of Technology, Manipal, Karnataka, India

Manipal Institute of Technology (*MIT*), Manipal is the largest constituent unit of India's top ranked private university, Manipal Academy of Higher Education (MAHE). MAHE was declared as an Institute of Eminence Deemed to be University (IoE) by the Government of India in the year 2020. Dr. T M A Pai was the visionary leader and the founder of the MAHE Manipal, whose commitment to the cause of quality education and social responsibility led to the establishment of world class institutions, health care facilities and banking institutions in this part of the country as early as 1925. These institutions transformed the lives of millions of people in and around Udupi district in the state of Karnataka and served as a model for many other institutions in this region.

Established in 1957, MIT Manipal has been a pioneering institution in engineering education and is one of the first self-financed engineering colleges in India. The campus is spread over an expanse of 313 acres. Long time ago, this land was a barren plateau of hard, laterite rock in the Southern Indian state of Karnataka's Udupi district. Today, MIT Manipal accommodates 18 academic departments, awarding undergraduate, graduate and post graduate degrees in this very land. It is imperative to share that it is not only a hub of excellence for engineering students but is proving livelihood to its adjoining community that provides the establishment with the necessary support system.

At the heart of MIT Manipal's educational philosophy lies the understanding that student success extends far beyond academic prowess. It encapsulates the holistic growth of an individual nurturing not only intellect but also character, leadership skills, and a profound sense of community. MIT has a pan student India representation as well as from non-resident Indians or the NRIs.

This section explores how MIT Manipal's campus ecosystem fosters holistic development among its students pursuing various disciplines of engineering. In the capacity of an Associate Director, Student Welfare, the author's[1] experiences have been intricately woven with the tapestry of student experiences. With over fifty technical and cultural clubs that pulse with creativity, to the dynamic student council co-steering the ship of governance, we explore the avenues that pave the way at MIT Manipal, for a comprehensive education; one that prepares students not just for careers but for life.

The unraveling of the strategies, initiatives, and collaborative endeavors that shape the MIT, Manipal experience, stand as a testament to the belief that rules and guidelines are not mere constraints but enablers of growth. As we navigate the realms of experiential learning, community building, and student leadership, we illuminate the path towards a campus ecosystem where every facet of a student's potential is acknowledged, nurtured, and allowed to flourish.

Building Twenty First Century Skills: **MIT Manipal's approach**. India continues to be a global hub for technology and innovation. The role of engineering colleges, both government-funded and private, remains pivotal in shaping the future of the nation's technical workforce. These evolving global conditions require future engineers to prepare and equip themselves with twenty-first century skills. These skills are divided into three categorized in Bernie Trilling (Figs. 3, 4 and 5).

Foundations of a Holistic Campus Ecosystem at MIT Manipal

Operationalizing Synergy between teachers, students, and peers. Any institute that is intent on enabling holistic student development requires a focused transformation of interaction between primary stakeholders, teachers, and students.

A few mechanisms in place here at MIT, Manipal:

1. *Teacher Guardian System*: All first-year engineering students are divided into batches of 15 to 20 students. They are assigned one of their teachers as their ***"Teacher Guardian"*** *or TG*. This system was developed for better coordination and communication with the first-year students since they follow a common

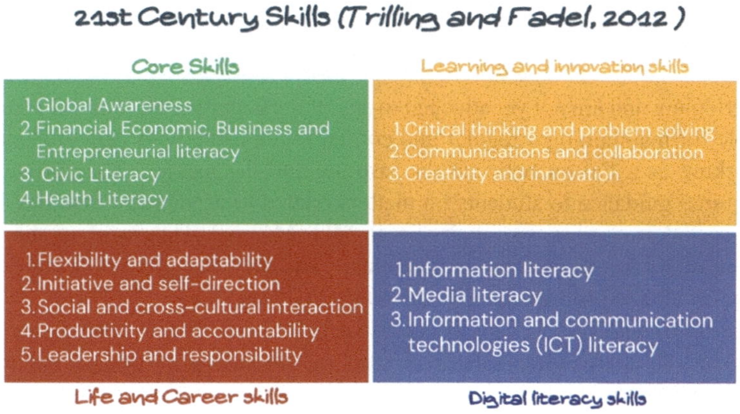

Fig. 3 The twenty first century skills explained [5]

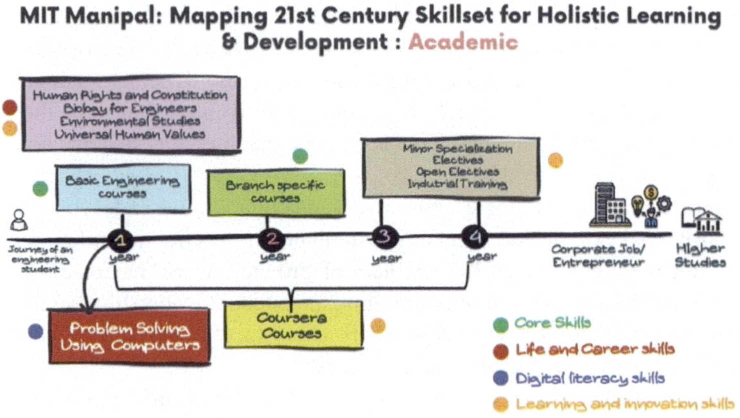

Fig. 4 MIT Manipal (academic development) mapped to twenty-first century skills

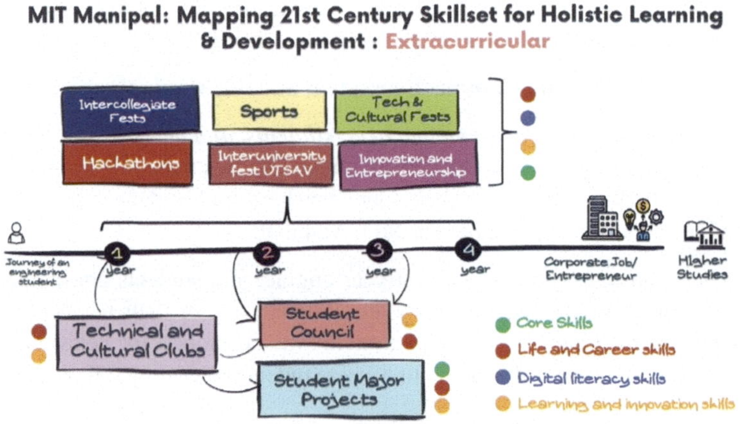

Fig. 5 MIT Manipal (extracurricular development) mapped to twenty-first century skills

curriculum and are not yet attached to any of the departments. The TG is responsible for maintaining a regular and cordial relationship with their students and working as a liaison between the institute and the parents. They provide the required guidance to students on matters related to academics/ institute/ hostel or any other issue to ease their stay in MIT, Manipal. They work as the single point of contact in case of medical issues, and they also advise the students on their academic progression and attendance in classes.

2. *Mentor Mentee System*: This system is very similar to the TG system except that the students are from the higher semesters and affiliated to a particular department of the Institute.
3. *Study Buddy System*: This is an institute initiative that pairs senior students, who are more experienced with newly joined undergraduates for guidance and

valuable advice that would help the juniors navigate the first year at MIT. So, the buddy system usually pairs second or third-year students with new undergraduate students. These "buddies" are volunteers. In essence, the buddy system is the easy transition system where the senior helps a junior decide on joining clubs, addressing challenges of adapting to the new environment and place.

Promoting Intellectual and Cultural Leverage of students: Role of Clubs & Projects. MIT Manipal's diverse array of more than fifty-eight clubs that cater to technical and cultural events and twenty-two student major projects. These serve as vital hubs for student interaction and support. These clubs not only contribute to academic and extracurricular growth but also act as communities where students find a sense of belonging, that help deal with the feeling of isolation. The cultural clubs are varied and includes clubs for music, dance, dramatics, debating and literature, manga anime, film making, art and fashion, social work, and photography, with the latest entrant being the radio club. The technical clubs cater to specific student chapters of IEEE, such as application development, astronomy, women empowerment, open-source platforms, data science and agriculture. Members of these clubs represent MIT at various national level fests, and there they exhibit the same enthusiasm as when MIT conducts its own fests.

Apart from these student clubs, MIT also has about twenty-two student major projects that propelled MIT Manipal into the international arena of competition. These projects include Formula Manipal (formula cars from Manipal), SolarMobile (solar powered cars), Project Manas (AI and Robotics), Mars Rover Manipal (next generation rovers), Robomanipal (Robotics), AeroMIT (Aeromodelling and Aerial Robotics), thrustMIT (rocketry) and many other such exciting clubs. Here, the students design, build, deploy and demonstrate real time engineering skills working beyond academic hours. Some of these clubs have represented MIT at international competitions and won laurels for the institute. ThrustMIT announced recently that their team was selected to compete in the coveted Spaceport America Cup 2024. Out of the 183 teams that applied, this was the first Indian team to compete in the 30 K Commercial Off-The-Shelf (COTS) category. The launch will be in June of 2024 at Spaceport America Cup, New Mexico, USA.

Research indicates that participation in clubs fosters a sense of purpose and community, positively influencing mental health and well-being [6]. By actively engaging with these clubs, students not only enhance their engineering skill sets but also cultivate life skills, teamwork ability, leadership skills, communication skills and also a support network that can prove crucial for their overall well-being.

MIT's Student Council and Leadership. The MIT Student Council [7] is the voice and the face of the MIT student community. It is an elected student body and is a vital component of the institution, playing a crucial role in shaping the student's experience and contributing to the overall campus environment. The responsibilities of a Student Council typically encompass a broad range of activities aiming to enhance student life, foster community engagement, and ensure the well-being of all students. Being the official representative voice of the student body, it acts as a liaison between students and the administration, conveying student perspectives on various issues, policies, and campus matters.

The elected members serve for a period of one year in their given capacities as Senior council members (if they belong to the fifth semester) or as Junior Council members (if they belong to third semester).

One of the roles of the Student Council is to organize and coordinate a variety of events and activities. These can include cultural festivals, academic seminars, sports events, and community service initiatives that contribute to a dynamic campus life.

They conduct the two major fests:

Revels, the cultural and sports fest and

Tech Tatva, the technical fest, which transform MIT's physical and cultural landscape with students working as organizers, participants, and volunteers. It is an experience of a lifetime for many of them.

The fests are the best examples of promoting a sense of community and camaraderie among students by encouraging interaction, collaboration, and inclusivity among students from diverse backgrounds. The Student Council also manages funds allocated for student activities. This responsibility involves transparent financial management, ensuring that funds are utilized effectively for the benefit of the student community. They are also responsible for identifying and pursuing sponsorship opportunities. This exercise builds marketing and negotiation skills in a student.

The Student Council acts as a platform for students to express their concerns and grievances. Council members work to address these issues by collaborating with the administration and facilitating open communication channels. A structured team meeting is the forum for these discussions where the management of MIT meet on a common date and time with the council members to address their concerns. Such meetings establish collaborative decision-making practices which involve inclusion of students in the actual decision-making processes related to institutional policies. During these meetings, the members of the student council advocate for policies that promote the welfare, interests, and rights of students. The council also serves as a platform for students to voice their academic and personal grievances within the institute.

Experiential Learning Initiatives. At MIT, Manipal, we believe that learning extends beyond written words in textbooks. The integration of experiential learning into the curriculum stands as a testament to our commitment to providing students with real-world applications of their academic knowledge (Emily, 2015) [8]. The learning spaces extend beyond the classroom to the library to the mechanical workshop to the student club meeting room, and to the student project discussion rooms. Every minute that a student engages in observing and communicating, learning becomes an integral part of the experiential learning at MIT Manipal. The numerous collaborative workshops, hackathons, or interdisciplinary projects create spaces for students to engage in experiential learning within a supportive community.

From the faculty's perspective, the integration of experiential learning within engineering curricula has been heralded as a transformative pedagogical approach

[9]. Faculty members have come to recognize that theoretical knowledge, while foundational, finds its true resonance when students can apply it to real-world challenges. This creates a paradigm shift in delivery mechanisms, with project-based learning [10] or case-based learning, where case-based learning being the preferred pedagogical modes. This transformative journey aligns with Dewey's philosophy of **'learning by doing,'** thereby fostering a dynamic educational environment (Dewey, 1938).

***Sports and Recreation*: *Building Team Spirit and Discipline*.** MIT Manipal has a staff sports advisor who is part of a larger University (MAHE) council. The University Sports Council organizes inter-college sports events and makes selections for the university team for various inter-university tournaments. There is an in-house MIT Campus Sports Facility as well. A student of MIT can freely avail these facilities which include indoor badminton courts, boxing ring, gymnasiums for both men and women, state of the art basketball courts, swimming pool, table tennis and lawn tennis courts, football, and hockey grounds. Several inter collegiate tournaments for men and women are organized throughout the year in athletics, basketball, badminton, best Physique, chess, cricket, cross country race, football, hockey, swimming, squash, table tennis, lawn tennis, throw ball and volleyball. The best players from each of these events are sent to Inter University tournaments, state level or national level events.

***Innovation Center & Entrepreneurship*: *Building Leadership*.** MIT Manipal has an inclusive "Innovation Center" in one of the academic blocks, which is the place where ideas come to life, inspiring students to build a culture of creativity and innovation that addresses larger community problems. The league of innovations includes Idea café where inter-institute networking is encouraged through ideation, MakerSpace with well-functioning Fablab Makerspace on the Technology campus and Co-Working that provide collaborative working spaces. These foster interpersonal and interprofessional collaborations in the spirit of nurturing overall development of students.

Entrepreneurship is amalgamated in MIT Manipal through E-Cell MIT (also a student body under the student council) which was established with a vision to empower entrepreneurship by promoting a vibrant startup ecosystem and facilitating budding students to develop their ideas. The mission of the E-Cell is to act as a one-stop destination for all students looking to convert their ideas into viable start-ups. This in turn inculcates the spirit of entrepreneurship among young minds through two major events called Manipal Entrepreneurship Summit (MES) held in February every year. It is heartening to see a good number of start-ups mushrooming under MIT's futuristic outlook. The prominent of these start-ups being Blackfrog Technologies that developed cold chain technologies for the transport and accountability of vaccines, serums, COVID-19 specimens, and other biologicals. Another one being Impact, a Fintech Startup that specializes in customer profiling and acquisition solutions for enterprises. These are just examples of the leaps MIT Manipal has taken in terms of moving ahead in harnessing student ideas and working on fostering an ecosystem where holistic learning prevails in the environment.

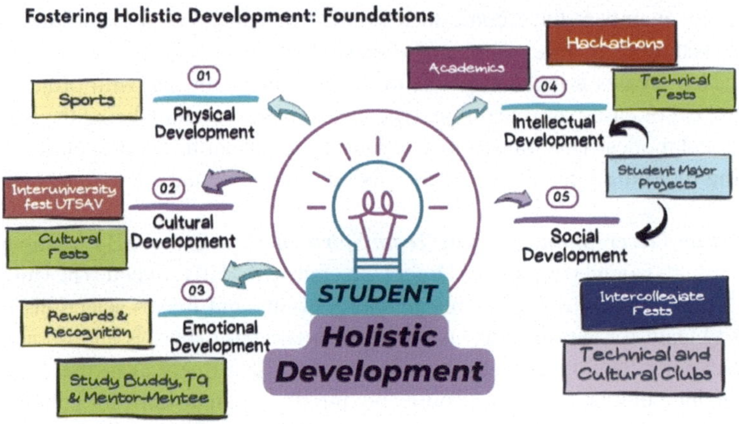

Fig. 6 Holistic development: foundations

MIT has been in the forefront of establishing industry-academic connect with the creation of Institution's Innovation Council (IIC) program that brings forward departmental activities leading to or driven as innovation and entrepreneurial activities, as directed by the Ministry of Education, Government of India.

Rewards & Recognition*: *Acknowledging Accomplishments. MIT Manipal believes in recognizing student achievements and honoring them with Certificates of recognition. The forum for saluting these student achievers is commemorated in a special event called the "Tea with the Director". The Director of MIT personally interacts with and congratulates the students on this august forum. These achievements are not academic in nature but are extra-curricular activities such as participation in national and international competitions or events, art, music, drama, theatre, or NCC cadets, NSS volunteers or sports. Such recognition boosts the confidence of students who perform better and excel in their chosen field of excellence–both academically and extracurricular activities (Fig. 6).

The Feedback mechanism*: *Rework on the existing plan, do not re-plan, saves time. MIT uses feedback as a powerful tool taken from our primary stakeholders' which are the students, faculty, and industry partners. They together partner to design a curriculum that fosters holistic development. Mechanisms is in place to collect feedback on experiential learning initiatives through the Student Life Cycle Management (SLCM) system that contributes to the iterative improvement of our on-going programs.

MIT Manipal's understanding of receiving **feedback** has been instrumental in backing the academic growth of our students. Beyond grades, constructive feedback provides students with insights into their academic strengths and areas that require improvement. This targeted guidance enables them to refine their learning strategies, encouraging a deeper understanding of engineering principles. Therefore, a feedback mechanism from peers, faculty/mentor serving as a mirror reflecting areas for improvement in these essential interpersonal skill sets, is a part of the institution's policy for continuous improvement.

Engineering projects often involve collaboration and leadership. Feedback on group projects is provided by mentors and faculty to attain valuable insights into how students navigate team dynamics, contribute to group goals, and exercise leadership skills. This is the hallmark of the overall development arch of an engineering graduate.

Feedback, when delivered with empathy and encouragement, contributes to the mental and emotional well-being of students. Recognizing their efforts and providing constructive criticism in a supportive manner fosters a positive learning environment. Faculty and mentors at MIT Manipal are encouraged to be empathetic towards students. To this end, faculty are sensitized through various programs organized by the university.

Feedback from industry experts or alumni can bridge the gap between academic learning and real-world expectations. Understanding industry requirements allows engineering institutes to align their curriculum and experiences with the demands of the professional landscape. This is facilitated by including Industry personnel in our Board of Studies and Department Curriculum Committees.

Feedback is not merely an evaluative tool but a catalyst for holistic development in engineering institutes and MIT Manipal is aware of the significance of this tool.

3 Case Study 2: Student Support: A New Vision for the Work at Milwaukee School of Engineering, Wisconsin, USA

The emergence of engineering and technological knowledge as subjects of advanced study occurred during the eighteenth century [11]. Previously, this knowledge was largely attained through apprenticeship work in various technical settings and was still largely the conventional model. The formalization of engineering curricula was inspired through the French military. Faced with a shortage of military engineers, the monarchy established engineering institutions rooted in military instruction, ultimately leading to the influence of American adaptation of the model. The French model, the polytechnic, first came into existence in the United States with the founding of West Point, that treated engineering education separate from the rest of the university. Changes by the Morrill Act of 1862 led to the introduction of engineering and technical education in formal higher education along with established disciplines.

Moving forward to 1903 and the establishment of the Milwaukee School of Engineering, (MSOE), Oscar Werwath, MSOE's founder, and president, saw the need for a blended approach:—one that converged the historical apprenticeship model featuring hands-on, applications-based technical education but strongly supported by academic instruction. The purposive history of the technical institution had an explicit, albeit crucial job—to train individuals to solve technical problems and support industry.

But in the modern world of higher education, this vision and scope has become broader. Institutions of higher education are called, and at some level, expected to develop the whole person, not only a trained, technical employee.

MSOE found itself presented with new opportunities as its fifth president, Dr. John Walz, began his tenure as president in 2016. Operations based on a history of centralized, effective leadership during past presidencies provided the stability and the opportunity for advancement beyond the boundaries of establishing MSOE as a major contender of technical education and into the business of curating an institution that found itself equally qualified in caring for and developing the whole student, without losing the strong history on which it was founded. In short, MSOE learned it could be just as impactful at teaching the student outside of the classroom and laboratory as it could be inside.

In concert with this newfound purpose is the recognition of a demographic shift in the collegiate student. With an increased need for care and support of their mental health to supporting students in the aftermath of COVID, colleges are faced with the challenge of taking care of their students in a fundamentally different way.

This case study aims to outline and describe the reformation of support services for students to meet their evolving needs. It is fair to recognize and presume that engineers have been taught in a very particular way for decades. And given demographic shifts and the changing makeup of the college student, conceptualizing a new framework for the delivery of that education seems overdue. The university's responsibility in developing the whole student is not a new idea but recognizing the need for the change in operations of traditional offices, as well as the importance of involvement of all campus constituents, requires a significant effort. At MSOE, we have begun this work.

More than a Data Point: Retention as a Medium for Student Satisfaction. Like all colleges and universities, MSOE places great value on achieving the highest possible first-year retention. There are many benefits of high retention rates, not the least of which is the health and success of the first-year cohort, which paves a smooth road toward persistence and graduation. With a goal of 90% retention, MSOE realized a need for creativity when it came to student support. It was clear that we could not abandon traditional methods entirely, such as academic advising, counseling services, and tutoring. But it was just as clear that we would not reach this goal without change. It had become evident that there were needs students had that could not be met through the traditional service of academic advising. The said need in turn was also placing undue burden on the mental health and counseling services of the institution. The lack of initiative to address these needs had direct impact on the satisfaction, or lack of satisfaction, students were experiencing. Thus, a new role of student support professionals was created:–**the student Success Coach**.

Designed to be an academic support rooted in student development theory, the Success Coach can serve as a filter for counseling with respect to the mental health and emotional needs of the developing college student that do not require a licensed provider. The Success Coach additionally serves as a college mentor, whose role extends beyond approving class registration and answering general questions about career readiness and internships. The role of the Success Coach is vast. According to the MSOE website, a Raider Success Coach is a collaborative relationship between a professional staff member from the MSOE community and an incoming student, to support their transition to the university, recognize and maximize their potential, connect them to MSOE resources, and empower personal accountability in their learning and development. Your coach will work one-on-one with you to achieve your personal and academic goals, such as: Time management, Procrastination/ Motivation, Test preparation, Note-taking and Reading strategies, Study Strategies, Communication with Faculty and Staff and Personal WellTH (well-being & health).

Students have access to their Success Coaches on need-basis. Each freshman student is assigned a Success Coach to follow them through the first two years of their college transition and then are called on as needed. While the role of the Success Coach places a large focus on academic logistics (time management, study skills etc.), the coaching staff uses a relationship-based approach to build a rapport that curates trust within the student. Once trust is built, the student moves from perceiving an employee giving them the same advice their parents have probably been giving them, to a new person, outside of the parental role, in whom they can confide.

The Office of the Dean of Students: Not Just a Place for the Misbehaved. There is a common misconception that the Dean of Students is a role that exists in higher education for no other reason than to enforce rules and hold students accountable when they inevitably break them. While it is true that the dean is traditionally the chief conduct officer, reducing their role to a mere disciplinarian is a poor description of its actual function. At MSOE, the Dean of Students office diligently and intentionally works to ensure the student body, as well as faculty, staff, and parents, understand the office's role to be a service provider rather than merely an enforcer of policy and rule. This allows for several productive initiatives to be realized. It gives a way for the dean's office to function as a central office for concerns from all campus constituents, which also facilitates a centralized place for data on student issues and concerns that inevitably arise. It helps to support data-driven decision making about how to best support students through programs and interventions. Presenting the dean's office as a service provider also encourages students and parents to reach out sooner than they may have done in the past. It often culminates in students connecting to services and intervention sooner, thus supporting the retention efforts of the institution, which in turn aid in positive outcomes for student satisfaction. The Dean of Students office also aims to support faculty and staff in their work of supporting students. In a student-centered approach to caring, it helps to have a student-centered office to assist the community with their work, which fosters an ecosystem of learning and support.

A New Approach to College Mental Health Care. The mission of mental health care and counseling services within the university has remained mostly unchanged since its inception at Princeton University in 1910.

In the modern-day university, counseling centers function mostly to support students in their emotional and psychosocial development, with the primary goal of removing barriers to academic success.

There is a variety of services offered with counseling centers, which is largely influenced by size of the campus, as well as access to resources and budget. But the scope and purpose should reflect the growing needs and changes of students.

At MSOE, we have taken an updated approach to student mental health care. The growing demand for care cannot be met with the current resources most colleges have, and many schools may lack the ability to add to their operating budgets an expensive, in-demand professionals who are qualified to provide therapy. Rather than add more staff to support the need, we have begun to engage employees who are already working with students. It is worth noting that only licensed mental health professionals on campus are authorized to conduct therapy, author, and execute treatment plans, and conduct official psychological assessments and diagnostics. However, many students in college who approach with mental health struggles, need not necessarily be referred to, or taken care by a licensed mental health provider. College students face a wide range of acute stressors that fall under the banner of mental health, for example–homesickness, school stress or family problems. These can be appropriately managed by a trusted and willing-to-listen student development professional within the university.

Initiatives to lead training and development within this scope have been met with great eagerness and success as many employees in the student affairs division (and even some professors) are excited about the opportunity to support students outside of their normal professional role and see the true value in being able to contribute to the mental health care of the student body. Training is also given to them to empower and educate them in being alert to signs of significant distress so they can make a referral to a campus mental health provider. This approach to mental health care in the campus setting contributes to good resource management, but more importantly, brings a wider range of campus professionals to the table for supporting students.

At MSOE, we are preparing students to solve the complex problems of today and tomorrow. But in the finite amount of time, we must do that, we must focus on developing the student as a whole person, rather than focusing exclusively on their skill development as an engineer. STEM institutions in the twenty-first century must not only be in the business of training competent, high-functioning engineers and STEM professionals. Instead, we must be in the business of training competent and high-functioning emerging adults who are capable of being high-functioning engineers and STEM professionals. This is best accomplished through an ecosystem of holistic learning and development.

4 Key Takeaways

Global Engineering Education Trends: The article highlights the dynamic response of engineering education to technological advancements, interdisciplinary learning, and challenges related to inclusivity, diversity, and industry needs.

21st Century Skills: Future engineers need to develop skills such as critical thinking, digital literacy, adaptability, and leadership to thrive in the evolving global landscape.

Challenges and Opportunities: Engineering education faces challenges like maintaining quality and aligning with industry needs, but these challenges also present opportunities for innovation and reform.

A Case Study on Holistic Learning at MIT Manipal: MIT Manipal focuses on holistic development through initiatives like the Teacher Guardian System, Mentor Mentee System, and Study Buddy System to foster academic and personal growth, structure of Student Council and Leadership, role of clubs and student projects, along with sports and recreation practices, and integration of experiential learning into the curriculum through workshops, hackathons, and interdisciplinary projects to bridge theory and practice. The push given to innovation and entrepreneurship at MIT Manipal brings in a futuristic vision. The importance attached to rewards and recognition of student and faculty achievements boosting confidence and motivation. The drive to retrospect through a strong feedback mechanism to continuously improve programs, foster self-awareness, and align with industry needs.

A Case Study on New Vision at MSOE, Wisconsin: A paradigm shift from apprenticeship-based learning paved the way for the establishment of institutions like the Milwaukee School of Engineering (MSOE), blending hands-on technical education with academic instruction. With demographic shifts in the collegiate student body, institutions like MSOE recognize the imperative to support holistic student development, beyond technical training. Through innovative approaches like the Raider Success Coach program and reframing the role of the Dean of Students office, MSOE aims to address evolving student needs and enhance retention efforts. Furthermore, MSOE adopts a novel approach to college mental health care, leveraging existing campus professionals to provide support, thereby expanding resources, and fostering a campus-wide ecosystem of care and development. In navigating the complexities of modern engineering education, institutions like MSOE recognize the importance of nurturing well-rounded individuals equipped to tackle the challenges of the twenty-first century.

5 Future Directions and Continuous Improvement

As education evolves, so does the approach to experiential learning. Engineering institutes across the globe have set into place numerous ingoing measures to adapt to the changing landscape of education, incorporating new technologies, interdisciplinary collaboration, and industry partnerships to enhance the experiential learning curves of engineering students. Apart from the academic rigor, engineering education must also focus on the overall holistic development of students. The vision of future engineers should extend beyond technology and should embrace a versatile skill set of leadership, technical proficiency with adaptability, creativity and effective communication along with global citizenship skills.

In conclusion, the two case studies discussed in this chapter emphasize the importance of continuous improvement by prioritizing welfare and wellbeing of students, fostering active engagement, promoting shared governance, implementing feedback loops as in the case of MIT Manipal. The second case study at MSOE believes that a proactive approach to student support, adapting to evolving needs of students and emphasis on comprehensive support mechanisms would be beneficial to all stakeholders in the long run.

Both institutions, however, strongly highlight the need for building a strong sense of community, recognizing its significance in overall holistic student development.

References

1. Paul M (2023) John Dewey's theory. https://www.structural-learning.com/post/john-deweys-theory. Accessed 10 October 2023
2. Bartle E (2015) Experiential learning: an overview. The University of Queensland, Institute for Teaching and Learning Innovation
3. Zhu X (2014) On the theoretical model construction of teachers' professional development. Educ Res 35(06):81–90
4. Zhu X, Li J (2019) Being a teacher of holistic profession: a perspective of transformative learning. Beijing Int Rev Educ 1:472–487. https://doi.org/10.1163/25902539-00102004
5. Fadel C, Trilling B (2012) Twenty-first century skills and competencies. In: Seel NM (eds) Encyclopedia of the sciences of learning. Springer, Boston, MA. https://doi.org/10.1007/978-1-4419-1428-6_763
6. Buckley P, Lee P (2021) The impact of extra-curricular activity on the student experience. Act Learn High Educ 22(1):37–48. https://doi.org/10.1177/1469787418808988. Accessed 10 October 2023
7. Manipal Institute of Technology (2023) Student Welfare. https://manipal.edu/mit/mit-experience/student-welfare.html. Accessed 23 September 2023
8. Solar Mobil (2023) https://www.solarmobilmanipal.org/
9. Kolb D (2015) Experiential learning: experience as the source of learning and development. 2nd Edn
10. Andrés Díaz L (2022) Engineering education 5.0: strategies for a successful transformative project-based learning. IntechOpen. https://doi.org/10.5772/intechopen.102844
11. Geschwind L, Broström A, Larsen K (2020) Technical universities: past, present and future. Springer Nature, p 242

Diverse and Inclusive Learning and Working Environments

Ashish Agrawal, Brianna Benedict McIntyre, Chloe Agg, and Yi-Xin Zou

Abstract This chapter aims to equip leaders in engineering with the tools to create a more diverse, equitable, and inclusive education and industry. This chapter first highlights issues experienced by various marginalized groups due to curricular designs and tangential academic work environments. The topics covered includes creating a more inclusive environment for women through FemTech in the United Kingdom, decolonizing the curriculum to focus on problems relevant to students' local context in South Africa, and advancing Diversity, Equity, and Inclusion (DEI) alongside professional organizations through inclusive leadership in the United States of America. Next, the chapter documents three cases that aim to provide examples of strategies to manage DEI problems by tackling students and educator barriers to belonging and continuously building more diverse and inclusive systems and teams. Overall, this chapter shares ways by which educators and leaders can learn from each other and create a more sustainable and inclusive setting for learning and working environments.

Keywords Curriculum design · Problem-based learning · Decolonization · FemTech · Advancing DEI through inclusive leadership

1 Introduction

> Leaders need to design organizations [and classrooms] that equitably meet the needs of their talent. And this exciting movement toward building tomorrow's equitable organization is dependent on a new type of leader: the inclusive leader [1, p. 6].

A. Agrawal (✉)
Rochester Institute of Technology, Rochester, NY 14623, USA
e-mail: atacet@rit.edu

B. B. McIntyre
NACME, Alexandria, VA 22314, USA

C. Agg · Y.-X. Zou
Imperial College London, London SW7 2AZ, UK

R. Kandakatla et al. (eds.), *Academic Leadership in Engineering Education*, Lecture Notes in Networks and Systems 1097, https://doi.org/10.1007/978-3-031-68282-7_18

When seeking excellence through diversity, equity, and inclusion (DEI) in engineering education, we face several challenges presented by structural issues that materialize as institutional and interpersonal messages that are normalized in our policies, practices, programs, and cultures [2]. However, we must confront these challenges through the global exchange of insights to inform who we teach, how we teach, and what we teach in engineering education [3]. For example, actionable dialogues may involve diversifying the composition of students and faculty, building inclusive learning, and working environments wherein students and faculty can thrive, implementing curricula that students can relate to, building and implementing equitable assessments, and teaching the importance of inclusive design and the social responsibility of engineering solutions.

Consequently, it describes multiple efforts to cultivate diversity and inclusion in learning and working environments in engineering education.

This chapter aims to equip leaders with new perspectives that will, in some areas, help them work toward building an inclusive classroom and team. Engineering educators, practitioners, and leaders will benefit from extending their understanding of some of the rewards and challenges faced by students, educators, and professional development leaders in engineering education, as well as gaining inspiration on how to resolve them. Likewise, those seeking to influence policymakers or funding agencies will find themselves armed with evidence-based insights to lead change in their learning and working environments through the perspectives of a diverse group of authors and leaders in their respective positions and communities.

2 Literature Review

This literature review situates each case study's issue and suggests some alternative approaches to address barriers to prioritizing inclusion in learning and working environments. We describe how FemTech, and decolonization have been used to create and foster inclusive and equitable learning environments. We then explore how professional associations leverage their position as national and international bodies to advance DEI in STEM fields.

2.1 Inclusive Learning Environments

Integrating FemTech into Curricula to Enhance Women's Sense of Belonging in Engineering. Traditionally, engineering education and practice have focused on engineers as problem solvers [4], lending it well to current trends in authentic, contextual, and project-based learning. For students to succeed within higher education, we cannot solely focus on what they are taught, but also on how they learn. Male and Bennett's research finds that students need to understand what engineering is and to envision themselves as future engineers [5]. This recognition is particularly challenging for marginalized students who lack role models for framing that future and relatable context within their learning.

The field of engineering struggles with recruiting and retaining marginalized students due to a lack of belonging within education and industry. One group historically impacted is women of all backgrounds. Consequently, this lack of representation leads to many issues with accepted standard designs and processes that are not suitable for women, contributing to a greater gender disparity [6, 7]. For example, snow clearance schedules in Sweden prioritized roads over pavements, which ultimately benefited male commuter routes, leaving women at higher risk of hospitalization due to using pavements for their commutes and 'school-runs' [7]. While this is not always the case, many products addressing this gap are often related to the rising FemTech industry.

FemTech focuses on products designed for healthcare related to female physiology; design solutions ranging from menstrual product manufacture and fertility trackers to breast pumps and vibrators are considered FemTech. It targets the needs of cisgender and transgender women, some transgender men, asexual and non-binary individuals, as well as pre-teens through to the post-menopause age range.

In an education setting, FemTech enables students to engage in authentic, contextual project-based learning using user-centered design, where the focus is on the spectrum of people as described above. This approach allows students, especially female students, develop technical skills within a relatable context. The focus on the development of products related to women's health and wellbeing generates connectedness and belonging to the study of engineering and to the otherwise perceived foreign environment. It helps women in engineering cross the threshold of the dogma of 'engineering being an all-men domain', into thinking of themselves as engineers.

Decolonizing the Engineering Curricula. Colonialism, which refers to the geographical, political, judicial, and economic subjugation of different parts of the world by the European imperial powers, ended with nations gaining independence. However, coloniality outlived the end of colonization, i.e., the hierarchies imposed in the minds and systems of the then-colonized populations. As such, remnants of colonialism manifested in their governance structures, educational institutions, cultural patterns, and other day-to-day experiences [9].

Decolonization signifies the end of the imperial power's territorial control over the erstwhile colonized territories, and the various racial, gender, sexual, religious, epistemic, and other hierarchies imposed on the previously colonized people [10]. As Smith [11] notes, it is "a long-term process involving the bureaucratic, cultural, linguistic and psychological divesting of colonial power" (p. 112).

Various scholars and activists worldwide have worked on decolonization over the last several decades. In the context of the academic curriculum, decolonization has been seen as a practice that challenges the Western conceptualizations of knowledge production and sharing, while legitimizing other epistemic traditions [12]. Thus, the goal is to create a dialogue between different ways of knowing [13].

Decolonizing engineering education goes beyond creating a level playing field for diverse epistemic traditions. Decolonizing engineering involves considering the developmental needs of erstwhile colonized societies and acknowledge the need to design locally-relevant curricula to address these needs [14]. Moreover, decolonization also involves developing critical awareness among students and empowering them to work towards them, such as food security, healthcare, sanitation, water, etc. Furthermore, decolonization democratizes the curriculum by critically questioning the actors who are involved in producing knowledge [11], who is disseminating that knowledge [14], and who gets a say in implementing that knowledge for the betterment of people [15].

Additionally, scholarship on decolonization has also called for understanding the experiences of students from marginalized backgrounds and creating academic environments that are welcoming to all students. For example, in the South African context, Soudien [16] argues that the university setup can alienate students with forms of social and cultural capital that differ from middle-class and White students, which are often not recognized as valuable in traditional learning environments.

Dube [17] and Motala et al. [18] highlight the linguistic barriers in learning when the language of instruction is not the same as the students' primary language of communication. This linguistic barrier in teaching and learning poses a key challenge for students in several erstwhile colonies in Africa, Latin America, and the Indian subcontinent where languages such as English, French, Portuguese, and Spanish have replaced the local languages in academic settings.

These issues highlight the need for incorporating strategies that are inclusive of alternate ways of knowing and students' diverse linguistic backgrounds.

2.2 Inclusive Working Environments

Advancing Diversity, Equity, and Inclusion Through Professional Organizations National and international professional associations are membership networks that help advance the profession and empower professionals through various member services, such as professional development programs, education and outreach, and the distribution and exchange of scientific knowledge through publications and conferences [19]. Likewise, these associations are also situated to inform accreditation boards [20], and influence policy and legislation through advocacy [19, 20]. At the surface, some STEM professional associations have made strides to promote diversity, equity, and inclusion [20]. However, STEM professional associations are also plagued with issues pertaining to bias, discrimination, exclusion, and harassment of scientists and engineers based on their race and ethnicity, gender identity and sexual orientation [21], and disability. This resulted in the direct and indirect reinforced exclusionary disciplinary norms, cultures, and values by ignoring misconduct and condoning the exclusion of people, through inaction and narrow framings of what counts and who belongs in STEM.

However, in recent decades and, for some, expedited in the wake of the #MeToo Movement and George Floyd's murder, association staff, leaders, and volunteers have recognized how professional associations are in a prime position to transform the profession through shifts to policies and practices that center and intentionally include those historically and systemically excluded in STEM fields [20, 22]. For example, George Floyd's murder made a global impact on the need for police reform and racial justice [23]. Likewise, the #MeToo movement served as a catalyst to address gender bias and sexual harassment in STEM. These collective actions to create a more diverse, equitable, inclusive, and just STEM profession often materialize through multiple structures, such as diversity focused conferences (e.g., CoNECD), diversity task forces [24], strategic STEM alliances [20], committees [25], commissions and divisions [26], amongst others. Similar to higher education institutions, these approaches are not mutually exclusive, meaning professional organizations may simultaneously implement multiple structures to support DEI work [27]. One recent contribution by a STEM alliance in the United States involved the development of an inclusive professional framework to provide professional associations with a lens to examine how they "support and hinder DEI aspirations and help set a foundation for lasting organizational change" [20, p. 2]. Otherwise, the sustainability of these professional associations is at risk of failure [28]. Together, these insights help ground a conversation about what it means to build and sustain diverse and inclusive teams charged with advancing diversity, equity, and inclusion in partnership with and through professional organizations.

3 Positionality

The authors of this chapter represent multiple facets of the engineering education ecosystem within the United Kingdom (UK) and the United States of America (USA). As such, we chose to disclose our positionalities as educators, researchers, students, and leaders to help the reader understand how we are positioned in relation to the chapter's context. Each author reflected on a personal story to impart their commitment to advancing diversity, equity, and inclusion in our respective environments, core elements of our identity, and experiences of privilege and marginalization, and its influences on our work as educators and leaders. Lastly, we describe our connection to the case and how our positionality influences the selected cases presented in this chapter.

3.1 Agg

Agg was brought up in the UK and recalls a close and influential relationship with her grandfather, as an engineering hobbyist. He did not permit his granddaughter into his workshop due to tradition and fear of her being hurt. Agg's decision to become an engineer came from a determination to stand beside him while he overhauled steam engines together. The journey has not been an easy one, though, experiencing a range of prejudicial barriers, such as being told by her grandfather's acquaintances that she should not be studying engineering, as a woman could never amount to anything in that field. Working in construction, Agg's line manager told her to crawl under a desk to plug his laptop in because she was wearing a skirt. She would be mistaken for the secretary in meetings with external stakeholders in their forties. On-site visits, she would walk into the shared office space of men in their twenties to find calendars of topless women on the wall. Things are improving, but the journey through engineering has provided Agg with little sense of belonging. Thus, she has endeavored to create one for the next generation of women entering the engineering industry. She has done this through school outreach, creating training and support schemes in industry, and now as a student experience-focused lecturer in academia. Coming from a minoritized gender within engineering, she reflects upon her own experiences to develop more inclusive practices, especially for women in the field. Examples of inclusive practice include using anonymized live forums to create in-session discussions and questioning opportunities for large class teaching and teaching about methods for fostering inclusive teamwork prior to group assessments. This approach, though is limited by her cis-gender, middle-class, white, euro-centric life experience, and thus she strongly values the diverse range of students, trainees, and colleagues she has had the privilege to work with through these engineering projects.

3.2 Agrawal

Agrawal identifies as an upper-caste queer academic from India. He works at a university in the United States (US). He has previously worked in diverse national academic contexts, including South Africa and India. He has spent a significant part of his academic and professional journey studying curricular and pedagogical practices and the resulting student learning experiences. Additionally, he has worked with faculty members to help them adopt evidence-based and equitable teaching approaches in their classrooms.

Moving across national academic cultures has allowed him to experience privileges and disadvantages due to shifting identities. For example, being in a relatively privileged position in India, he first realized the importance of individual backgrounds based on caste, class, English competency, and rural versus urban upbringings in influencing academic success during his undergraduate studies. As a graduate student in the US, he was exposed to other facets of diversity and the systemic disadvantages experienced by different groups due to their identities through his first-hand experiences of being an immigrant.

As Agrawal moved to South Africa amid the #FeesMustFall movement aimed at improving students' access to higher education, he realized that educators must design and facilitate inclusive learning environments for the success of underrepresented students. In addition to policy changes to reduce tuition fees and to broaden access for students from low socioeconomic backgrounds, efforts need to be made at the curricular and pedagogical levels to increase the relevance of the course material for students, which can lead to a deeper and more critical engagement with the concepts. As a result, Agrawal became interested in the ongoing debate on curricular transformation through decolonization. For Agrawal, decolonization of curricula refers to curricular changes to make learning more relevant to individuals from diverse backgrounds while developing their ability to critically reflect on their learning and life experiences. Additionally, decolonization allows for a greater involvement of the community through its engagement in the teaching and learning process.

3.3 McIntyre

McIntyre is a Black woman who grew up in a blended, middle-class family in eastern North Carolina. She is a proud Historically Black College and University (HBCU) alumna. Among her research interests, she is passionate about civic engagement in her local and professional community. Through service and mentorship, she has dedicated time to advancing undergraduate and graduate scholars in higher education and K-12 settings (e.g., primary and secondary education). Throughout her upbringing, she was aware of the educational and economic disparities for African American and Hispanic families because her mother and aunts were educators and leaders at Title I schools (e.g., a United States federal education program to support low-income students), so it was instilled at an early age to help and serve others, which laid the foundation of her commitment to educational equity and civic engagement. However, as a child, she did not make the connection or understand the implications of these inequities on the national and global underrepresentation of Black engineers. As she progressed in higher education and transitioned to a historically white institution, she became more conscious of systemic issues influencing what counts and who belongs in engineering education and the workforce.

Within engineering education, she has experienced how engineering education is siloed, in terms of where diversity, equity, and inclusion fit within the field. She values the importance of creating a culture where her colleagues and students feel empowered to contribute without negotiating who they are to establish a sense of belonging. She believes engineering education should recognize diversity, equity, and inclusion as central to its purpose and mission rather than sprinkled in as an afterthought. However, this agenda must begin with strengthening the preparation of our engineering education professionals. These experiences drew her to join the ASEE Commission on Diversity, Equity, and Inclusion and collaborate with other dynamic educators and leaders to build a community of education professionals passionate about diversity, equity, and inclusion in engineering education.

3.4 Zou

Zou is a cisgender woman who is currently a final year student of Mechanical Engineering at Imperial College London. She grew up as a Chinese immigrant in Chile and has had the privilege to experience diverse educational systems and cultures in Chile, China, USA, and UK. She chose her current degree because it seemed like the logical choice to make, given her interest. At school, she always performed better at STEM subjects and was more interested in application rather than theory. However, it was not until university that she understood the importance of female representation in engineering.

At the start of her university studies, she viewed her degree as steppingstone to a stable work position. It was further enhanced while continuously being presented with traditional engineering examples, related to construction and motorsports throughout her degree. She saw this as a proof that she was not made to be an engineer. However, this perspective changed when she worked on a FemTech literature review and designed a menstrual testing rig for her third-year projects. These projects exposed the lack of inclusion and consideration in many objects and processes used in our daily lives. Therefore, working on creating more women-centered designs provided her with a sense of being needed and valued as an engineer. It highlighted the importance of having diverse engineers bring their lived experiences to the table and advocate for more inclusive designs.

These inclusive projects allowed her to recognize a gap in the engineering industry that she can fill by working on women-centered projects. This new sense of belonging in the engineering industry allowed Zou to see herself as an engineer in training, and not just a woman studying engineering. Consequently, she has become interested in promoting more opportunities for students who do not fit the usual mold of an engineer and help them find a sense of belonging in the industry. However, she also recognizes the immense privilege she has had from encouraging mentors and economic circumstances that have allowed her to pursue her interests freely. Knowing that many students do not have this privilege, drives her to take advantage of her fortunate position. She uses her platform to inform faculty and administrators about the importance of integrating inclusive projects into the curriculum, so that academics can offer them, and students can be empowered to ask for them.

4 Leading by Example

In the following three subsections, three cases are presented to demonstrate how engineering educators can begin to consider how to build diverse and inclusive learning and working environments.

These cases were selected based on previous and current experiences of the authors' leading efforts to advance diversity, equity, and inclusion (DEI) through curriculum reform and professional development efforts. Consequently, these cases uncover insights about integrating FemTech into a Mechanical Engineering curriculum in the UK to improve the retention of women in engineering, decolonizing an engineering curriculum in South Africa to address inequities in engineering education, and building and sustaining diverse and inclusive remote teams in the USA that is designed to support a community of engineering educators and practitioners committed to learning how to advance diversity, equity, and inclusion in their research, teaching, service, and leadership.

4.1 Case Study 1: Integrating FemTech into a Mechanical Engineering Curriculum

Background Building upon the findings of Bennett and Male's study of possible selves [5], engineering undergraduate students need to see themselves as engineers to achieve highly in their studies.

However, many engineering degrees are based on the science behind the engineering fundamentals, giving a 'pure', theoretical, and person-agnostic approach to education. Within this pedagogical approach, it's difficult for our students to visualize (or recognize) themselves as applied engineers. In the few cases where specific persons are mentioned, such as Bernoulli or Pythagoras, they are almost all white European men. To give context to the engineering science, mechanical engineering syllabi historically tend to focus on motorsports, aviation, and other modes of transportation. While for some students this content is highly engaging and exciting, for others it does not align with their values or with their ability to identify with a possible future self within engineering. This lack of relatable or value aligned content is therefore a potential reason why our female students progress into engineering careers at a lower rate than their male counterparts. It may also be a cause for the lack of recruitment and retention efforts designed to support diverse cohorts, particularly women, in engineering education and subsequent careers.

How can we, as academics, adapt our curricula to provide an inclusive environment, that enables a more diverse range of students to imagine themselves achieving self-actualization within the engineering sector?

Huge efforts are being made across the sector to decolonize engineering science, provide more examples of role models, and provide greater visibility of the science and research conducted beyond the traditional euro-centric teachings. However, while important, this work is constrained by the historical and current biases in access to education, funds, and publication. To remove these limitations, we must recognize the possibilities within teaching and learning so that we can transition from teaching pure theoretical concepts to pedagogical innovations that center on the applied science where we can situate learning within a human-centered context.

Consequently, this case study utilizes project-based learning opportunities, which already exist within the UK curriculum, and enhances their inclusivity by introducing

FemTech themed projects. FemTech is a relatively new term that describes technologies focused on women's health. Examples range from modern and high-tech, such as menstrual tracker apps and pelvic floor exercisers, to longer standing but underdeveloped designs, such as breast pumps and the speculum, used for gynecological procedures. These project-based modules are open-ended and offer both students and staff the opportunity to innovate.

This 'new space' in which projects exist within engineering education allows us to work on research and design areas that have not previously been explored in depth. Traditionally, this space has enabled students and their supervisors to extend knowledge, or identify knowledge gaps, in areas of long-standing research. On the FemTech inclusion program, however, we are enabling students to work on areas with little prior research. In addition to giving female-specific subject areas, this gives students the benefit of emulating the difficulties and excitement experienced by small startup businesses and inventors in new technologies. Lastly, the liminal nature of innovative project work aligns closely with the liminal space required for the transformational nature of threshold concepts, allowing students to see themselves and their education differently.

At Imperial College London we have introduced FemTech projects across three different project formats (Table 1).

The menstrual product testing rig 'DMT' and the biodegradable waterproof menstrual pad layer 'FYP' have won departmental and university-level awards within

Table 1 FemTech project examples

Project type	Type description	Project description	Participants
LRP	Literature Review Project 30-page report covering 30–50 references 3rd year students 5 ECTS/2.5 US Credits	'The Rise of FemTech'—student to define FemTech and analyze its growing usage and impact in education	1 F
		'Testing methods for menstrual products'—investigating publications on this topic and drawing comparisons to related research such as testing incontinence products	1 M
DMT	Design Make Test Group project with Sub-assemblies teams of ~ 4 contributing to an overall 'super-project' team 3rd year students 20 ECTS/10 US Credits	'Menstrual Product Testing Rig' students created a replica fluid and injection system, an artificial vagina with heating system and a biomimetic chassis with sensors	7 M 5 F
		'Manufacture of Banana Fiber based Menstrual Products' students created a fiber cutting system and a fiber carding system	5 M 3 F
FYP	Final Year Project 12,000-word report as main assessment 4th year students 25 ECTS/12.5 US Credits	'Banana Fiber-Based Sanitary Pad Project for Women in Rural Rwandan Villages' student researched manufacturing methods for a biodegradable plant-based waterproof layer for menstrual pads	1 F

one year of launching the project, demonstrating the individual and academic value of the projects.

Anecdotal outcomes suggest that students have gained a broader awareness of possible engineering roles, industries, and applications, such as menstrual product testing. In addition, these projects have provided students with a clear demonstration of the potential impact engineering can have on women's health in particular, and social justice more broadly. Many students indicated a renewed interest in remaining in the engineering sector, citing their projects as the reason for seeing themselves as future engineers. The projects have also catalyzed more widespread conversation and change; for example, staff members who second-assessed project reports stated an increased awareness of the need for gender-disaggregated data and specific female-centered projects within biomedical engineering research. Another example includes the development of a national menstrual tech collaborative platform for academics in the UK, given the lack of currently available published papers and other formal knowledge sharing on this topic.

At the department's end-of-year showcase, the products and posters generated by the FemTech projects were amongst the most popular stands at the exhibitions. Fellow students, staff, and the public were all highly interested in learning more about these new research areas, illuminating the positive impact we engineers can have in changing the conversation and generating new interest in the engineering sector. However, FemTech has limitations because most publications about FemTech projects and research focus on the experiences and needs of English-speaking, white cis women [29], though this has not been the experience within our setting where students on the projects had a roughly even split between male and female, and where white students were in the minority. Further work is required to understand the impact of FemTech projects in undergraduate engineering education. However, in this case, students have indicated positive outcomes, not solely for white female students.

Taking our Chinese-Chilean student author's own experience into account, she informs us how the FemTech project has impacted her engineering identity.

> The most impactful factor that has made me consider a future in engineering is feeling like I am able to make a difference and fill a space in the industry that is currently lacking. Seeing the lack of inclusivity for women in many designs and products has shown me that more diverse engineering groups are needed to create more inclusive products. And the only way to make a lasting change is to stay within the industry and push for more inclusivity. As an engineer in training, I am already in a privileged position to impact others' lives and create opportunities for more inclusivity. Having this realization has sparked a feeling of belonging and responsibility to stay in the industry and make the changes I want to see.

4.2 Case Study 2: Decolonizing the Engineering Curriculum in South Africa

Background The student-led protests starting in 2015 under the umbrellas of #RhodesMustFall and #FeesMustFall created a sense of urgency for universities to address the different forms of inequities in South African higher education [30]. These protests started at the University of Cape Town with demands to take down the statue of Cecil Rhodes on campus. Upon the success of the #RhodesMustFall campaign, the protests soon grew nationwide under the banner of #FeesMustFall. As the name suggests, the latter protests aimed to bring the attention of the government and the university authorities to the rising costs of higher education, making it exclusionary for people from lower economic strata. In addition, these protests also aimed to address other forms of inequities at the universities, including discrimination based on race, gender, and other factors, and to decolonize the curricula to make them relevant for a larger body of students [14].

While responding to this issue, several universities started transforming their curricula to make them more equitable [30]. One such initiative was adopted at the case site, a large, research-focused, and historically English-medium university. The Department of Chemical Engineering, that replaced its existing second-year undergraduate course project in response to the issue.

In the sections below, we first describe the new course project, highlighting the changes made to the existing course project. Then, we discuss how the course project met the goals of decolonization. Quotes from the interviews conducted with the course lecturer and a course tutor who was involved in designing and implementing the project, along with excerpts from project tasks given to students, are included as evidence. Interview data were also collected from students to capture their experience of the project. However, the analysis of the student interviews is beyond the scope of this chapter. These data were collected following the human subjects' approval from the Ethics-in-Research Committee at the data collection site. We encourage the readers to refer to other publications from this project [31, 32] for project design and implementation details.

Design of the new project. The undergraduate chemical engineering curriculum at the research site requires students to engage in a semester-long course project during the second year of the degree. Prior to 2018, the students were expected to complete a project that involved the design of a large petrochemical plant. Specifically, students were asked to create a plan to synthesize a precursor to produce polymers using refinery products.

The learning goal for students was to execute the different aspects of designing plants for large industrial processes and the key factor driving this design was the commercial concerns of making a profit for the manufacturing plant. There was little emphasis on the social or the environmental aspects of engineering except when students were asked to analyze the potential impacts of the petrochemical plant on the nearby surroundings in case of a catastrophic event. However, since the project was situated in the context of a large-scale industrial plant, the students were unfamiliar with the problem's context.

To rectify this issue concerning students' unfamiliarity with the problem's context, the department introduced a new course project in 2019. This new project required students to design an anaerobic digester in an informal urban setting located in the peripheries of the city where the university is located. The feed for the digester included waste generated from people's day-to-day activities, including waste from slaughtering animals, cooking food, gardening, and human excreta.

In addition to shifting the context, the new project incorporated an asset-based community-driven approach. An asset-based approach necessitates that the students consider the existing resources that a community already has and then work closely and collaboratively with community members to plan initiatives to develop the community [33]. This approach is a departure from a more commonly adopted needs-based engineering practice that looks at a community from a deficit standpoint to identify its needs and then designs solutions to fulfill them with little to no input from the community.

Besides adopting an asset-based mindset, the students were required to work with a community-based organization (CBO) that was seen as the driver of the project while the students acted as consultants. The students in the role of consultants sought feedback from the CBO on their progress, which was evaluated from the community's needs and its potential impact on the people. The CBO that the students engaged with was role-played by the course instructor and the course tutors. We used this approach due to the logistical and financial constraints of hiring a CBO on a long-term basis. However, the instructor and the tutors provided a reasonable substitute for CBO members, given their experiences of living in informal communities and being exposed to the literature on the lives of people in these communities.

It is important to note that the new project did consider the commercial concerns related to the economic gains. However, the commercial concerns were addressed simultaneously with the proposed design's social and environmental impacts, specifically the contributions by and the resulting benefits to the community.

New Project and Decolonization Goals. There were several ways in which this new community-driven course project met the goal to decolonize the curriculum. The first was to move the context of learning from large industrial plants to the local community. As the course tutor reflected on the experience of redesigning the project:

> The [project] scope remained the same in the case of the key result areas with the experience students do get, but I think the concept changed. [Thus, the new project addressed] that issue of decolonizing science and trying to introduce concepts to students which were Africa-based, or community based.

As the context of learning changed to a place and more familiar project, it was easier for students, especially those from underprivileged backgrounds to better relate to the course.

> To what extent ... is a portion of the class that comes from socially disadvantaged backgrounds, from the types of backgrounds we want to benefit from this, it seemed to me there were people in [the] group that had that relation, emotional or direct through family where they had personal experiences to that. (Course Lecturer)

Second, this project increased stakeholder engagement by having a deeper engagement of the community in engineering teaching and learning. Thus, the project modified "who gets to play" in the engineering space.

> The CBO thanks you for the poster you presented.... However, the poster session also made the community aware of two other issues that could potentially impact on the success of the project. The first is the cost of the storage vessels for the biogas and biomethane.... The second is the safety considerations associated with the handling and storage of biomethane.... Thus, the CBO would like your team to investigate and report on the storage and safety issues. (Project Task)

As the above excerpt exemplifies, the community as actualized through this project was not just passive. Rather, the community directed the work of students (in the role of engineering consultants) based on their own needs and the available resources.

Finally, the project provided an opportunity to critique the existing canons of engineering knowledge. As students were working on this small-scale community-focused project, they realized that the resources (i.e., textbook or the internet) available to them to solve the given problem were inadequate. As the course lecturer reflected:

> Because the plant was so much smaller than if it had been an industrial plant, all of the wonderful heuristics that are used were out of range.... And so indeed there's a vacuum of engineering methods at community scale work. And that's an important finding actually of this piece of work.

This finding calls for expanding the scope of engineering knowledge to include concepts and heuristics that can be applied to small-scale community driven projects.

4.3 Case Study 3: Building and Sustaining Diverse and Inclusive Teams

Background In 2009, the American Society for Engineering Education (ASEE) assembled a task force that aimed to create a strategic plan to increase diversity in engineering through various activities and policy changes [34]. Since then, the taskforce has evolved into a volunteer-led commission called the Commission on Diversity, Equity, and Inclusion (CDEI). The commission comprises six committees and five teams. The committees include Awards, Communications, Conference Planning, Policy and Letters, Professional Development, and Strategic Planning and Assessment. The teams include Spotlight, Outreach, Institutional Change, Encouragement, and Community Building. This case is based on a diverse team of engineering educators that advanced from the Virtual Workshop team to championing the Professional Development Committee. Above all, this team was selected as the basis of a case that was focused on diverse and inclusive working environments due to the commitment and growth of a volunteer-driven team for four consecutive years.

Before the pandemic, the executive leadership for CDEI decided to expand the professional development program to include virtual learning opportunities for the ASEE community that extend beyond the lifecycle of the annual conference held in June. As a result, the virtual workshop team was established to operate as a sub-team of the professional development committee. In Spring 2020, right on the cusp of the Coronavirus outbreak, we facilitated the first virtual workshop, **"Engineers Show Up: Let's Talk about Allyship,"** with approximately 20 participants. It was primarily in response and support of a campaign to 'Show up and Disrupt' in engineering education through a week of action [35, 36]. Drawing from lessons learned from the transition to the virtual conference and virtual workshops, we launched the Focus Fridays series in Fall 2020. This series is a virtual learning program for engineering education professionals specifically designed to offer the engineering education community an opportunity to translate their work into interactive virtual workshops to encourage professional development beyond the annual conference. Most importantly, we aim to promote sessions that provide participants with knowledge and skills readily applicable to their teaching, research, and leadership. Within six months of establishing the virtual workshop team, our team expanded from two to six dedicated members who are trained to review proposals and support facilitators from acceptance to implementation.

To date, our team has nearly doubled within the past four years. It includes eight women and one man representing nine universities in the US across multiple functions in the engineering education ecosystem (e.g., graduate student, postdoctoral research associate, engineering collections and research analyst, assistant professor, diversity or broadening participation in STEM office director, research manager, and learning scientist).

Over time, the Focus Fridays series has grown to offer 15 virtual workshops and three panels, impacting nearly 725 participants collectively across the sessions and panels. These virtual workshops cover various topics to ensure our community is equipped with the knowledge and skills to support the diversity, equity, and inclusion of marginalized or historically excluded groups in engineering. For example, as a community, we have discussed latent diversity, culturally relevant pedagogy, advancement of women in STEM, positionality statements, accessibility for faculty with disabilities, microaggressions, parenting during the pandemic, contemplative pedagogies, and inclusive leadership. Likewise, our panels have discussed the experiences of graduate students who identify as LGBTQ + and the mental health of Black graduate students and faculty.

Another significant point here is that, our facilitators represent a combination of faculty, student support staff, graduate students, administrators, and practitioners. As we have transitioned to lead and support the professional development (PD) committee, we lead the coordination and execution of in-person professional development programming for the national annual conference, the Focus Fridays virtual series, and Distinguished Topical Plenary Speaker. On par with the impact of the previous PD committee, we coordinated the logistics of 15 PD sessions offered at the 2023 annual conference in Baltimore, MD. These sessions focused on decolonization, cultural change, equity audits, harassment in work and school, stewardship over gatekeeping, and critical mentorship in engineering education.

We have built and sustained inclusive relationships and practices through transparency, accountability, communication, respect, support, and care. For example, outside the realm of the commission, we have also supported one another during various life experiences and milestones, such as dissertations, career transitions, and other life events. In addition, we provide space in our bi-weekly meetings to share various updates and exciting projects we are leading and contributing to in our respective positions. Notably, we cherish and protect the working environment we have created where everyone feels valued and heard.

As an inclusive leader, providing a supportive and responsive environment and leveraging the team's expertise is important.

Another key element is the intentional practice of shared responsibility. As the team's demands grew, we implemented a team structure encompassing multiple roles to accommodate our expanding submission requirements and community responsibility. These roles include assessment and evaluation specialists, digital communications specialists, reviewer specialists, and resource specialists. This work also involves championing the development of new processes to ensure we practice what we recommend regarding how we select proposals, panelists, and distinguished speakers. Together, we contribute to the overarching mission and strategic efforts of CDEI, recognizing that we all play a vital role in our respective positions.

Our central dilemma, influencing our operation, involves the reality of limited resources, which has substantial implications for the individuals and teams willing to share their insights in virtual and in-person formats. Likewise, as a byproduct of the limited resources, we could not provide our distinguished speaker with an honorarium. We acutely understand how soliciting an unpaid speaking engagement (re)produces inequities and are seeking ways to rectify this prevailing issue. However, we have received financial assistance (e.g., external donors and registration waivers) to support some program costs. Above all, we use insights to build inclusive systems to inform how we provide a platform for ourselves and others to advance diversity, equity, and inclusion in engineering education.

5 Discussion, Challenges, and Connections

Throughout the literature review and case studies, we have observed certain commonalities. In the FemTech and Decolonization topics, we see the use of project-based learning to facilitate the integration of DEI concepts into engineering education, and empowering students to form and validate their own engineering identities through ownership and relatable contexts, and for staff to share their own experiences. This strength in identity is also seen in our third case study, where the team is encouraged to work in specialist roles and sit within an environment where life experiences are shared and supported. This positioning of personal identity and 'otherness' as a strength enables the creation of welcoming environments and a sense of belonging, and the democratization of knowledge and skills.

Nevertheless, all this positive power of DEI work does not come easily. Within the case studies, we have identified multiple constraints imposed on the leaders, such as time, workload, and financial constraints, which ultimately limit the speed, quantity, and quality of progress made.

The shared experience of marginalized groups enables them to join in supporting one another, sharing knowledge and expertise, and addressing some of the gaps and workload issues raised above. Thus, adapting solutions to local or even individual contexts is the key, which requires DEI advocates to be simultaneously an international and a decentralized resource [27, 37]. This approach requires sufficient staff time, resources, and awareness of international networks, and opportunities to engage in local activism and research.

While academic staff, DEI advocates, and empowered students are a hugely valuable resource in creating inclusive learning and working environments, it is important to recognize that those people will not have a lived experience of all marginalized characteristics and that inclusion goes beyond visible issues. Many reasons for exclusion, such as disabilities or difficulties, sexuality, or socio-economic background, cannot be seen. As we develop curricula and professional development platforms, considerations must be taken to ensure that engineering educators are equipped with the knowledge and skills to prevent or mitigate experiences that further reinforce marginalization and exclusion. To achieve this, we must, as this chapter has reinforced, focus on creating environments where people can recognize their experiences as having value, and from there, we can co-create projects and platforms to enable change through a diverse lens.

6 Recommendations

The three cases discussed in this chapter highlight how fostering diversity and inclusivity is possible at both the curricular and the organizational levels. However, the first step to this process is to reflect on who is marginalized and how they are marginalized within a given context. Hence, it is important for stakeholders (e.g., faculty, staff, and administrators) across the engineering education ecosystem to critically evaluate who is represented and who is excluded while thinking about how to build and sustain diverse and inclusive learning environments for engineers [38]. For example, while a project, focused on teaching students' industrial processes and ways to enhance corporate financial gains is helpful for future employment, it may alienate several students, especially those who struggle to relate to the project context or are not represented as potential beneficiaries.

Integrating socially and locally relevant problems in engineering curricula and course design should not be considered an either-or debate. We acknowledge that industrial systems and manufacturing are necessary in the modern world to support the needs of people worldwide. Instead, the suggestion here is to incorporate curricular elements that make engineering more relatable for students while challenging the scope of engineering knowledge and work. This is exemplified through the FemTech and the decolonization case studies.

While it is important to design engineering curricula inclusive of different student backgrounds and epistemic traditions, we must also provide institutional support to individuals seeking to learn how to integrate diversity, equity, and inclusion in their work. The ASEE Commission on Diversity, Equity, and Inclusion initiative is an example in this direction, where their committee can be used as a vehicle to amplify practical examples of creating inclusive learning and working environments through various pedagogical and structural innovations, like the examples presented in this chapter.

The following recommendations reinforce case study insights and other scholars and practices in engineering education.

- **Recommendation #1:** Use intersectional and comprehensive approaches to identify the students' needs and leverage their interests and community needs to design meaningful learning experiences.
- **Recommendation #2:** Allocate resources (e.g., time, funding, and people) necessary to adequately address issues concerning diversity, equity, and inclusion in the learning and working environment.
- **Recommendation #3:** Intentionally collaborate with local stakeholders and national and international practitioners to develop best practice appropriate to your context.
- **Recommendation # 4:** Invest in continual learning and recognize that no one individual is the expert on all elements of diversity, equity, and inclusion.

Lastly, we recognize the constraints some individuals and organizations may have considering the federal and state anti-DEI legislation in the USA context and others. This reality may influence your ability to adapt some of the recommendations presented throughout this chapter.

Acknowledgements Collectively, we want to express our gratitude to the editors for the invitation to contribute to this book dedicated to sharing emerging global trends in engineering education. This chapter was only possible with the collective efforts of the individuals involved in developing these cases. Agg and Zou expressed how the FemTech projects could only have been embedded into the curriculum with the support of the relevant module leaders, particularly Marc Masen, who leads the Design and Projects teaching group in Mechanical Engineering at Imperial College London. Agrawal thanks his colleagues, Hilton Heydenrych, Harro von Boltnitz, and Jenni Case, for the support in conducting the study in decolonization of the course project. McIntyre deeply appreciates her team and past and present leadership for their time and effort in shaping the program presented in this chapter–Dr. Christina Alston, Dr. Sarah Over, Dr. Emilie Siverling, Dr. Kali Morgan, Dr. Katreena Thomas, Dr. Meagan Pollock, Dr. Alison Kerr, Peter DeCrescenzo, Cynthia Pickering, and Carolina Blassick. Lastly, a special thanks to Dr. Christina Alston, whose critical review significantly enhanced the quality of this work. Authors Agrawal, McIntyre, and Agg contributed equally to this chapter. The ideas and opinions expressed in this chapter are solely those of the author(s) and do not necessarily reflect the views or positions of any affiliated organizations.

References

1. Tapia A, Polonskaia A (2020) The 5 disciplines of inclusive leaders : unleashing the power of all of us. Berrett-Koehler Publishers, Incorporated, Oakland

2. Sellers V, Villanueva Alarcón I (2023) From message to strategy: a pathways approach to characterize the hidden curriculum in engineering education. Stud Engin 4(2):176–200.https://doi.org/10.21061/see.113
3. Delaine DA, Williams D, Sigamoney R, Tull R (2016) Global diversity and inclusion in engineering education: developing platforms toward global alignment. Int J Eng Pedagogy 6(1)
4. Williamson JM, Lounsbury JW, Han LD (2013) Key personality traits of engineers for innovation and technology development. J Eng Technol Manag 30(2):157–168
5. Male SA, Bennett D (2015) Threshold concepts in undergraduate engineering: exploring engineering roles and value of learning. Australas J Eng Educ 20(1):59–69
6. McMillan C (2022) Monitoring female fertility through 'Femtech': the need for a whole-system approach to regulation. Med Law Rev 30(3):410–433
7. Perez CC (2019) Invisible women: data bias in a world designed for men. Abrams
8. Almeida T, Balaam M, Comber R (2020) Woman-centered design through humanity, activism, and inclusion. ACM Trans Comput-Hum Interact TOCHI 27(4):1–30
9. Grosfoguel R (2007) The epistemic decolonial turn: beyond political-economy paradigms. Cult Stud 21(2–3):211–223
10. Maldonado-Torres N (2007) On the coloniality of being: contributions to the development of a concept. Cult Stud 21(2–3):240–270
11. Smith LT (2021) Decolonizing methodologies: research and indigenous peoples, 3rd edn. Zed Books
12. Mignolo W (2011) The darker side of western modernity: global futures, decolonial options. Duke University Press
13. Mbembe AJ (2016) Decolonizing the university: new directions. Arts Humanit. High. Educ. 15(1):29–45
14. CHE, "Decolonising the curriculum: stimulating debate," Briefly Speaking, vol 3 (2017)
15. Riley D, Pawley AL, Tucker J, Catalano GD (2009) Feminisms in engineering education: transformative possibilities. NWSA J 21(2):21–40
16. Soudien C (2020) The learning challenge in south African higher education. In: Transforming universities in South Africa, Brill (2020), pp 131–155
17. Dube B (2017) Afrikaans must fall and English must rise-ironies and contradictions in protests by South African university students. Afr. Insight 47(2):13–27
18. Motala S, Sayed Y, de Kock T (2021) Epistemic decolonisation in reconstituting higher education pedagogy in South Africa: the student perspective. Teach High Educ 26(7–8):1002–1018
19. Bugher RD (1983) Historians in professional associations. Public Hist 5(3):77–83
20. Leibnitz GM et al (2022) The inclusive professional framework for societies: changing mental models to promote diverse, equitable, and inclusive STEM systems change. Front Sociol 6:784399
21. Marín-Spiotta E (2018) Harassment should count as scientific misconduct. Nature 557(7706):141–142
22. Madzima TF, MacIntosh GC (2021) Equity, diversity, and inclusion efforts in professional societies: intention versus reaction. Plant Cell 33(10):3189–3193
23. Silverstein J (2021) The global impact of George Floyd: how black lives matter protests shaped movements around the World, CBS News, Jun 04, 2021. https://www.cbsnews.com/news/george-floyd-black-lives-matter-impact/
24. Smith-Jackson T, Pak R, Johnson K, McLaughlin A, Rovira E (2016) The HFES diversity task force: advancing the HFES vision of inclusion. In Proceedings of the human factors and ergonomics society annual meeting. SAGE Publications, Sage, pp 422–424
25. Segarra VA et al (2020) Scientific societies advancing STEM workforce diversity: lessons and outcomes from the Minorities Affairs Committee of the American Society for Cell Biology. J. Microbiol. Biol. Educ. 21(1):10–1128
26. American Society for Engineering Education, what's the difference between an ASEE division and commission?. https://diversity.asee.org/deicommittee/2023/05/12/whats-the-difference-between-an-asee-division-and-commission/

27. Holcombe EM, Paul J, Kezar AJ, Vigil D (2024) Organizing shared approaches to equity work. J Higher Educat 1–29. https://doi.org/10.1080/00221546.2024.2301913
28. Morris VR, Washington TM (2017) The role of professional societies in STEM diversity. Am Math Soc 65
29. Krishnamurti T, Birru Talabi M, Callegari LS, Kazmerski TM, Borrero S (2022) A framework for Femtech: guiding principles for developing digital reproductive health tools in the United States. J Med Internet Res 24(4):e36338
30. Koopman O (2019) Is the decolonisation of the South African university curriculum possible in a neoliberal culture? Alternation 24:48–69
31. Agrawal A, Heydenrych H, Case JM, Guni FA, Brown S (2021) Leaving the 20th century behind: exploring how a community-focused design project responds to the call for decolonising an engineering curriculum. In: Presented at the South African chemical engineering congress virtual conference, 2021, pp 361–370
32. Agrawal A, Heydenrych H, Harding G (2022) Application of a decolonisation framework to a second-year chemical engineering design project. South J Eng Educ 1:39–60
33. Mathews S (2013) Asset-based, community-driven development (ABCD) in South Africa: Rebuilding communities from the inside out. In: Presented at the 4th annual Soweto conference, centre for small business development conference, University of Johannesburg
34. American Society for Engineering Education, Engaging a community in diversity, advocacy, and inclusive practices. https://diversity.asee.org/deicommittee/
35. Foster E, Riley DM (2020) Campaign among engineering educators. In: 2020 ASEE virtual annual conference content access
36. Riley D, Foster EK, Karlin J (2020) Show up and disrupt. J Engin Educat 109(1). Wiley Online Library, pp 7–10
37. Kezar A, Eckel P, Contreras-McGavin M, Quaye SJ (2007) Creating a web of support: an important leadership strategy for advancing campus diversity. High Educ 55(1):69–92. https://doi.org/10.1007/s10734-007-9068-2
38. Pawley AL (2017) Shifting the 'default': the case for making diversity the expected condition for engineering education and making whiteness and maleness visible. J Eng Educ 106(4):531–533

Support Systems and Welfare Measures to Encourage Physical and Mental Wellbeing

Jo-Anne Tait, Rosa-María Rodríguez-Jiménez, Sarah A. Wilson, and Karin Jensen

Abstract The focus on engineering students' health and wellbeing has increased due to a global rise in mental health issues and the increased need for more engineers. The engineering industry grapples with recruitment and retention challenges, partly attributable to a high-stress work environment. Further, entering university can be a significant period of change and overlaps with the age when adolescents undergo profound cognitive, emotional, physical, behavioral, and social development. While University students are often considered privileged in many ways, the transition to higher education can result in health and wellbeing challenges. Universities worldwide offer social and psychological support and health promotion services to address these issues but there is need for improvement to meet the increasing demand for mental healthcare services. Studies have shown that engineering students are less likely to seek professional mental healthcare when compared to students from other subjects. Studies also reveal a high prevalence of mental disorders in university students, even among those without diagnosed mental illnesses. These issues are frequently accompanied by physical symptoms, underscoring the importance of physical activity for wellbeing. This chapter highlights the work being carried out internationally in this area, explores relevant case studies and provides recommendations for engineering education leaders to consider in their own institutions. Strategies for enhancing wellbeing include exercise, mindfulness techniques, creative movement, time management, professional skills, and person-centered approaches.

Keywords Engineering · Health · Wellbeing · Higher education · Mental health

---

J.-A. Tait (✉)
Robert Gordon University, Aberdeen AB10 7GJ, UK
e-mail: j.e.tait@rgu.ac.uk

R.-M. Rodríguez-Jiménez
Francisco de Vitoria University, 28223 Madrid, Spain

S. A. Wilson
University of Kentucky, Lexington 40506, USA

K. Jensen
University of Michigan, Ann Arbor 48109, USA

R. Kandakatla et al. (eds.), *Academic Leadership in Engineering Education*, Lecture Notes in Networks and Systems 1097, https://doi.org/10.1007/978-3-031-68282-7_19

We must recognize that the burden for change should not be placed on the individual student. Instead, we must focus on structural changes at higher education institutions aimed at providing students with the support systems necessary to support their wellbeing. Efforts must be taken to shift the culture of engineering from one that normalizes and often glorifies stress, to one that prioritizes wellbeing for all engineering stakeholders.

1 Overview of Mental Health and Wellbeing in University Students

1.1 Introduction

Population health and wellbeing have become increasingly prominent concerns in recent decades due to the high prevalence and severity of mental health-related problems globally [1] in parallel with the increase in life expectancy of the world's population [2].

However, health and wellbeing problems are not exclusively due to an increasingly ageing population. For instance, global data show that suicide is the fourth leading cause of death for individuals between the ages of 15 and 29 [3]. This highlights the critical importance of mental health efforts for individuals within this age range.

The significant mental health challenges found in school-aged adolescents could be due to the significant cognitive, emotional, physical, behavioral and social development that occurs during this stage of life [4]. Additionally, most mental health disorders emerge during this age, with the onset of 48% of mental health disorders by age 18 and 63% by age 25 [5]. In 1996, the World Health Organization (WHO) created a framework called Health Promoting Schools (HPS) to address these challenges. It focuses on implementing health improvement strategies in six key areas: healthy policies, physical environment, social environment, competencies for healthy living, health care and promotion services, and links to the community [6]. This framework continues to be applied in primary and secondary school settings [7, 8] and has also been expanded by the WHO to university settings through the Health Promoting Universities framework [9].

Within a university setting, students are entering a key transitional time where they must face new challenges and make new choices. Often, they have to leave their homes and support networks for the first time and move into a new and challenging environment with people from different backgrounds, cultures and values. It is a period of increasing autonomy and also of decision making at different levels (companions, new experiences, life habits) that will have an impact on their adult life. That being said, students enrolled in a university setting are often considered a privileged group. For instance, while efforts in the United States have been made to increase the enrollment of historically excluded populations in post-secondary education, there are still significant enrollment gaps between people of color and White Non-Hispanics, and individuals from lower socioeconomic status [10].

Despite the layers of privilege that many students within a university setting experience, there has been a significant increase in health and wellbeing problems in university students, especially since 2010 [11]. These levels further increased globally due to the COVID-19 pandemic [12]. Therefore, while obtaining a university degree offers many positive lifelong benefits, it does not protect students from varying health problems.

For instance, in a large self-report study of 13,984 students by the WHO, 35% reported symptoms consistent with at least one diagnosable mental health disorder [13]. Within the United States, studies have shown a 50% increase in the prevalence of mental health conditions in college students between 2013 and 2021, with 60% of students having one or more mental health challenges in 2021 [14]. Further, among college students who do not suffer from a mental illness, studies show that there are still high levels of anxiety, stress, and life dissatisfaction [15–17]. Associated with this are a wide range of physical symptoms, such as lack of sleep and recurrent back, head, and stomach pains [18, 19]. It has been observed that this symptomatology is greater in students who do little physical activity, which clearly shows how movement and a physically active life are strong contributors to greater wellbeing [20, 21].

In response to this perceived mental health crisis, there has been an increase in global awareness on the importance of mental health and wellbeing [11]. To this end, many universities offer social and psychological support and health promotion services. Many campus services are resorting to waitlists to keep up with student demand and this delay has been shown to significantly impact student engagement in treatment [22]. Though strategies are being implemented to address this high demand for psychological services, more research is needed to understand the impact of these strategies on treatment outcomes [23]. Further, there has been an increase in proactive efforts aimed at supporting student wellbeing including stress management [24] and mindfulness [25] interventions.

There has been a progressive shift from a paradigm of medical models based on the psychopathology of mental health (diagnosed mental illness) towards models that focus on people's capacities and attributes to cope with difficulties. This has resulted in the term ***'wellbeing'*** being introduced into the field of mental health.

WHO itself defines mental health as:

> A state of wellbeing in which every individual realizes [their] own potential, can cope with the normal stresses of life, can work productively and fruitfully, and is able to make a contribution to [their] community [26].

In terms of wellbeing, there are often two ways of defining it in the literature: subjective wellbeing and psychological wellbeing.

Subjective wellbeing refers to one's perception of oneself in terms of positive affect and high life satisfaction, that is closely connected to the idea of happiness [27].

Psychological wellbeing is defined as a state of positive psychological functioning of the human being, considering a variety of dimensions such as self-acceptance, autonomy, purpose in life, personal growth, mastery and positive relationships [28].

Complete mental health results from the absence of mental illness and high levels of wellbeing [11].

1.2 The Case for Encouraging Good Health and Wellbeing in Engineering

Engineers are directly involved in solving some of the world's biggest challenges, and, as a result, are in high demand in the international workforce. Despite this, the engineering industry is experiencing significant challenges in recruiting and retaining talent. In the UK, the increasing age of the engineering workforce highlights that retirements will lead to further engineering shortages [29]. This issue is compounded by declining interest in engineering degrees among students and a decrease in pupils pursuing subjects necessary for engineering entry [30]. Further, technological changes and evolving job prospects divert talent away from engineering; for example, in the US, students are increasingly favoring computer science over electrical engineering [31]. As a result, the engineering sector is grappling with a massive skills crisis, evident in the UK's annual shortfall of around 20,000 engineers [29, 30]. This shortage is a global concern, with reported deficits in China (Kaur, 2021), Europe [32], India [33], the USA [34, 35], and Australia [36].

Challenges in both the recruitment and retaining of engineering talent could be, at least in part, due to a perceived high-stress education environment. Because many perceive the engineering education system as a stressful environment [37, 38], this could dissuade potential students from enrolling in engineering degree programs. Further, the experiences that individuals have as they navigate their engineering education could result in them choosing to leave engineering prior to graduation. As a result, shifting the culture of engineering to be more supportive of and conducive to positive mental health and wellbeing could contribute to an overall increase in the number of engineers entering and remaining in the profession [39].

While there is limited research on the mental health of engineering professionals, findings do shed light on some of the challenges faced by engineers in industry that could result in individuals choosing to leave the engineering sector. In the UK, over half of engineers report experiencing negative effects of workplace stress on their mental health or wellbeing, and two thirds describing presenteeism (going to work despite feeling mentally unwell) [40]. Internationally, engineers are leaving the workforce due to burnout [41], which can differentially impact women engineering professionals [42]. Additionally, there are challenges in retention of a diverse range of engineers with high rates of career move out from engineering for disabled people, ethnic minorities, and women [43, 44].

Universities have the unique opportunity to address challenges in the recruitment and retention of engineers as they pursue their engineering degrees. Through developing a culture that is supportive of mental health and wellbeing, universities can introduce engineers into the workforce that recognize the importance of prioritizing mental health. Further, incorporating the development of stress management and coping skills can help new engineers navigate the stressful environment that they might face within the engineering workforce.

2 Health and Wellbeing in Engineering Students

To create an engineering culture that is supportive of student mental health and wellbeing, it is critical to examine the current experiences of engineering students as well as perceived norms and expectations around mental health.

In a narrative analysis of how engineering is portrayed in public forums, Sochacka and colleagues describe engineers as "heroes" who struggle through the curriculum and are rewarded on the other side with high-paying and stable employment [45]. The engineering education environment has been described as a "meritocracy of difficulty" [46] with themes of hardness and continuous struggle [37]. Further, the engineering student workload has been described as the defining stressor for students [38], with an ethnographic study of engineering describing the workload as "horrific" [37]. This extreme workload is described by students to leave little time, if any, to care for their wellbeing, and students feel that they must prioritize their academics over their mental health [47]. Students also feel pressure from not only themselves, but also faculty, to achieve high academic performance regardless. In fact, students described a normalization of high stress and poor self-care as necessary for success as an engineering student [38]. Not only do these cultural factors have the potential to influence student mental health and wellbeing, but they must be taken into account when developing proactive strategies to support engineering students and their development of the strategies necessary to support their own wellbeing.

Research within engineering has identified significant rates of mental illness and poor mental wellbeing in students [48–52]. While stress is the most commonly experienced mental health concern reported in the literature, engineering students also experience high rates of anxiety, depression and post-traumatic stress disorder [53]. This stress is most commonly linked to high academic workloads experienced by students, as well as sleep deprivation [53]. While there is minimal work to date looking at the physical health of engineering students, there have been reports of students experiencing increased musculoskeletal problems potentially due to high levels of computer use [54, 55].

Research also shows that mental health disproportionately impacts students who are minoritized within engineering. For instance, women engineering students report even poorer mental health and wellbeing than their male counterparts [48, 56–59]. Embedded sexism has been reported to be experienced by women engineering students, who often experience being talked over and ignored in group assignments and treated differently by engineering staff [60]. This can increase risk of poor mental health and wellbeing and contributes further to academic engineering culture where stress is being labeled as "common" and conflated with "unproblematic" [61]. Danowitz and Beddoes (2022) further report that engineering students with a physical disability were more likely to screen positively for a mental health disorder and Hispanic engineering students were significantly more likely to suffer from depressive disorders and PTSD compared to their White peers [59].

Gender expansive [57] and first-generation students [58] have also been shown to have higher rates of mental health distress when compared to majority populations within engineering. Importantly, these studies highlight the differential impact that mental health distress can have on students who are minoritized within engineering. As efforts continue to increase the recruitment of historically excluded student populations into engineering, it will be important to ensure that these students have the structural support necessary to maintain their mental health and wellbeing.

Engineering students experiencing mental health distress are also less likely to seek professional help for their mental health when compared to students from other majors, even when controlling for other factors such as race/ethnicity and gender identity [62]. This is true not only for undergraduate students but also engineering doctoral students and Masters students [62, 63]. Because professional help seeking is associated with decreased risk for progression to more chronic or severe mental illness [64], this puts engineering students at risk for increased mental health distress. Further, mental health distress is linked to negative academic outcomes such as dropout and poor academic performanc [65–68], increasing the the need for students to access professional support.

The historically high proportion of men within engineering has resulted in an engineering culture that adheres to hegemonic masculine norms [69–72], which have the potential to negatively influence help seeking within this population. For instance, there are gender differences in mental health help seekers where men are significantly less likely to seek professional help for their mental health [73–75]. This could be due to masculine norms such as independence, control of emotions and prioritization of work over relationships. These norms have been identified within engineering, where engineers have been described as "hard" and relying on logic and reason over emotions [37].

Within the undergraduate student population, studies have found that only a small fraction of those students who are self-reporting symptoms of a diagnosable mental health condition have actually received a mental health diagnosis [59]. This further highlights the low rates of help seeking within this student population. Quantitative studies have looked at the help seeking beliefs of engineering students. For instance, students believe that seeking help is a positive behavior, that they are in control of their ability to seek help and that they have the skills and ability to seek help [80]. Despite this, students had lower mean scores for their intention to seek help if they were struggling with their mental health. There is also the beliefs they harbor about their acceptability within their community for mental health help seekers [80]. Lastly, help seeking beliefs were found to be more favorable in women, gender expansive and non-heterosexual student groups [57, 80, 81].

Engineering students have identified challenges in finding ~~a~~ mental health professional who understands their experiences and their barrier to seeking help [47, 61], with one student reporting that the mental health professional felt that their stress should be expected based on their enrollment within engineering [61].

Students also often seek informal sources of help, such as their engineering peers [38, 82]. Further, students perceive lack of flexibility and empathy from their professors as a sign that they are not supportive of mental health [61, 83]. This perception makes them more reluctant to prioritize their mental health. This plays right into the engineering cultural norms where the focus is on productivity than over personal wellbeing [47], and treat high stress environments as both normal and necessary in engineering [38].

Within the graduate student population, the low rates of help seekers for Masters students might be attributed to the shorter course duration (typically 1–2 years) rather than doctorate or undergraduate degrees which are expected to take longer.

Within undergraduate engineering student populations, qualitative work has identified that students feel that they can just "tough it out" and that their education is "just four years" and then they can prioritize their mental health and wellbeing [47]. This could apply more strongly to Masters programs, where the shortened timeline puts additional pressure on students to just get through rather than seek professional healthcare.

There is also the established scenario where students in higher degrees have adopted alternative coping strategies throughout their undergraduate degree which do not include professional help seeking. For instance, qualitative work has identified that students use stress relief strategies such as exercise, mindfulness and maintaining spiritual health [82]. They are also likely to rely on their engineering peers for support when stressed [38].

3 Case Studies from Engineering Institutions

Across the literature, there are references to different strategies that students adopt to support their mental health and wellbeing, such as physical activity, mental relaxation, meditation, and others. To date, two key papers, reviewing interventions supporting the mental health and wellbeing of undergraduate [84] and graduate [63] engineering students have been published.

Interventions include:

(1) **Psychological interventions**—such as mindfulness training, creative movement, enhanced counselling support and listening to music,
(2) **Physiological interventions**—such as breathing and body awareness exercises,
(3) **Educational interventions**—related to mental health and wellbeing awareness, changes to teaching approaches, and changes to curriculum scheduling [84].

As this field of research is relatively young, there are some key challenges in identifying which strategies are effective, including the wide variety of interventions that have been tested and the variability in assessed outcome measures [84]. The following case studies highlight work undertaken by the authors to support good health and wellbeing in engineering students.

3.1 Case Study 1 Mental Health and Wellbeing Training in Engineering Staff

Several recent studies have highlighted a lack of conversations around mental health occurring in engineering spaces [47, 53, 83, 85], highlighting a key opportunity for faculty to influence the way that engineering students perceive the engineering environment. In particular, students have highlighted that the faculty and staff that they feel are most supportive of their mental health and wellbeing are those who explicitly talk about the importance of mental health [83]. While many engineering faculty and staff care about student wellbeing and feel that supporting student wellness is a part of their job responsibilities [86], they also feel underprepared to engage in conversations with students about their mental health [86, 87]. Faculty also recognize the energy that it takes to support students' mental wellbeing and have reported negative impacts on their own mental wellbeing as a result of providing students with such support [86]. Therefore, it is important that faculty and staff are provided with the resources necessary to support students through referring them to appropriate resources [88], while also protecting their own mental health and wellbeing.

In Robert Gordon University, Scotland, there was a drive to improve pastoral support in the School of Engineering.

Dr Jo-Anne Tait developed an enhancement plan where key staff (personal tutors, technicians, administrators, and course leaders) undertook Mental Health First Aid training.

A staff support network within the School of Engineering was also developed. This resulted in increased capability and confidence of staff to support students suffering from poor mental health and wellbeing. The School also witnessed an increase in proactive help-seeking from students as there was now internal School support. Further to this, an online tailored resource was developed in collaboration with engineering staff and students, health scientists and student counsellors to provide information in a way that would encourage engineering students to take ownership of their mental and physical wellbeing. In addition to fostering good staff/student communication, these enhancements had the added impact of reducing academic administration relating to assessment deferrals, mitigations, and appeals.

At the European University of Madrid and Spain, interventions have been carried out for more than twenty years with teachers from the School of Engineering, Architecture and Design, which started with the creation of the European Higher Education Area. Curricular changes from a structure based on technical content, to one based on the development of generic and specific competencies required specific training for teachers in a variety of soft skills. Training was delivered by the Human Resources department and covered the following topics: communication, leadership, emotional intelligence, and new teaching methodologies, among others.

The research group led by Dr. Rodríguez-Jiménez incorporated training courses based on non-verbal communication, creative movement, and dance to promote the development of intra and interpersonal competencies among engineering teachers. The training lasted between 6 and 20 hours, spread over different days and adapted to the teachers' schedules. All of them were fundamentally experiential, working on the content through proposals for play, movement, and body awareness with a time for reflection and verbal integration at the end of each session. In the longer training sessions, focus groups were held in which the learning achieved and the connection with the classroom were discussed. Teachers increased their level of self-knowledge and self-awareness in their movement and relational patterns, highlighting how this learning had a positive impact on their relationships with coworkers and students [89]. As one teacher said: "if I enter the classroom as if I were going into a battle, the students will perceive this, and they will feel restlessness and distant." [89, pp. 233]. There were workshops that focused on the wellbeing, life satisfaction and stress reduction of engineering teachers, as highlighted in the literature [90, 91]. Generating healthy educational environments includes enhancing the wellbeing and health of teachers and their work environment, since this will undoubtedly affect the relationship with students.

3.2 *Case Study 2 Embedding Wellbeing into the Engineering Classroom*

Faculty can also help to further shift the cultural expectations around mental health in engineering through introduction of the concept in the classroom. This can be done through modelling behavior, developing positive relationships with students and including wellness activities within the classroom [92]. Emotionally positive environments have been found to positively influence student wellbeing, motivation and academic performance [93, 94]. Therefore, faculty engagement in strategies to improve classroom climate, communication between teachers and students and teachers' wellbeing, can also contribute to the wellbeing of students [90, 91].

There has been an increase in studies about using mindfulness techniques to improve the mental wellbeing and positive outlook of engineering students [95–98].

For instance, introduction of mindfulness training for first-year engineering students resulted in improved intrapersonal competencies such as self-regulation and wellbeing, as well as interpersonal competencies such as empathy and communication [96].

For graduate students, an eight week mindfulness training course resulted in positive changes such as improved emotional health and satisfaction with their research work [98]. Breathing exercises have also been found to be particularly effective and regarded positively by engineering students [99]. Ashgar et al. (2022) described several perceived factors that support subjective wellbeing in engineering students, including faculty support, academic satisfaction, and a supportive environment [100]. Their study is one of the few that looks at wellbeing from a positive stance rather than a deficit approach. A deficit approach is one that considers the weaknesses of an individual with a view to "make them better," which would position the engineering student as someone in need of improvement, which could potentially discourage help seeking [101]. On the other hand, a positive stance is that which advocates 'wellbeing for all' that is aimed to enhance all participants' experience of life.

Some mental health and wellbeing strategies have been incorporated in classrooms, for instance, Miller et al. integrated biometrics and mental wellbeing activities into classroom activities, allowing students to measure the physiological impact of stress management techniques on their body [102]. Other authors have incorporated listening to the body and body awareness as a daily routine in the classroom [103], including the incorporation of small body-motion breaks. Short-term activities such as attention to one's own breathing, movement and joint activation or creative games in movement, can not only improve physical health but also contribute to the improvement of cognitive functions, including attention [103, 104].

Listening to students is key in these activities in order to carry out those that can improve the emotional and cognitive state of the students, and in some cases to generate a state of greater relaxation, in others to activate energy and attention.

Time management has also been identified as a stressor for engineering students [82], who face significant and often unmanageable academic workloads [38]. These challenges in time management can result in students feeling as though they do not have time to prioritize their wellbeing [82]. Further, time management can also be an effective coping strategy that allows students to utilize their time more effectively. For instance, one study found that students used time management to support self-discipline and boundaries [105], improving opportunities for maintaining mental health and wellbeing [105, 106] and academic performance [106–108]. Because time management could be a significant difficulty for many engineering students, given the heavy workloads and is something that doesn't come naturally to everyone [108], there are opportunities to integrate time management skills into the engineering curriculum. Furthermore, soft skills learning from a person-centred approach contribute to the wellbeing of future engineers by increasing their self-awareness and self-knowledge, and thus their capacity for self-care [109].

3.3 Case Study 3 Embodied Wellbeing

Because of the high stress environment of engineering, there is an important need to investigate novel and creative strategies for stress management. For instance, the WHO has collated evidence on the influence of art on health and wellbeing [110] and have recently announced a forthcoming Lancet Global Series on the health benefits of arts.

In Europe, pioneering research was carried out during the last 25 years by introducing interventions with creative movement and Dance Movement Therapy (DMT) in higher education contexts [89–91, 111, 112]. The studies focused on the development of intra- and interpersonal skills as well as on self-care and stress reduction. DMT is part of the group of so-called creative therapies, those that incorporate creative resources in a therapeutic context. There is proven evidence about the health benefits in different populations and problems [111, 112].

The fundamental and differentiating principles of DMT are the concept of embodiment and bodyfulness, the existence of a creative process, the symbolic and metaphorical meaning of movement and the use of kinesthetic empathy in a relational context.

Traditionally used in the clinical setting, DMT has also been introduced for prevention and health promotion in randomized controlled studies proving the increase of wellbeing and life satisfaction and stress reduction in engineering students [91]. These studies incorporated for the first time the analysis of both psychological and physiological variables to analyse the impact of the use of creative movement on reducing stress and increasing life satisfaction.

One of these studies focused specifically on engineering students and was carried out in the European University of Madrid.

Through a mixed methodology, an embodied learning program based on DMT was designed, implemented and evaluated [113]. The objective was to measure psychological variables such as stress level, life satisfaction and wellbeing using both psychometric tests and physiological measures. Additionally, it was intended to reflect on the introduction of embodiment to increase the self-knowledge and awareness of students, thus providing them with resources to deal with stress.

A total of 37 students randomly divided into two experimental and control groups participated. The experimental group carried out a total of ten body work sessions lasting 90 minutes each. The development of the sessions was based on the results obtained from previous projects. All the sessions had the same structure and addressed different themes including 'listening to my body and that of others', limits, rhythms, movement patterns, tonic states and emotions, modes of relationship and presence. Each of the sessions began with a check-in, a body warm-up structured according to Hatha Yoga principles, a development phase in which the indicated themes were worked on through specific dynamics with creative movement, play, dance, and a closure. The closing phase gave time to put words and meaning to the metaphorical aspects that emerged through the movement, and therefore supported integration of the work. Both psychometric and physiological measures showed evidence of decreased stress and improved life satisfaction [113]. The qualitative analysis of the students' perceptions offered relevant information about their needs, increased self-awareness and self-knowledge, and the impact of the training on their daily lives.

The engineering students expressed their desire for this type of training to be maintained on an ongoing basis. They considered that the work environment and the quality of health of the entire engineering school would benefit [113]. Integration of wellbeing through bodily listening and creativity across institutions could result in benefits for both teachers and students, generating healthier educational environments in engineering and creating a culture of self-care and care for others.

4 Key Takeaways for Academic Leaders

We must recognize that the burden for change should not be placed on the individual student.

We must focus on structural changes at higher education institutions aimed at providing students with the support systems necessary to support their wellbeing.

Efforts must be taken to shift the culture of engineering from one that normalizes and often glorifies stress, to one that prioritizes wellbeing for all engineering stakeholders.

Therefore, the below section will focus on recommendations for interventions and policy changes that can be implemented at the institution level, as well as within engineering programs (school, department or faculty level) to improve student wellbeing outcomes.

4.1 Institution Level

Institutions (Table 1) and Schools or Departments (Table 2) must work collaboratively with staff and students to identify barriers to help seeking within the institution and co-produce support systems that reduce or fully dismantle these barriers. Recommendations have been split into the high-level themes of the Curriculum, The Student, and The Culture.

4.2 Recommendations for Schools or Departments

Supporting staff in departments or schools of engineering should be trained to understand that encouraging good mental health and wellbeing in engineering students is critical. These recommendations arise from identifying aspects that are likely to be in the department's or school's control.

5 Conclusion

It is vital to acknowledge that engineering students are not simply passive receivers of support and knowledge. Engineering students hold the power to support curriculum development as well as themselves, and also drive a positive cultural shift within the engineering profession. Staff, institutions, professional, statutory, and regulatory bodies, and researchers can go a long way to enhancing the student experience and improve mental health and wellbeing, but it is the engineering students who are integral to any success in this area. As such, it is imperative that academic leadership

Table 1 Recommendations for mental health interventions and policies that could be developed at the institutional level

	Intervention	Motivation
The curriculum	Engage with professional and national engineering bodies to evolve and enhance engineering learning outcomes to include and prioritize health and wellbeing	Regulatory bodies have significant impact over the curriculum and learning outcomes within engineering. These changes can have broad impact across institutions
	Develop supportive mechanisms to reduce the impact of heavy test loads that might be associated with assessment requirements to meet engineering regulatory standards	Some engineering programs require major assessments at the end of a term that result in high student stress levels and test anxiety
The student	Increase student access to resources that support health and wellbeing. Ensure that university staff receive appropriate training about these resources and how to serve as effective referral agents for students [88]	Literature shows that many students do not feel they have sufficient access to mental health resources [38, 47]. University staff also feel unprepared to support student wellbeing [86, 87, 91]
	Support staff and students by providing clear pathways and signposting to existing support. Collaborate with programs to develop clear, manageable processes and procedures to support positive health and wellbeing	There is evidence supporting the importance of creating spaces where students can interact and learn about themselves and about resources for wellbeing [113]
The culture	Develop programs that encourage students to advocate for and support the mental health and wellness of their peers through both informal (in-class training on resources [114]) and formal programs (peer support programs [115])	Engineering students feel that they would be more likely to access professional help if they received support from a friend [47]. Bringing training into the classroom can help to destigmatize conversations around mental health
	Provide pathways to encourage staff to implement mental health interventions within their programs. Provide funding to help support data collection to measure the impact of these programs	It requires significant energy to implement mental health programs. University staff should receive recognition and support for their efforts
	Develop and implement a debriefing network for those who are supporting mental health and wellbeing	It is important for staff members who are providing mental health support to also feel supporte. Engineering staff have indicated concerns about supporting students resulting in decreases in their own wellbeing [86]

Table 2 Recommendations for mental health interventions and policies that could be developed at School or Department level

	Intervention	Motivation
The curriculum	Ensure courses are well organized and follow a sensible progression of topics and academic difficulty	Workload is a significant stressor for students. Ensuring the curriculum is distributed across a term can reduce student stress levels
	Encourage faculty to implement mental health in the classroom by incentivizing efforts in teaching evaluations. Ensure there is a plan for assessing the value work aimed at supporting student wellbeing	Expectations are high and it is difficult to implement initiatives that are not incentivized. This is especially true for minoritized staff (women and racial/ethnic minorities)
The student	Develop strategies to support student organizational and time management skills. Incorporate activities for building a sense of community within their program	Time management and organizational skills are key challenges for students particularly in courses with a full curriculum [82]
	Develop mental health training that is specific to the engineering student experience. Incorporate literature on mental health in engineering into the training	Tailoring training to engineering students empowers them to see their own experiences when talking about mental health [114]
	Work within your program to develop policies for supporting student wellbeing. Include in these policies how mental health will be handled in terms of missing assignments and class absences. Make sure these policies are clearly communicated to all faculty and students	Student experiences with engineering staff can result in them feeling that they do not care about their wellbeing [83]. Having policies that are consistent across the program can ease student stress
The culture	Consider ways in which a culture of trust within engineering Schools or departments can be built	Students will repeat patterns and behaviors observed during their studies in their adult lives. If they learn the value of attention to wellbeing and care, they will transfer this to their future work environments
	Recognize and acknowledge diverse cultures and backgrounds of the engineering student cohort. Develop a debriefing network for those supporting the mental health and wellbeing of engineering students. Facilitate more social opportunities for students and staff to be together. Recognize the importance of incorporating emotional and care aspects in a transversal way	Generating healthy environments is a transversal task that must encompass all levels in the institution. The evidence shows the suitability of generating spaces of wellbeing and care for all members [90, 91, 115]

actively engage with the student body to become enhancement partners to ensure success.

References

1. Organization WH (2022) World mental health report: transforming mental health for all
2. Rudnicka E et al (2020) The World Health Organization (WHO) approach to healthy ageing. Maturitas 139:6–11
3. Organization WH (2021) Suicide worldwide in 2019: global health estimates
4. Kutcher S, Wei Y, Weist MD (2015) School mental health: global challenges and opportunities
5. Solmi M et al (2022) Age at onset of mental disorders worldwide: large-scale meta-analysis of 192 epidemiological studies. Mol Psychiatry 27(1):281–295
6. Organization WH (1996) Regional guidelines: development of health-promoting schools-a framework for action
7. Langford R et al (2014) The WHO Health Promoting School framework for improving the health and well-being of students and their academic achievement. Cochrane Database Syst Rev 2014(4)
8. Lee A et al (2020) Health promoting schools: an update. Appl Health Econ Health Policy 18:605–623
9. Tsouros AD et al (1998) Health promoting universities: concept, experience and framework for action. 1998, World Health Organization. Regional Office for Europe, Copenhagen
10. Ayyad F (2015) Higher education and privilege: 21st century issues. Coll Stud Affairs Leadership 2(2):2
11. Hernández-Torrano D et al (2020) Mental health and well-being of university students: a bibliometric mapping of the literature. Front Psychol 11:1226
12. Li Y et al (2021) Prevalence of depression, anxiety and post-traumatic stress disorder in health care workers during the COVID-19 pandemic: a systematic review and meta-analysis. PLoS ONE 16(3):e0246454
13. Auerbach RP et al (2018) WHO world mental health surveys international college student project: prevalence and distribution of mental disorders. J Abnorm Psychol 127(7):623
14. Lipson SK et al (2022) Trends in college student mental health and help-seeking by race/ ethnicity: findings from the national healthy minds study, 2013–2021. J Affect Disord 306:138–147
15. Beiter R et al (2015) The prevalence and correlates of depression, anxiety, and stress in a sample of college students. J Affect Disord 173:90–96
16. Ibrahim AK et al (2013) A systematic review of studies of depression prevalence in university students. J Psychiatr Res 47(3):391–400
17. Mahmoud JSR et al (2012) The relationship among young adult college students' depression, anxiety, stress, demographics, life satisfaction, and coping styles. Issues Ment Health Nurs 33(3):149–156
18. Lund HG et al (2010) Sleep patterns and predictors of disturbed sleep in a large population of college students. J Adolesc Health 46(2):124–132
19. Schneiderman N, Ironson G, Siegel SD (2005) Stress and health: psychological, behavioral, and biological determinants. Annu Rev Clin Psychol 1(1):607–628
20. Johnston SA et al (2021) Effects of team sports on anxiety, depression, perceived stress, and sleep quality in college students. J Am Coll Health 69(7):791–797
21. Kędra A et al (2017) Back pain in physically inactive students compared to physical education students with a high and average level of physical activity studying in Poland. BMC Musculoskelet Disord 18:1–8
22. Krendl AC, Lorenzo-Luaces L (2021) Identifying peaks in attrition after clients initiate mental health treatment in a university training clinic. Psychol Servic

23. Thomas KA, Schroder AM, Rickwood DJ (2021) A systematic review of current approaches to managing demand and waitlists for mental health services. Ment Health Rev J 26(1):1–17
24. Amanvermez Y et al (2023) Stress management interventions for college students: a systematic review and meta-analysis. Clin Psychol Sci Pract 30(4):423–444
25. Dawson AF et al (2020) Mindfulness-based interventions for university students: a systematic review and meta-analysis of randomised controlled trials. Appl Psychol Health Well Being 12(2):384–410
26. World Health Organization, Promoting mental health: concepts, emerging evidence, practice: summary report. 2004: World Health Organization
27. Diener E (2012) New findings and future directions for subjective well-being research. Am Psychol 67(8):590
28. Trudel-Fitzgerald C et al (2019) Psychological well-being as part of the public health debate? Insight into dimensions, interventions, and policy. BMC Public Health 19(1):1712
29. Quantum Marketing. The UK's engineer shortage – what you need to know 2022 [cited 2023]
30. EngineeringUK, Engineering UK 2020: Educational pathways into engineering 2020: https://www.engineeringuk.com/research-policy/educational-pathways-into-engineering/educational-pathways-into-engineering/
31. Atwell C (2022) Electronics and electrical engineering jobs on the decline—can they be saved?, In: ElectronicDesign 2022. https://www.electronicdesign.com/technologies/embedded/article/21255051/electronic-design-electronics-and-electrical-engineering-jobs-on-the-declinecan-they-be-saved
32. Bergman A, Ogunshakin N (2022) Shortage of engineers is threatening climate action, in Euractiv. 2022. https://www.euractiv.com/section/energy-environment/opinion/shortage-of-engineers-is-threatening-climate-action/
33. Raj A (2022) Is India about to experience a huge talent shortage for tech developers? In: Tech Wire Asia 2022. https://techwireasia.com/06/2022/is-india-about-to-experience-a-huge-talent-shortage-in-tech-developers/
34. Jamali MM et al (2023) Role of undergraduate summer research in improving retention of engineering students. In: 2023 IEEE international symposium on circuits and systems (ISCAS). IEEE
35. Hoske MT (2023) Control engineering career and salary survey, 2023. In: Control engineering. https://www.controleng.com
36. Yu H, Chai H, Ravishankar J (2023) Identifying the needs of electric power industry through online job ads: a mixed-methods approach. In: 2023 American society for engineering education annual conference and exposition. Baltimore, MD (2023)
37. Godfrey E, Parker L (2010) Mapping the cultural landscape in engineering education. J Eng Educ 99(1):5–22
38. Jensen KJ et al (2023) Undergraduate student perceptions of stress and mental health in engineering culture. Int J STEM Educ 10(1):30
39. Jensen K (2021) The time is now to build a culture of wellness in engineering. Stud Engin Educ 2(2)
40. Flaig J (2022) Feature: pandemic pushes already stressed engineers to breaking point. Institution of Mechanical Engineers
41. Phillips T (2022) Is there a shortage of developers? Developer shortage statistics in 2022. In: CodeSubmit. 2022
42. Ronen S, Pines AM (2008) Gender differences in engineers' burnout. Equal Oppor Int 27(8):677–691
43. O'Connor R (2021) UK engineering firms struggling to hold onto diverse range of talent, new report reveals
44. Lavalin S (2021) Career deflection: exploring diversity, progression and retention in engineering
45. Sochacka NW et al (2021) A narrative analysis of stories told about engineering in the public discourse: implications for equity and inclusion in engineering. Stud Engin Educ 2(2)

46. Stevens R et al (2007) Engineering as lifestyle and a meritocracy of difficulty: two pervasive beliefs among engineering students and their possible effects. In: American society for engineering education conference and exposition. 2007. Honolulu, Hawaii
47. Wright CJ et al (2023) Mental health in undergraduate engineering students: Identifying facilitators and barriers to seeking help. J Engin Educ
48. Deziel M et al (2013) Analyzing the mental health of engineering students using classification and regression. 2013. Citeseer.
49. Vats N, Sharma A (2017) Stress and anxiety among two groups of engineering students. J Adv Schol Res Allied Educ 12(2):223–227
50. Danowitz A, Beddoes K (2018) Characterizing mental health and wellness in students across engineering disciplines
51. Reta Y, Samuel T, Mekonnen M (2020) Mental distress and associated factors among undergraduate engineering students of Hawassa University, Ethiopia. J Multidiscip Healthcare 13:99–107
52. Siddiqui N et al (2021) A standardized analysis of regularity of depression among the engineering students. World J Phamacy Pharmaceut Sci 10(5)
53. Asghar M, Minichiello A, Ahmed S (2023) Mental health and wellbeing of undergraduate students in engineering: a systematic literature review. J Engin Educ
54. Menéndez CC et al (2009) Upper extremity pain and computer use among engineering graduate students: a replication study. Am J Ind Med 52(2):113–123
55. Sial S et al (2020) Prevalence of low back pain among engineering students of Isra University. J Physiotherapy Res 4(1)
56. Negi AS, Khanna A, Aggarwal R (2019) Psychological health, stressors and coping mechanism of engineering students. Int J Adolesc Youth 24(4):511–520
57. Hargis LE et al (2021) Relationship between mental health distress and help-seeking behaviors among engineering students. In: American society for engineering education annual conference
58. Jensen KJ, Cross KJ (2021) Engineering stress culture: relationships among mental health, engineering identity, and sense of inclusion. J Eng Educ 110(2):371–392
59. Danowitz A, Beddoes K (2022) Mental health in engineering education: identifying population and intersectional variation. IEEE Trans Educ 65(3):257–266
60. Ravijolan U (2020) How engineering students are enduring more than a mental health crisis
61. Beddoes K, Danowitz A (2022) In their own words: how aspects of engineering education undermine students' mental health
62. Lipson SK et al (2016) Major differences: variations in undergraduate and graduate student mental health and treatment utilization across academic disciplines. J Coll Stud Psychother 30(1):23–41
63. Bork SJ, Mondisa JL (2022) Engineering graduate students' mental health: a scoping literature review. J Eng Educ 111(3):665–702
64. Mitchell C, McMillan B, Hagan T (2017) Mental health help-seeking behaviours in young adults. British J General Practice 8–9
65. Deberard MS, Spielmans GI, Julka DL (2004) Predictors of academic achievement and retention among college freshmen: a longitudinal study. Coll Stud J 38(1)
66. Eisenberg D, Golberstein E, Hunt Justin B (2009) Mental health and academic success in college. B.E. J Econ Anal Policy
67. Bruffaerts R et al (2018) Mental health problems in college freshmen: prevalence and academic functioning. J Affect Disord 225:97–103
68. Hartley MT (2011) Examining the relationships between resilience, mental health, and academic persistence in undergraduate college students. J Am Coll Health 59(7):596–604
69. De Pillis E, De Pillis L (2008) Are engineering schools masculine and authoritarian? The mission statements say yes. J Divers Higher Educ 1(1):33
70. Frehill LM (2004) The gendered construction of the engineering profession in the United States, 1893–1920. Men Masculinit 6(4):383–403

71. Akpanudo UM et al (2017) Hidden in plain sight: Masculine social norms in engineering education. In: 2017 IEEE frontiers in education conference (FIE). IEEE
72. Akpanudo UM, Huff JL, Godwin A (2018) Exploration of relationships between conformity to masculine social norms and demographic characteristics. In: 2018 IEEE frontiers in education conference (FIE). IEEE
73. Möller-Leimkühler AM (2002) Barriers to help-seeking by men: a review of sociocultural and clinical literature with particular reference to depression. J Affect Disord 71(1):1–9
74. Galdas PM, Cheater F, Marshall P (2005) Men and health help-seeking behaviour: literature review. J Adv Nurs 49(6):616–623
75. Yousaf O, Grunfeld EA, Hunter MS (2015) A systematic review of the factors associated with delays in medical and psychological help-seeking among men. Health Psychol Rev 9(2):264–276
76. Levant RF, Richmond K (2016) The gender role strain paradigm and masculinity ideologies
77. Juvrud J, Rennels JL (2017) "I don't need help": gender differences in how gender stereotypes predict help-seeking. Sex Roles 76(1):27–39
78. Mahalik JR, Di Bianca M (2021) Help-seeking for depression as a stigmatized threat to masculinity. In: Professional psychology: research and practice
79. McDermott RC et al (2018) College students' conformity to masculine role norms and help-seeking intentions for suicidal thoughts. Psychol Men Masculin 19(3):340
80. Armstrong I et al (2022) Investigating mental health distress and help-seeking perceptions in first-year engineering students. In: American society for engineering education annual conference, Minneapolis, MN
81. Sánchez-Peña, M.L. and S.A. Kamal. A comparative analysis of mental health conditions prevalence and help seeking attitudes of engineering students at two institutions in the U.S.A. in 2023 World Engineering Education Forum - Global Engineering Deans Council (WEEF-GEDC). 2023.
82. Ban N et al (2022) Identifying common perceived stressors and stress-relief strategies among undergraduate engineering students. In: ASEE annual conference exposition
83. Ban N et al (2023) It's very important to my professors…at least most of them": How messages from engineering faculty and staff influence student beliefs around seeking help for their mental health. In: ASEE annual conference exposition
84. Tait JE et al (2024) Interventions to support the mental health and wellbeing of engineering students: a scoping review. Eur J Eng Educ 49(1):45–69
85. McAlister AM et al (2023) Stigma of mental health conditions within engineering culture and its relation to help-seeking attitudes: Insights from the first year of a longitudinal study. In: American society for engineering education annual conference and exposition
86. Sanders J et al (2024) "Not a therapist": why engineering faculty and staff don't engage in supporting student mental health and wellbeing. Int J Engin Educ 40(1)
87. Wilson SA, Hammer JH, Usher EL (2021) Faculty experiences with undergraduate engineering student mental health. In: 2021 ASEE virtual annual conference content access
88. Wilson SA, Goldberg DS (2023) Strategies for supporting engineering student mental health. Chem Engin Educ 57
89. Terrón-López MJ et al (2013) Non-verbal teacher training program: emotional intelligence, body awareness and communication. Literacy Inf Comput Educ J (LICEJ) 4(3):1058–1065
90. Rodríguez-Jiménez R-M, Carmona M (2021) A rationale for teacher change from a bodyfulness paradigm: an experience in higher education. Educ Sci 11(9):460
91. Rodríguez-Jiménez R-M et al (2022) Stress, subjective wellbeing and self-knowledge in higher education teachers: a pilot study through bodyfulness approaches. PLoS ONE 17(12):e0278372
92. Wilson SA, Jensen KJ (2023) Strategies to integrate wellness into the engineering classroom. Chem Engin Educ 57(4)
93. Upadyaya K, Salmela-Aro K (2013) Development of school engagement in association with academic success and well-being in varying social contexts. Eur Psychol 18(2):136–147

94. Davidson M, Lickona T, Khmelkov V (2014) Smart and good schools: a new paradigm for high school character education. In: Handbook of moral and character education
95. Miller I, Jensen K (2020) Introduction of mindfulness in an online engineering core course during the COVID-19 pandemic. Adv Engin Educ 8(4)
96. Huerta MV et al (2021) Inner engineering: evaluating the utility of mindfulness training to cultivate intrapersonal and interpersonal competencies among first-year engineering students. J Eng Educ 110(3):636–670
97. Dalton ED, Estrada T (2023) Implementation and feasibility of a group mindfulness intervention for undergraduate engineering students. Adv Engin Educ 11(1)
98. Crone WC et al (2023) Cultivating well-being in engineering graduate students through mindfulness training. PLoS ONE 18(3):e0281994
99. Joshi A, Kiran R (2020) Gauging the effectiveness of music and yoga for reducing stress among engineering students: an investigation based on Galvanic Skin Response. Work 65(3):671–678
100. Asghar M, Minichiello A, Iqbal A (2022) Perceived factors contributing to the subjective wellbeing of undergraduate engineering students: an exploratory study. Int J Environ Res Pub Health 19. https://doi.org/10.3390/ijerph192316284
101. Meyer D, Strevens C (2022) Is using the term wellbeing with students a mistake? Evaluating a wellbeing intervention in a UK Law School-suggestions for repositioning strategies to address the impact of deficit-discourse. Law Teach 56(1):54–66
102. Miller I et al (2022) Development and implementation of a biometrics device design project in an introductory BME course to support student wellness. Biomed Engin Educ 2(1):75–82
103. Rodríguez-Jiménez R-M, García-Merino S (2017) Enactive and embodied learning in higher education. J. Funct. Neurol. Rehabil. Ergon 4:5–9
104. Merriman W, González-Toro CM, Cherubini J (2020) Physical activity in the classroom. Kappa Delta Pi Record 56(4):164–169
105. Gelles LA et al (2020) Compassionate flexibility and self-discipline: student adaptation to emergency remote teaching in an integrated engineering energy course during COVID-19. Educ Sci 10(11):304
106. Aeon B, Faber A, Panaccio A (2021) Does time management work? A meta-analysis. PLoS ONE 16(1):e0245066
107. Razali S et al (2018) The impact of time management on students' academic achievement. In: Journal of Physics: Conference Series. IOP Publishing
108. Adams RV, Blair E (2019) Impact of time management behaviors on undergraduate engineering students' performance. SAGE Open 9(1):2158244018824506
109. Rodríguez-Jiménez R-M, Lara-Bercial PJ, Terrón-López M-J (2021) Training freshmen engineers as managers to develop soft skills: a person-centred approach. Sustainability 13(9):4921
110. Fancourt D, Finn S (2019) What is the evidence on the role of the arts in improving health and well-being? A scoping review. In: 2019: World Health Organization. Regional Office for Europe
111. Koch S et al (2014) Effects of dance movement therapy and dance on health-related psychological outcomes: a meta-analysis. Arts Psychother 41(1):46–64
112. Karkou V et al (2019) Effectiveness of dance movement therapy in the treatment of adults with depression: a systematic review with meta-analyses. Front Psychol 10:936
113. Rodríguez-Jiménez R-M et al (2022) Embodied learning for well-being, self-awareness, and stress regulation: a randomized trial with engineering students using a mixed-method approach. Educ Sci 12(2):111
114. Wilson S et al (2023) Delivery of mental health training across a college of engineering. In: European engineering education conference. Dublin, IR
115. Berry A, Friend J (2023) Advocacy and engagement on a college campus: implementation of a peer support program. Coll Stud J 57(1):64–68

Models for International Collaborations to Nurture Students into Global Engineers

Renetta Garrison Tull, Jamie Gurganus, Dirk Reith, Libis Valdez Cervantes, Xu Lihui, Iris Groß, Michael Malschützky, Sundiata Jangha, Catalina Aranzazu-Suescun, Luis Alberto Cruz Salazar, Jose Texier, Jusmeidy Zambrano, Maria Mercedes Larrondo Petrie, Li Fan, L. I. Shuangshou, Qiao Weifeng, Wang Sunyu, Robin Cresiski, Yarazeth Medina, Krishna Vedula, Veena Kumar, and Rovani Sigamoney

Abstract Engineering schools and colleges around the world have developed a variety of innovative models to train their undergraduate (post-secondary, tertiary) students and graduate students (post-baccalaureate, masters, doctoral) for global engagement. This chapter shares global student mobility models from Africa, China, Germany, India, Latin America, and the United States that are both internal to the

R. G. Tull (✉)
University of California Davis, Davis, CA 95616, USA
e-mail: rgtull@ucdavis.edu

J. Gurganus · S. Jangha · R. Cresiski · Y. Medina
University of Maryland, Baltimore County (UMBC), Baltimore, MD 21250, USA
e-mail: jgurganus@umbc.edu

S. Jangha
e-mail: sjangha@umbc.edu

R. Cresiski
e-mail: rcresiski@umbc.edu

Y. Medina
e-mail: yarazeth@umbc.edu

D. Reith · I. Groß · M. Malschützky
Hochschule Bonn-Rhein-Sieg (H-BRS), Grantham-Allee 20, 53757 Sankt Augustin, Germany
e-mail: dirk.reith@h-brs.de

I. Groß
e-mail: iris.gross@h-brs.de

M. Malschützky
e-mail: michael.malschuetzky@h-brs.de

L. V. Cervantes · J. Texier · M. M. L. Petrie
Latin and Caribbean Consortium of Engineering Institutions (LACCEI), Colombia, USA
e-mail: decano.fadi@unitecnar.edu.co

R. Kandakatla et al. (eds.), *Academic Leadership in Engineering Education*, Lecture Notes in Networks and Systems 1097, https://doi.org/10.1007/978-3-031-68282-7_20

curriculum, and external to the campuses. Collaborations between the leaders of the models throughout these featured regions have been fostered by engineering education organizations such as the Latin and Caribbean Consortium of Engineering Institutions, and the Global Engineering Deans Council, which connect through the

J. Texier
e-mail: texier@laccei.org

M. M. L. Petrie
e-mail: petrie@fau.edu

L. V. Cervantes
Unitecnar, Cartagena, Colombia

X. Lihui · L. Fan · L. I. Shuangshou · Q. Weifeng · W. Sunyu
Tsinghua University, Beijing, China
e-mail: xulihui@tsinghua.edu.cn

L. I. Shuangshou
e-mail: icee@tsinghua.edu.cn

Q. Weifeng
e-mail: lss@mail.tsinghua.edu.cn

X. Lihui · Q. Weifeng · W. Sunyu
International Centre for Engineering Education Under UNESCO (ICEE), Beijing, China

C. Aranzazu-Suescun
Embry-Riddle Aeronautical University, Daytona Beach, USA
e-mail: aranzazc@erau.edu

L. A. C. Salazar
Universidad Antonio Nariño, Bogotá, Colombia

Technical University of Munich, Arcisstraße 21, 80333 Munich, Germany

L. A. C. Salazar
e-mail: luis.cruz@tum.de

J. Zambrano
Universidad Nacional de Chilecito, Chilecito, Argentina
e-mail: jzambrano@undec.edu.ar

M. M. L. Petrie
Florida Atlantic University, Boca Raton, FL 33431, USA

K. Vedula
IUCEE (Indo Universal Collaboration for Engineering Education), Hyderabad, India
e-mail: krishna@iucee.org

University of Massachusetts Lowell, Lowell, MA 01854, USA

V. Kumar
University of Maryland Global Campus, Adelphi, MD 20783, USA
e-mail: krishna@iucee.org

R. Sigamoney
UNESCO Office, Harare, Zimbabwe
e-mail: r.sigamoney@unesco.org

International Federation of Engineering Education Societies and convene annually at locations across the globe. Collaborations formed during the convenings yielded promising practices that can be utilized by professors around the world as they create their own sustainable programs and learning opportunities to develop new generations of global engineers.

Keywords Global · Engineering · Competencies · Collaboration · Intercultural

Every student deserves an opportunity to globally engage.

1 Introduction

As technologies advance, engineering societies are taking a closer look at ways to facilitate international communication and collaboration opportunities for engineering students so that they will be prepared to be the next generation of technological innovators and leaders. International engineering education organizations such as the Latin and Caribbean Consortium of Engineering Institutions (LACCEI), and the Indo Universal Collaboration for Engineering Education (IUCEE) in India, share and elevate research developed by scholars from their member universities, but these organizations also facilitate networking opportunities for engineering educators from around the world. This chapter can serve as a guide for professors, instructors, and academic administrators in colleges and universities who want to develop their students so that they can become global engineers.

The University of California Davis in the USA is one of the universities around the world calling for students broadly to receive global education and global competency, through their "Global Education for All" campaign, led by the Division of Global Affairs. That model is considered within this chapter, which shares that global competencies can be developed in a variety of ways. Competencies can be taught within a curriculum and through co-curricular activities outside of the classroom. Training for global engagement can also be developed within a country—without a "study abroad" component or international location exchange, as funding and sponsorship for costly travel may not always be readily available. Of course, traditional ways of engaging globally remain with the utilization of opportunities outside of one's own country for an external global exchange or education experience. To facilitate planning for training students to be ready for participation in global contexts, whether the training takes place within country or externally, this chapter offers a collaborative "guide"—**Global Universities Investing in Developing Engineers (GUIDE),** to share information that will serve students around the world. Focus areas include—future faculty development, global competencies, internationalization, Engineering Education, and cultural awareness. Many of the programs have taken care to align with the United Nations Sustainable Development Goals (SDGs) which represent a "call to action" by the global community to partner and address the world's challenges [1, 2].

2 Engaging Global Training Models from Around the World

This chapter highlights six models that prepare engineering students for global engagement. These models come out of engineering programming from universities and organizations around the world: Africa, China, Germany, India, Latin America, and the United States.

Programs from Africa and China that are highlighted in this chapter are connected to centers within UNESCO, the United Nations Educational, Scientific and Cultural Organization. UNESCO's "Category 2" centers develop capacity (including training students and professionals) through cooperation on both regional and global levels.

The centers in Ghana and Beijing that are featured in the chapter have the bulk of their activities within their respective regions, but they build out a knowledge base that is applicable to international technical advancement.

The programs from Germany and the United States that are discussed here have training for global competencies within their respective universities and regions, but they also leverage connections to international conferences or student competitions.

The constructs in this "GUIDE" for India and Latin America showcase international organizations that comprise several universities, primarily within Asia and South America, but which also connect to communities of practice for global education within the Global Engineering Deans Council (GEDC) and the World Engineering Education Forum (WEEF), which host meetings and conferences around the world.

2.1 *Model 1: Connect to International Center*—Africa—Kwame Nkrumah University of Science and Technology (KNUST)

The Kwame Nkrumah University of Science and Technology (KNUST) in Ghana, houses the Technology Consultancy Centre, a Category 2 Centre of Excellence for Engineering Innovations, Manufacturing and Technology Transfer under the auspices of UNESCO.

The UNESCO Center is developing opportunities for students to gain global competencies in the following areas: *Knowledge Production, Capacity Building, Technical Service, Modernizing Indigenous Technologies, Policy Advocacy and Information Sharing*, and *Building Partnerships.* The Centre of Excellence focuses on adaptive research and development, technology innovation and transfer, manufacturing skills development and entrepreneurship development—particularly sustainable development of the West Africa Sub-region. The Centre helped countries in West Africa, especially Burkina Faso, Ivory Coast, and Benin to develop and grow their metal goods sector, in a sustainable way. This is a new center that is starting its operations in partnership with the government of Ghana, KNUST, and UNESCO. While there are other UNESCO category 2 centers in Africa in sciences, the new center at KNUST will be the only UNESCO category 2 center specific to "Engineering."

2.2 *Model 2: Transform University Project to Include Internationalization*—China—Transcultural Engineering Practice Education from a Global Perspective

In Beijing, China, Tsinghua University has a *"Transcultural engineering practice education from an international perspective"* project that engages 30 undergraduate students per year. The project includes students in mechanical, industrial, automation, and electrical fields of study, from respective verticals. There is also Intelligent manufacturing, and collaboration with the Academy of Arts and Design, School of Science at Tsinghua. The program gives students opportunities to experience personalized innovative product development with student-centered, interest-oriented, and product-driven principles. Tsinghua University's program connects to the International Center for Engineering Education under the Auspices of UNESCO (ICEE), in collaboration with the Chinese Academy of Engineering (CAE). The center provides high quality technical education, while fostering an "equal, inclusive, developmental and win–win engineering education community." This combined focus on promotion of both quality and equity in engineering education connects to the UN's SDGs [2–6].

2.3 *Model 3: Connect to International Competition—Germany—Formula Student: International Competitions as Catalysts for Student's Global Cooperation*

In Sankt Augustin, Germany, Hochschule Bonn-Rhein-Sieg (H-BRS) hosts the *Formula Student: International Student Competitions* as catalysts for student's global cooperation, and a program of impact that trains students for global competency. The program brings together approximately 70 students annually from Electrical Engineering, Mechanical, and Environmental Engineering. Other departments include Computer Science, and Management Sciences (e.g., Business). The project areas include: Sustainable Materials, Virtual Materials Design, Optimization of Energy Networks, Efficient Mobility and Transportation Alternatives, Life-Cycle Assessment, Reception of Technology in Society, and Gender Diversity Aspects of Technological Innovations. *Formula Student* is an international design contest for students following regulations of SAE International, a global standards development organization. There are more than 1000 active teams in one of the three competition classes: Combustion, Electric, or Driverless to manufacture a single-seat formula race car. *Formula Student* offers students a motivating problem- based learning setting (developing and building a race car from scratch) and forms a powerful, self-reliant, interdisciplinary group to solve problems. Focus areas include connection to an interdisciplinary project, complex tasks, problem-based learning, peer-assisted learning, knowledge transfer, and utilizing professors in a mentoring capacity.

2.4 *Model 4: Connect to Continental Organization*—India—Indo Universal Collaboration for Engineering Education

The Indo Universal Collaboration for Engineering Education (IUCEE) is another organization that promotes global training for students through its organizational structure. Experts from around the world offer webinars, workshops and mini-courses to students and faculty in engineering colleges in India. Led by India and supported by several countries all over the world, IUCEE includes more than 100 universities/colleges in India, serving more than 50,000 students. IUCEE's programming is geared toward "Improving the quality and effectiveness of teaching and learning processes for undergraduate education in India, so that the graduates have the skills relevant to industry and societal needs." IUCEE aims to improve the quality and global relevance of engineering education in colleges in India, by emphasizing problem-based learning approaches. Engineering graduates should have the skills to be valuable to industry and society anywhere in the world, including communications, working in teams, and technical writing. The globally based emphasis is to help produce engineers who can contribute to projects that address the UN's SDGs. The National Education Policy 2020 for India has emphasized a need for higher education to be holistic and multidisciplinary. IUCEE focuses on faculty development, student development and academic leadership development by strengthening the ecosystem for fostering the employability, entrepreneurship and leadership skills of engineering graduates. More than 100 global professors and experts connect through this ecosystem.

2.5 *Model 5: Connect to Hemispheric Initiatives—Latin America—the Latin American and Caribbean Consortium of Engineering Institutions Bootcamp*

The Latin American focus, "*Bootcamp LACCEI,*" covers universities in South America and the Caribbean, sponsored by LACCEI, the Latin American and Caribbean Consortium of Engineering Institutions. The organization connects to North and South America, and western hemisphere activities through partnership with the Organization of American States (OAS) [7, 8]. LACCEI has over 250 institutions from 24 countries. The Bootcamp addresses applied problems, and helps develop participants' professional, general, and life competencies with special emphasis on project development, social entrepreneurship, and innovation. Participants engage collaboratively in project proposal development, evaluated by an international jury. The Bootcamp accesses LACCEI's student chapters, competitions, resources, and network. The curriculum includes modules on applied AI, design thinking, technological surveillance, innovation management, Scrum methodology, SDG analysis, and communication, with visits to relevant universities and industries. The initiative significantly contributes to participants' professional and life competencies and the broader engineering education landscape in the region.

Bootcamp LACCEI facilitates the transition from conceptual understanding to practical application, embodying the principle of learning by doing. This immersive experience unfolds over 8-days where participants engage with international peers [9–14].

2.6 *Model 6: Connect to International Conferences*—United States—The PROMISE Engineering Institute (PEI)—Global

There are several US-based programs that train students for global engagement. The PROMISE Engineering Institute's "PEI-Global" initiative, just one highlighted here, combines in-country mentorship at local university levels with international conferences. PEI-Global expands opportunities, access, and mentoring to diverse groups of engineering students, with a deep connection to fostering racial and gender diversity. PEI-Global is an outgrowth of National Science Foundation funded programs such as the Alliances for Graduate Education and the Professoriate (AGEP), and the Louis Stokes Alliance for Minority Participation (LSAMP). Practices from these programs along with incorporation of a certificate connected to the nationally networked Center for the Integration of Research, Teaching, and Learning (CIRTL), have given PEI-Global a structure which involves the University of California Davis (UC Davis) and the University of Maryland, Baltimore County (UMBC) as leads, with partners from other universities. The foray into international engagement has focused on participation in international conferences sponsored by WEEF, GEDC, and the International Federation of Engineering Education Societies (IFEES) [15–18].

In 2022, PEI-Global collaborated with Hochschule Bonn-Rhein-Sieg (H-BRS) in Germany to offer cultural competency workshops to students who were attending the WEEF in South Africa. The collaboration also invited students who were part of the Student Platform for Engineering Education Development (SPEED). The collaboration between PEI, H- BRS, and SPEED continued in 2023 at the WEEF conference in Mexico to provide mentoring and feedback support for student research presentations. Students presented research to international audiences of engineers, and they received feedback from members of the GEDC who came from universities around the world.

3 How Should Professors Prepare?

For global competency, engineers should be prepared and trained to solve real-life problems based on scientific knowledge [19, 20]. Engineering professors can help students to bridge the gap between theory and practice. To facilitate this, professors *themselves* may need some additional preparation and attention toward their own professional development. This GUIDE provides a few points for such preparation: pay attention to content, collaboration, current state-of-the art, and context.

- **Content—Courses and Curriculum:** Attention to the content in the engineering students' curriculum and courses should be a focus. To be globally competitive, be sure that students are receiving updated content in the classroom. The engineering professors should also consider enhancing their own teaching practices with new instructional pedagogies. For example, the faculty can learn how to connect and integrate knowledge from external activities into regular study courses. Cooperating with a student initiative can be fruitful—if professional support is given reliably and constantly, and if the knowledge that the students gain on their own can be transferred back into the curriculum. Teaching content should also be combined with the latest industry developments and topics of global concern.
- **Collaboration—Universities, Research organizations, Industry:** Collaboration in successful global engagement recognizes that projects can be enhanced when they are not done in isolation. Collaborations can be between universities, with universities and other types of research institutions, or for-profit entities. In-depth collaborations among universities and industrial enterprises can be the basis for a successful program. Consider collaborative research that will be connected to developing and transferring innovative engineering solutions.
- **Current State-of-the-Art—Innovation:** UNESCO suggests making sure that one facilitates space for product development, dissemination and knowledge uptake, and continuous professional development in relevant areas of engineering

and technology. Continuous learning about current state-of-the art and technological advances is important and should be taught using modern digital and smart technologies.

- **Context—International:** It's important to have some knowledge about global contexts and policies. If a new project is going to have connections to another country or region beyond the home country, professors and administrators can learn from and develop policy briefs to inform (and influence) national and sub-regional policy direction on engineering innovation, manufacturing, and technology transfer. Consider seeking and promoting collaboration with regional and international partners to achieve targets that are specific to the sub-areas and are connected to each of the United Nation's 17 SDGs. Several of the guiding points above have been established by UNESCO as areas that need to be considered by faculty and administrators who are training students to be global engineers. The centers in Africa and China described in Sects. 2.1 and 2.2 are among the programs that have utilized these guidelines, which were among the attributes that allowed them to be qualified as UNESCO Category 2 Centers.

4 Competencies that Students Are Receiving to Facilitate Global Engagement

The programs representing different models from various parts of the world that are highlighted in this chapter are undertaking several practices to facilitate global competency for their students. This section will highlight some of the competencies that students are receiving to prepare for global engagement.

4.1 Multilingual Communication

The project at Tsinghua University in Beijing, China and the ICEE promote multilingual proficiency. Their program includes course material in English as a focus for communications, and trains students using common and well-known tools that will facilitate mobility (e.g. industrial software, manufacturing equipment). Similarly, in Germany, the "Formula Student" competition is executed world-wide with the same set of rules (introduced by the SAE), and each event is held in English. This is a huge advantage when it comes to exchange of knowledge between teams from different countries or regions of the world. In principle, everybody speaks the same language, faces the same problems and, consequently, develops similar ideas—summarized in the development of a "Formula Student spirit." Pre-cursors to the PEI-Global program included encouraging students to take time to become familiar with the language of the country that was hosting the international conference. For example, when the World Engineering Education Forum was held in Seoul, South Korea, some of the participants took evening classes outside of the university to learn Korean. A more direct facilitated approach included sponsoring pre-conference workshops at UMBC in the U.S. to learn basic Spanish prior to traveling to conferences in Costa Rica and Peru, which were among the locations of LACCEI conferences.

4.2 Socio-Cultural Exchange

PEI-Global gives students pre-sessions for the international conference, both before entering the new country and within that country, but, prior to the start of the conference. These pre-sessions spend time on information such as language, cultural norms, and customs. The students are also required to take cultural awareness workshops at the beginning of the conference to prepare them to engage internationally. In Germany, attention to facilitating cultural connections is embedded in the training related to promoting the understanding of different traditions and learning the language of the host nation for the competition. In the case of "Formula Student," the students from Germany learn English which also includes fostering an attitude of mutual support of other teams, especially in situations of crises. Nearly every team is befriended by one or more other teams in the world. Often, such friendships start by accident, e.g. because they shared the same pit lane box at a competition event. It often leads to mutual visits and cooperation in a very organic way. This automatically includes social activities, cultural exchange, and personal insights.

4.3 Skill-Building in State-of-the-Art Technologies and Professionalism

Building skills in State-of-the-Art technologies, in addition to attending to professional skills are part of all the models discussed in this chapter. In India, the IUCEE is strict about the quality control of the instruction, and the students have short courses and projects guided by global experts. Bootcamp-LACCEI in Latin America meticulously incorporates a wide range of academic modules aimed at providing participants with both depth and breadth in their learning experiences and relevant soft skills. Combining knowledge of technical skills with professional skills is a strategy that has motivated engineering students [21, 22]. Bootcamp-LACCEI's academic and technical skills modules include Artificial Intelligence Applied to Project Formulation, Design Thinking, Technological Surveillance, Innovation Management, Change Management, and Application of Scrum Methodology. These modules are carefully selected and crafted to equip students with cutting-edge skills and knowledge, enabling them to navigate and respond adeptly to the complex, ever-changing landscapes of engineering and technology. Participants are exposed to modules like Prospective Analysis of the SDGs, Assertive and Effective Communication, and Soft Skills. These components are integrated into the curriculum to ensure that students are not only technically proficient but also socially aware and effective communicators. UNESCO's SDG and engineering report underscored the importance of engineering education, the development of new engineering competencies, and the ability to accredit engineering training programs internationally. It emphasized the following competencies: multidisciplinary and interdisciplinary skills, adaptability to change, creativity and flexibility, practical experience, leadership, sustainable development, and employability [1, 2].

5 Signature Activities

Some of the programs have signature activities with aspects that can be considered for scaling up, scaling down, or direct replication. While professors and administrators who are developing new programs find that funding streams, capacity, resources, and even cooperation may vary, segments of the programs below can also be utilized and adapted for a given context.

5.1 *State-of-the-Art Knowledge Through the* Unmanned Aerial Vehicle Project

China has focused on the design and prototyping of Unmanned Aerial Vehicle as a project that has relevance around the world. Students from Tsinghua University and foreign institutes were taught, inspired, and guided from multiple disciplines, so that they were familiar with the application of specific industrial software in a short time, completed the UAV design in a mixed team, and made a prototype.

5.2 *Collaborating on Professionalism by Leveraging Co-Located Conferences*

One genre of activities that has been successful is the construct of co-located conferences, or the "conference within a conference." The PROMISE Engineering Institute (PEI) prior to adding the official "global" component, heavily leveraged the LACCEI conferences as international locations and launching points for their respective, specific meetings. The LACCEI Bootcamp does this as well and includes virtual components throughout the year. A key aspect of garnering global competency will always include having the opportunity to travel to a different country for the international experience. There are also "in-country" and virtual trainings that complement the global experience. Often, students in the PEI program are connected to the in-country (US- based) conference first.

5.2.1 Structured Mentoring at Conferences

Programming connected to the PEI includes mentoring components that take place at the various conferences. For example, 9 months prior to attending the WEEF conference in Cape Town, South Africa in December 2022, PEI students participated in an enhanced convention experience in the US, at the National Society of Black Engineers (NSBE) National Convention in California with mentors. African-American graduate students from all the participating campuses in the PEI (from Maryland and California) came to the NSBE convention to participate in mentoring activities with academic leaders of color, attend the convention, and recruit for their respective graduate programs and institutions. The PEI initiative had the added aspect of structured mentoring interaction with STEM academic leaders of color. The PEI graduate students were mentored by university presidents and deans who attended the conference. Discussions included ways to construct career trajectories, developing an entrepreneurial approach to research, and transitioning into administration following a record of scholarly excellence. PEI facilitated formal and informal interactions between the academic leaders and the graduate students. The insight provided by academic leaders whose social identities mirror those of the students was considered to be invaluable. PEI participants are also afforded the opportunity to participate in the Academic & Research Leadership Symposium (ARL Network) activities and sessions within the NSBE conference itself. The ARL Network is a subconference within NSBE for postdoctoral fellows and early faculty members. The recruiting element provides opportunities for light touch near-peer mentoring and cements a sense of community and belonging as the student scholars act as advocates for graduate school in general. The PEI programming includes career exploration & identity formation, essential skills, e.g., public speaking or networking, awareness & commitment to diversity, equity & inclusion.

The *PEI-Global* initiative was based on the general PEI, as a special activity inside the framework helping students to acclimate with international conferences, foreign cultures, and more. Students are encouraged to participate actively at an international conference with a scientific contribution and they are offered various training sessions to optimize their time at the conference. PEI-Global provides students with a unique opportunity to not only develop their networking and professional skills, but to increase their global and cultural competencies as they engage with the international community. PEI-Global extends the mentoring component to include international mentors who engage with the students at the international conferences. This kind of international exchange of mentors provides the students with an expanded perspective on the training that they are receiving regarding cultural competencies, as well as feedback that they receive when the international mentors have discussions with them about their research projects. Highlights of PEI-Global which engaged international mentors included: providing students with an opportunity to earn a new certification that focuses on building international connections, cultural interchanges within conferences, and sessions for students to share research with mentors during short, timed rounds.

5.2.2 Cross-Country Collaboration for International Certificate

In 2022, PEI-Global in the US and H-BRS in Germany collaborated to focus attention on training students at international conferences, and they launched a certificate for student global engagement at the WEEF conference in South Africa. The collaboration with H-BRS' Center for Teaching Development and Innovation (ZIEL) focused on cultural competency workshops, and an "international" *Center for the Integration for Research, Teaching and Learning International (ICIRTL) Certification* that was aimed to develop and facilitate global competency. ICIRTL, the new certification, was introduced through use of the learning outcomes of CIRTL, a network of more than 40 universities in the US that aims to develop future faculty and foster teaching excellence. Students who came to the WEEF in South Africa were provided with requirements that leveraged the conference and developed their scholarship in teaching and learning in the international space. Requirements for this certification included items such as: Attend workshops on cultural awareness and diversity, attend requisite technical sessions, connect with peers and mentors from different countries, and submit summary of learnings.

PEI-Global included Lightning Round Sessions, developed in coordination with SPEED, IFEES, and Petrus, a global company with offices in France, Romania, and the United Kingdom. The collaborative team worked together to facilitate a session for doctoral students to present their engineering research in front of global deans and administrators in a friendly competition. Petrus facilitated a workshop for the students to develop their public speaking and storytelling skills. The workshop helped students to overcome shyness, connect the story of their research to current issues or events, and engage the audience with real-time discussions and feedback. The cross-cultural awareness workshop, a requirement of PEI-Global and ICIRTL, focuses on ways to avoid cultural errors, e.g., showing a behavior that is appropriate in one's own cultural area but offensive in another. Students are coached to understand and be exposed to negative effects of such errors. They have sessions that allow them to experience sources of miscommunication through an experiential approach. After the workshop, the participants were able to (a) recognize the sources of miscommunication that can occur in a cross-cultural setting, (b) reflect on the occurrence of miscommunication using a constructivist perspective, (c) self-reflect their own reactions to "cultural triggers," and (d) apply techniques to overcome stumbling blocks in cross-cultural communication.

In the first phase of the cultural awareness training, an active learning, interactive group approach is used. This is based on an interactive exercise in a group setting, actively engaging the audience. It brings learning content to life and ensures a sustainable practical transfer.

Participants are confronted with rules deviating between groups (between departments in the same company or foreign cultures). Each participant experiences (a) the effects of explicit and implicit rules and norms for an appropriate behavior, (b) one's own and others' reaction to rules and norms that deviate from one's own, and (c) one's own and others' reaction when it comes to developing and enforcing common rules and norms.

The colliding rules create an experience of a culture shock. While experiencing authentic group processes, the participants were able to act in a protected space without negative consequences. They can try out and understand themselves better, while having time and support from the trainers and mentors to draw their own conclusions.

In the second phase, a visual presentation containing provocative "cultural triggers" is presented and discussed, allowing participants (a) to experience one's own reactions to "cultural triggers," (b) to experience ambiguity of "cultural triggers," and (c) to become aware of "stumbling blocks" in cross-cultural communication in a protective environment with empathy and respect for participants and some humor. The final result allowed the participants to be guided through a reflection of stumbling blocks and "cultural triggers" in a safe environment. By providing a toolbox for overcoming stumbling blocks, participants are encouraged to overcome fear of first contact with new cultures. An example of a powerful "cultural trigger" is a picture from members of confraternities of penitents wearing habits and capirotes, covering their faces, during Holy Week in Spain. This content is quite familiar for European students and shows culturally appropriate content from their perspective. However, US students react initially with strong negative emotions as they identify the penitents as members of a racist organization. As the instructor encourages the participants to exchange perception and reaction to the same stimulus, they realize how differently it can be interpreted.

5.2.3 Skill-Building in Teams Across Countries

The Formula Student project in Germany, highlights that success in such competitions includes an important factor: interdisciplinary teams. To participate successfully, engineering students from various disciplines need to team up with students from other STEM disciplines, e.g., computer science, and other fields, e.g., business management. As the team is faced with very complex, long-term issues, unique teaching, learning and utilization-oriented approaches, e.g., peer teaching [23, 24], are demanded. The H-BRS team, founded in 2007, reached a professional level which allows the university to include it within several other funded projects. This includes research-driven projects (e.g. on sustainable materials, aerodynamics, efficient drivetrain aspects) as well as international cooperation projects with partner universities. Specifically, in recent years, BRS Motorsport was invited to give one-week tutorials in Jordan and China. Currently, regular international buddy teams also include the Netherlands and Taiwan.

Student chapters of LACCEI in countries across South America and the Caribbean, bring together an international family comprised of peer students, academics, and professionals affiliated with the institutions of the LACCEI consortium. The intent is to support students throughout their careers. Participating in the chapter's activities will allow members to become familiar with a professional setup. It will also help develop their leadership skills and other soft skills [25].

Opportunities arising from joining the activities of a student chapter included: expanding social relationships with colleagues, academic leaders, and professionals, participating in in-person training and webinars, accessing mentors, participating in contests and competitions, visiting industrial facilities and companies, and accessing LACCEI publications. Additionally, participation encourages the creation of joint programs and projects that enable the implementation of professional practices abroad, student exchanges, and the development of open chairs at institutions that have effectively established chapters.

6 Lessons Learned from the Programs

The models from the different regions of the world, and development and implementation of the signature projects, demonstrate lessons learned related to awareness, agency, advancement, and awards.

6.1 Awareness of Global Perspectives

PEI-Global has seen an impact of participation on both the students and mentors. Over the last two years, PEI Institutions have sponsored graduate student participation at the NSBE National Convention In a case study, students indicated how they were more aware of the cultural and global differences among their expanded base of colleagues. The students also learned that they were able to engage in a way that looked beyond their typical, traditional environments. It introduced them to a fresh perspective of what they can do and who they can be, Further, they valued having an expanded view of engineering and the profession. As future educators, they noted that there was a need to consider global perspectives as a competency, and that they would incorporate it into their own future faculty philosophies in research and teaching.

6.2 Agency in Program Development

In Germany, the students have a level of agency as they have opportunities to contribute to the continuous improvement of the Formula Student program. The Formula Student team formed roughly 15 years ago, and it developed in all kinds of dimensions and reached a significant level of professionality [23, 24, 26]. The product (race car) evolves to higher levels every year because the leaders managed to implement effective means of knowledge gathering and transfer from one generation to the next one [27–31]. In terms of technical achievements, this includes the development of a fully adjustable double-wishbone suspension, a four wheel-hub drive train with 1,5 stage planetary gears, an advanced aerodynamic package that can generate downforce of roughly 100 kg at typical race speeds, or a self-made battery management system (BMS). This resulted in several wins at competition events (among others: Engineering Design in 2017, 2021 and 2023; overall 1st place FS Spain 2021) within the last 8 years, including top positions in the world ranking.

Interestingly, these wins are not the most important achievements of the project. In the view of the authors from Germany, the most important achievement of the project is rather the creation and implementation of effective instruments to guarantee continued high-quality output in partnership of university and student initiative.

Doing so, the professors augmented the overall progress of the team and formalized work that is done by the students, i.e., valuing their extraordinary engagement. In detail, that work led to (i) founding a "Race Academy" [27] to support the management of the project and to organize the knowledge transfer, (ii) offering customized project course modules [28, 29, 31, 32], (iii) introducing a special "racing engineer" certificate [30] additional to the regular degree, and (iv) hosting special sponsoring and review events.

In India, IUCEE is pleased to have been sustained for 15 years. IUCEE created a consortium and engaged more than 100 engineering colleges and universities, trained more than 1000 faculty in the IIEECP Teacher Certification Program, engaged more than 10,000 faculty in various other Faculty Development initiatives, and assisted more than 300,000 students to develop leadership and employability skills.

6.3 Advancement of Knowledge

In China, the project highlighted in this chapter considers its success, and notes that students from different countries experienced the integration of cultures, the collision of ideas, the collaboration of technologies, and the realization of products. It is characterized by promoting technical exchanges between Chinese and foreign university students, based on innovative fields such as intelligent manufacturing, transportation logistics, and public services, and combines product functional analysis with cutting-edge technology to integrate social value and commercial value.

Latin America's LACCEI Bootcamp counts the following as successful outcomes: Focused, in-depth learning; efficient use of time; enhanced understanding and retention; practical skills development; diverse, relevant skillset acquisition; preparedness for various engineering challenges; and multifaceted professional growth. The program also highlights entrepreneurial mindset cultivation, global perspective and networking, exposure to international standards and practices, expanded professional network, mentorship and guidance opportunities, acknowledgement of excellence and effort, and financial incentives for top-performing participants.

6.4 *Awards for Achievement*

Many of the programs described here have reward structures built into the respective initiatives. Global engagement initiatives that are part of courses within a curriculum have grading structures that serve as the reward. For programs that function as extra- or co-curricular programs outside of the classroom, there are different kinds of structures that mark participation and achievement. The PEI-Global program includes a certificate that connects to a national network that provides certifications connected to preparation for faculty positions. The new ICIRTL certificate connects to the national CIRTL network. ICIRTL was developed through the collaboration of UMBC in the US and H-BRS in Germany for PEI-Global. Formula Student in Germany is part of an international racing competition, which has awards ceremonies as part of the structure. The LACCEI Bootcamp in Latin America provides monetary awards as well as award packages that cover registration costs to other international conferences.

7 Challenges

Very few programs have successes without challenges, and it will be helpful to faculty and administrators who are developing their own programs to hear some of the issues that were experienced by the leaders of the programs highlighted here.

Challenge 1: Sustained Engagement: In the US, PEI-Global found that there could be difficulty in recruiting students who were hesitant to participate in traveling. In India, there was difficulty in engaging governmental organizations in India, issues with faculty quality and faculty motivation, and lack of quality awareness among institutional leadership.

Germany's Formula Student initiative is a large-scale project-based student off-curriculum initiative [30, 33–35]. The initiative's sustained success demands various expertise in technical, methodological, and social skills to ensure effective teamwork of students of all levels and various disciplines [35, 36]. Core team workshops or retreats for the active members are ideal to address these challenges. Input is delivered by the team members themselves and by supporting faculty and staff of the university. A faculty advisor, who is both part of the engineering department and the team's leadership council, plays an important role as mediator between the team and the university. A close and stable relationship between the team and the faculty advisor is desirable.

Another decisive factor is balancing the fluctuation of members. Not only must departing members be replaced, e.g., due to graduation, but the transfer of knowledge is also crucial. Annual recruitment events at the beginning of each new season enable the recruitment of new members and create university-wide awareness of the initiative itself.

Challenge 2: Sustained Funding and Time: Most of the programs highlighted here worked within the constraints of their funding. Ensuring repetitive and sustained funding streams can stall progress, especially when a large portion of the programming depends on federal grant awards and financial support from the institution. PEI-Global has segments of funding from the National Science Foundation, and the Dean of the College of Engineering and Information Technology at UMBC, but the funding is subject to proposal awards and availability of the dean's funding. The program in China found that their greatest challenge was effective organization of multi-national corporations in a short time.

8 Start-Up, Supports, Structures, and Sustainability

This section continues the GUIDE and focuses on building strong programs through focus on start-up, supports, structures, and attention to sustaining the initiative. Despite challenges, commitment to educating and preparing the students for global engagement, and seeing the impact is the best reward. Taking time to consider structure and supports can facilitate success.

Start-up and Support—China's program advocates for strong collaborations such as engaging a curriculum committee, international teaching team, and student affairs team; having appropriate international exchange policy, exchanges among international institutes, supports from industries; and ensuring that there is in-depth collaboration among international institutes. They also advocate for strong university-industry cooperation. The budget affects available teaching hours, number of students, duration and type of experiments, and cost of experimental material. University-industry cooperation is helpful toward providing the necessary financial support. PEI-Global in the US notes that it is important to have committed people with a commitment of resources, e.g., financial. Identifying potential students who could benefit from these projects may seem easy at the early stages as students can be identified through institutional databases. However, engaging those students and getting them to engage in substantive ways requires additional layers of faculty and administrative mentoring that can't be overlooked.

Structure: Governance and Goals—Formula Student in Germany is cognizant of budgetary constraints and notes that money is needed to build a race car, and close connection with sponsors from industry was essential.

If a professor wanted to have a project like Formula Student, but did not yet have the budget, they could follow a "bottom-up" approach with an initial group of students to start a Formula Student team. Early activities could also include connecting with teams in the international network to invoke the "Formula Student spirit" for assistance with the start-up activities. The authors in Germany leading a Formula Student team have assisted in this way. Formula Student is one of the oldest and largest projects for which universities worldwide can participate. However, it is just one example of a series of large-scale student competitions, which can be found in many technical disciplines. The authors from H-BRS advocate for two ways to successfully participate in this kind of project: ***Top-down, and Bottom-up.***

Top-down: The curriculum must include a lot of credit points, and hence, there must be options to work at least 50% of the student's full time (e.g. internship terms as part of the curriculum.) At the same time, personnel need enough dedicated time to organize and manage the project by the faculty. This is done by some universities in Europe, but it is not the standard option, because it demands a lot of commitment from the university to customize a whole study plan in terms of problem-based learning.

Bottom-up: The driving-force, management, and development, is governed by a student initiative, that can be legally organized by an independent club. Faculty serve as partners and negotiate for common goals, e.g., recognition of extra-curricular activities, course credit etc. This approach can be easier, and it consumes fewer university resources.

Sustainability—To start a project like *Formula Student*, all team activities need strong support from various levels at the university so that the racing team students can focus entirely on their extra-curricular core activities. On the single faculty and staff level, the students need volunteers serving as faculty advisors or short-term exchange and discussion partners. On the department level, there is a solid commitment to the sustainability of the racing team activity needed, starting with access to premises and facilities (especially during the start-up phase of the team.) It is also beneficial if the curriculum and the resulting workload leave room for extra-curricular activities. Additionally, a willingness to validate certain team activities curricular (e.g., internships) would also be helpful. As the interdisciplinary team set-up requires capabilities from students outside the engineering department (e.g., computer science and business management), arrangements and agreements between each affected department are needed. Furthermore, for specific activities, such as program evaluations, the support of central institutions is required. To sustain the project, all team activities must be consistent with the goals and values of the university. On the operational level, the team needs to regularly recruit new talent from the student pool, as well as win faculty and staff supporters. Knowledge transfer to new generations of team members is crucial. India's IUCEE encourages a network of motivated and dedicated people. India has a larger budget: $100,000 annually, but with more resources they would increase the number of institutions that would be involved, and also increase the number and type of programs offered. However, if the budget was smaller, it would mean that the number of programs offered would need to be reduced. The recommendation for those who are in start-up mode, if they don't yet have a budget, is to start working with a team of like- minded people and work with one or two colleges.

9 Recommendations for Faculty

This "GUIDE" shares some recommendations that can assist faculty with development of their own programs. The list below includes advice along with recommendations.

- **Collaboration:** ***Collaborate and share knowledge.*** In-depth collaborations among universities and industrial enterprises are the basis for a successful program. Teach other faculty to implement student-centric effective teaching strategies. Network with others having similar passion.
- **Content:** ***Be well-versed in the state-of the art technologies.*** Combine teaching content with the latest industry developments and topics of global concern.
- **Competencies:** ***Enhance your own global competencies and professional practice.*** Faculty should be equipped with language skills, professional knowledge in related fields of the program and international competence. Practice implementing "Effective Teaching and Learning Strategies" in one's own classroom. Professors can learn about the effectiveness of problem-based learning and gain motivation for developing their own projects that catch the students' interest. Professors should also be sure that they understand indigenous technologies, global policies, and socio-cultural competencies.
- **Continuation:** ***Plan for the long-term:*** Large-scale projects are demanding. Professors should be ready to support the project with strong efforts. If the project is going to be a long-term program, there should be planning for sustaining the program during its inception and development, so that once it enters the implementation phase, there is knowledge regarding how that program will continue from year to year. When a program is meant to be "permanent," the university and faculty should make an initial up-front decision to support the project for at least 3–5 years. A level of institutional commitment should connect to funding from sponsors or grants. Institutional funding can come in the form of committed human resources where a faculty member or a member of staff has the success of the project as part of their position and compensation.

10 Summary

This chapter provided a GUIDE for professors who are considering ways to prepare engineering students for global engagement. With highlights from programs in Africa, China, Germany, India, Latin America, and the US, the chapter shares different ways to engage students, and areas to consider for a professor's own professional development.

Collaboration and partnership are key themes that emerge. The programs highlighted depended on their connections to in-country resources such as those at their respective universities or local industry. However, each program also had deep engagement and collaboration with an international organization, where the international organization's reach and resources were leveraged to facilitate the goals of some of the more localized programs. For example, consider the case of a professor from an individual department who partnered with other faculty members, the academic department, centers on campus, and the college at the local university. After the local connections were solidified, the professor with the initiative, along with team of collaborators, then made connections to tie the activity into an initiative with an international organization. This strategy can provide an opportunity to have the initiative become part of an international center structure, membership in an international society, benefits and activities for associated students, and participation in an international competition.

In short, great success was noted with strategic collaboration or connection. No individual professor nor administrator in our examples launched a full program in isolation as a singular member of the faculty.

Regarding student competencies, technical knowledge is extremely important. However, programs should also build opportunities for students to learn professional skills and understanding of different cultures and languages so that the students will be prepared to work with other engineers who come from different countries. This training allows them to properly prepare to spend time in other countries.

Leaders of the model programs highlighted in this chapter reflected on their successes and identified themes for lessons learned. These themes of awareness, agency, advancement of knowledge, and awards, revealed that program leaders and administrators should take on the responsibility of developing well-prepared segments of programming that include opportunities for the students to participate in the program's improvement, as well as ways to keep them engaged with structures that reward their time and effort.

Planning successful programs should also include funding mechanisms, infrastructure, and long-term continuation plans, all of which should be considered in the early stages of program development.

A summary list of the themes identified throughout the chapter can be found below.

- **Models**: Connect to an *International Center*, transform a *University Project* to include internationalization, Connect to an *International Competition*, Connect to a *Continental Organization*, Connect to *Hemispheric Initiatives*, Connect to *International Conferences*
- **Areas for Professors to Prepare**: *Content*—Courses and Curriculum, *Collaboration* with industry and research organizations, *State-of-the-art* teaching methods and knowledge of innovations, Understand International *Contexts*
- **Key Competencies**: Multilingual Communication, Socio-cultural exchange, Skill- building in state-of-the-art technologies and professionalism
- **Lessons Learned**: *Awareness* of Global Perspectives, *Agency* in Program Development, *Advancement* of Knowledge, *Awards* for Achievement
- **Key Challenges**: Sustained *Engagement*, Sustained *Funding and Time*
- **Areas for Attention to Build Strong Programs**: *Start-up and Support*, *Structure:*

Governance and Goals, *Sustainability*

Recommendations: Be *collaborative*, present updated *content*, enhance one's own *competencies*, and in turn develop them in the students. Plan for *continuation* and sustainability of the program. The final recommendations highlighting collaboration, competencies, and planning for program continuation, bring the themes full circle as all the authors recommend developing programs in collaboration with others and finding connections to organizations that match the global mission that is sought. In the same way that professors study to master their respective engineering disciplines prior to teaching students, program leaders need to take time to develop their own globally-based proficiencies and competencies prior to engaging student participation in a new program that will prepare them for internationalization.

Consideration of the information put forth by this GUIDE has the potential to yield program success, whether newly developed programs are small—with a few students, or larger adaptations of programs that will serve large numbers of students.

There is hope for greater collaboration between countries and regions and enhanced cultural understanding between and among engineers from around the world, as professors and program leaders prepare students to contribute to innovation and globally engage.

Acknowledgements The authors acknowledge: *Peggy Oti-Boateng*, (The African Academy of Sciences, formerly at UNESCO); *Martina Grein* (Center for Teaching Development and Innovation—ZIEL); *Jan Behrendt, Dennis Günther, Elena Haedecke, Tobias Haedecke, Daniel Röthgen, Alexander Rundau* (BRS Motorsport Student Initiative); *Sampada Pachaury, Ravi Salagame, Jayant Sathe, Sushma Kulkarni, Gopalkrishna Joshi* (IUCEE) *María Laura Polo González* (SPEED); *Janet Rutledge, Keith Bowman, Anupam Joshi* (UMBC); *Cindy Rubio González* (UC Davis); *Rachel Jean-Baptiste* (Stanford University, formerly at UC Davis), National Science Foundation (Award #s 1309290, 1619676, 1734741, 2051599), Office of Academic Affairs of Tsinghua University and the Ministry of Education; Dassault Systèmes; Airbus; H-BRS: Department of Engineering and Communication, Office of the Vice-President Teaching, Learning, and Further Education, Formula Student Team and Alumni; MATLAB; Siemens NX; Altair; UC Davis Division for DEI; IUCEE Consortium; ABET; IGIP, Dassault Systemes; MathWorks; Ionidea; IEEE; and EPICS.

References

1. World Federation of Engineering Organizations (2010) Engineering Issues, Challenges and Opportunities for Development; UNESCO Report https://unesdoc.unesco.org/ark:/48223/pf0000189753. Accessed 2 July 2023
2. UNESCO International Center for Engineering Education (2021) Engineering for sustainable development delivering on the sustainable development goal https://unesdoc.unesco.org/ark:/48223/pf0000375644. Accessed 2 July 2023
3. Falchikov N (2001) Learning together: peer tutoring in higher education. Routledge, New York
4. Li F, Yang J, Wang J, Li S, Li Z (2019) Integration of digitization trends in learning factories. Procedia Manufact 31:343–348. https://doi.org/10.1016/j.promfg.2019.03.054
5. Guo Z, Xu L, Wang S (2022) Demand characteristics and training trend of engineering science and technology personnel based on education for sustainable development. Strat Study Chinese Acad Engin 24(2):179–188
6. Xu L, Wang S (2020) New mode of engineering talents training based on interdisciplinary cooperation: an exploratory multi-case study of engineering education. Tsinghua J Educ 2020(5):107–117. http://tje.ioe.tsinghua.edu.cn/oa/DArticle.aspx?type=view&id=20200513
7. Latin American and Caribbean Consortium of Engineering Institutions (2023) LACCEI. https://laccei.org. Accessed 7 Dec 2023
8. Latin American and Caribbean Consortium of Engineering Institutions (2023) LACCEI 2023. https://laccei.org/mission-and-goals/. Accessed 7 Dec. 2023

9. Novick P, Gadura N (2020) A summer biotechnology boot camp to increase STEM interest among minority high school students. Commun Coll Enterprise 26(2):9–23. https://eric.ed.gov/?id=EJ1281749
10. Ivkovic I, Willett T, Borland M, Gorbet M (2018) Design days boot camp: Enhancing student motivation to start thinking in engineering design terms in the first year. In: Proceedings of the Canadian engineering education association (CEEA), pp 1–7
11. Cico O (2019) The impact of IT bootcamp on student learning—experience from ICT enabled experiential-based course. In: International conference on software business 2019: software business, pp 430–435. https://doi.org/10.1007/978-3-030-33742-1_3
12. Smith J, Stoop C, Young M, Belou R, HELD S, (2017) Grant-writing bootcamp: an intervention to enhance the research capacity of academic women in STEM. Bioscience 67:638–645. https://doi.org/10.1093/biosci/bix050
13. Thayer K, Ko A (2017) Barriers faced by coding bootcamp students. In: Proceedings of the 2017 ACM conference on international computing education research, pp 245–253
14. Johnson T, Wisniewski M, Kuhlemeyer G, Isaacs G, Krzykowski J. (2012) Technology adoption in higher education: Overcoming anxiety through faculty bootcamp. J Asynchron Learn Netw 16(2):63–72. https://eric.ed.gov/?id=EJ971045
15. Gurganus J, Cresiski R, Tull RG, Rutledge, J (2021) Leveraging an EAST-WEST partnership to broaden participation of faculty from traditionally marginalized Groups. In: 2021 World engineering education forum/global engineering deans council (WEEF/GEDC), pp 441–446. https://doi.org/10.1109/WEEF/GEDC53299.2021.9657221
16. Tull RG, Rutledge JC, Carter FD, Warnick JE (2012) PROMISE: Maryland's alliance for graduate education and the professoriate enhances recruitment and retention of underrepresented minority graduate students. Acad Med 87(11):1562–1569
17. Tull RG, Jangha S, Medina Y, Bell T, Parker R (2018) Sharing peace engineering with US-based minority students, through the UN's sustainable development goals, in Peru. In: 2018 World engineering education forum-global engineering deans council (WEEF-GEDC), pp 1–6. https://doi.org/10.1109/WEEF-GEDC.2018.8629764
18. Delaine DA, Sigamoney R, Tull R, Williams D (2015) Global diversity in engineering education: An exploratory analysis. In: 2015 international conference on interactive collaborative learning (ICL), pp 378–388. https://doi.org/10.1109/ICL.2015.7318058
19. Reith D, Bachmeier C, Groß I (2015) Be an engineer right from the start. In: 2015 IEEE global engineering education conference (EDUCON), pp 249–254
20. Kitchen JA, Sonnert G, Sadler PM (2018) The impact of college- and university-run high school summer programs on students' end of high school STEM career aspirations. Sci Educ 102(3):529–547. https://doi.org/10.1002/sce.21332
21. Panaia M (2014) Abandonar la Universidad con o Sin Título, 1st ed. Miño y Dávila, Buenos Aires
22. Castaño Corvo MB, Huesca Ramírez MGE (2019) Causas de deserción de alumnos de los primeros semestres de una universidad privada. Revista Mexicana de Orientación Educativa 5(12):34–39. https://core.ac.uk/reader/268933307
23. Groß I, Malschützky M, Held T, Behrendt J, Röthgen D, Grein M, Reith D (2022) Self- reliant peer teaching as an additional dimension of university-level education for high potentials. In: 2022 IEEE global engineering education conference (EDUCON), pp 333–338
24. Reith D, Grein M, Röthgen D. Behrendt J, Held T, Malschützky M, Groß I (2022) How self-reliant peer teaching can be set up to augment learning outcomes for university learners. In: 2022 IEEE global engineering education conference (EDUCON), pp 57–62
25. Algarra Lopez LS, Cruz Salazar LA, Arbeláez-Soto A, Olguín E (2021) Competencias blandas para la mejora de la educación y el desarrollo social sostenible. "Prospective and Trends in Technology and Skills for Sustainable Social Development" Leveraging Emerging Technologies to Construct the Future: Proceedings of the 19th LACCEI International multi-conference for engineering, education and technology 2021, pp 1–8
26. Peitzmeier F, Mockenhaupt J, Reith D (2022) Optimum real time estimation and interfacing of braking points for racing car. IEEE Trans Veh Technol 72:1710–1719

27. Reith D, Haedecke T, Schulz E, Langel L, Gemein L, Groß I (2016) The BRSU Race Academy: A tutored peer-teaching learning approach. In: Auor M, Guralnick D, Uhomoibhi J (eds) Interactive Collaborative Learning. ICL 2016, Advances in intelligent systems and computing, vol 545. Springer, Cham, pp 166–174. https://doi.org/10.1007/978-3-319-50340-0_13
28. Reith D, Haedecke T, Schulz E, Langel L, Gemein L, Groß I (2017) Practical undergraduate projects within the BRSU race academy: an example for conducting a new tutored peer-teaching learning approach. IEEE Global Engineering Education Conference (EDUCON) 2017:146–151. https://doi.org/10.1109/EDUCON.2017.7942839
29. Gross I, Reith D, Grein M, Groß I (2019) Deeper theoretical understanding by means of practical experience in electric and electronic circuits for freshmen. In: 2019 IEEE global engineering education conference (EDUCON), pp 781–786
30. Reith D, Blume C, Grein M, Haedecke T, Könemann D, Malschützky M (2020) Sustaining complex projects by linking in- and off-curriculum elements: the BRSU racing engineer certificate. In: 2020 IEEE global engineering education conference (EDUCON):905–912.
31. Rundau A, Günther D, Behrendt J, Röthgen D, Grein M, Reith D (2023) Attracting engineering students to non-technical seminars by including international student competitions into the regular engineering curriculum. In: 2023 IEEE global engineering education conference (EDUCON), pp 1–7. https://doi.org/10.1109/EDUCON54358.2023.10125178
32. Franke D, Meilinger S, Groß I, Reith D (2022) Bachelor's program "Sustainable Engineering" at Bonn-Rhein-Sieg University of Applied Sciences: Development, concept and practical experience. In: Leal Filho W, Portela de Vasconcelos CR (eds) Handbook of best practices in sustainable development at university level. World sustainability series.
33. Renninger K, Hidi S (2016) The power of interest for motivation and engagement. Routledge, New York
34. Hmelo-Silver CE (2004) Problem-based learning: what and how do students learn? Educ Psychol Rev 16:235–266. https://doi.org/10.1023/B:EDPR.0000034022.16470.f3
35. Smith MK (2008) Informal learning. In: The encyclopedia of informal education. http://infed.org/mobi/informal-learning-theory-practice-and-experience/
36. Ryan R, Deci E (2016) Facilitating and hindering motivation, learning, and well-being in schools: research and observations from Self-Determination Theory. In: Wentzel K, Miele DB (eds) Handbook of motivation at school, 2nd edn. Routledge, New York, pp 96–119

Enhancing Engineering Education: Fostering Social Skills Through Peace Engineering Minor

Tania Cristina D'Agostini Bueno and Ramiro Jordan

Abstract The significant challenge facing engineers and engineering students today lies in their ability to deeply comprehend the social landscape in which they operate and to effectively address society's most urgent issues. This imperative is underscored by the context of the Fourth Industrial Revolution, which demands solutions to the social problems arising from technological advancements. While engineering education traditionally focuses on preparing students for employment, there is a growing urgency to introduce programs that broaden this perspective. One such critical initiative is the introduction of programs like the minor in Peace Engineering presented in this work. This minor program is pivotal as it equips engineering students with the knowledge and skills necessary to navigate and contribute meaningfully to a complex societal framework. By integrating principles of peace, ethics, and social responsibility into their education, students can harness their technical expertise to not only innovate but also to address the ethical and social implications of their work. Therefore, amidst the transformative landscape of the Fourth Industrial Revolution, the integration of programs such as the minor in Peace Engineering becomes not just relevant but essential in shaping a future where technology serves the betterment of society as a whole. The synergy achieved through the amalgamation of technology, principles of social justice, and emotional intelligence not only prepares individuals for the challenges of our evolving technological epoch but also imparts a comprehensive and inclusive perspective that transcends disciplinary boundaries. Understanding peace within this interdisciplinary framework significantly enhances the effectiveness of addressing this intricate and transformative issue within engineering education. The Peace Engineering Minor (PENG Minor), developed by New Mexico University, delves deeply into both the intended and unintended consequences of engineering solutions, relying on verifiable and trusted data to generate Actionable Knowledge. This approach forms the foundation of a new mindset for engineering students, empowering them to design innovative solutions while combatting disinformation.

T. C. D. Bueno (✉)
i3G Institute, Florianópolis, Brazil
e-mail: tania.bueno@institutoi3g.org

R. Jordan
ISTEC, Inc, and the University of New Mexico, Albuquerque, USA
e-mail: rjordan@istec.org; rjordan@unm.edu

R. Kandakatla et al. (eds.), *Academic Leadership in Engineering Education*, Lecture Notes in Networks and Systems 1097, https://doi.org/10.1007/978-3-031-68282-7_21

Keywords Peace engineering · Peace · Fourth industrial revolution · Engineering education · Verifiable trusted data · Actionable knowledge

1 Introduction

1.1 Peace and Sustainable Societies for Engineering Curricula Evolution

Peace, anchored in a collective social agreement and nurtured through the acknowledgment of diverse social roles, relies on the bedrock of moral and ethical principles essential to our societal fabric. However, the globalization surge has imposed a standardized behavioral template globally, risking the erosion of core values [1]. This shift is particularly noticeable in the evolving landscape of engineering education, marked by uncertainties arising from rapid technological advancements, environmental challenges, privacy breaches, social inequity and inequalities, and conflicts on the rise.

Calls for reform in engineering curricula are intensifying, advocating a multidisciplinary approach that integrates evidence-based pedagogies and outcome-driven learning mechanisms [2]. As engineering institutions endeavor to adapt, their primary focus is on crafting quality learning experiences that empower students to cultivate essential skills such as critical thinking, problem-solving, life-long learning, and effective teamwork [3].

Despite the surge in information, notably facilitated by the Internet and cell phones, an increase in wisdom has not necessarily followed. Instead, collective anxieties have heightened, particularly evident among the younger demographic, posing a pressing concern for global health. The aftermath of the pandemic has underscored this issue, revealing a society not only physically unwell but also grappling with a diminished capacity for creativity [4]. This prompts a crucial question: what trajectory are we on? While technology should act as a catalyst for human advancement, it paradoxically appears to be distancing us from our cherished and foundational values. The imperative now lies in recalibrating our relationship with technology and the planet to ensure it serves as an ally in our collective growth, rather than a force that distances us from the essence of our humanity, this was the conclusion of the World Economic Forum—WEF, in its global risks report [4].

This transformative shift aims to nurture engineers educated to reshape their world for peace and justice [3]. This paradigmatic shift not only acknowledges current disruptions but anticipates the future implications of technologies like artificial intelligence, blockchain and quantum computing, ensuring engineers are not just adaptive but well-prepared for the demands of this dynamic era.

Acknowledging that peace extends beyond conscious perception, this paradigm advocates for an exploration of vibrational energies through studies that transcend conventional experiences of peace and social justice. This endeavor aims to enrich methods and practices in the dynamic field of Peace Engineering.

Presented as a framework for the educational environment, Peace Engineering Minor emerges as a vital educational mindset in the midst of unprecedented global challenges, forging a connection between technological advancements and social consciousness. The proposed Minor not only tackles existing complexities but also establishes a foundation for engineers and other professions to navigate future uncertainties, ensuring their contributions lead to the establishment of a harmonious and sustainable global society. This evolution responds not only to changing trends in industry requirements but also to escalating concerns regarding sustainable development and climate change. Therefore, important premises for Peace Engineering are "One planet, One Environment, One Chance", Verifiable Trusted Data, and Actionable Knowledge [5].

In this context, the Peace Engineering Minor emphasizes fostering a harmonious relationship between technology, society, and the environment [6–9]. Introducing engineering students to these concepts, weaving justice into technology's effects, shifts us from a war-centric socio-tech setup to a global peace engineering agenda.

1.2 Peace Engineering as an Emerging Mindset

The values such as environmental sustainability and social responsibility become increasingly crucial in the context of the Fourth Industrial Revolution. By framing it as a revolution of values, there is a call to integrate sustainable practices into technological advancements, ensuring that progress is not only efficient but also environmentally and socially responsible. The landscape of academic professional training is undergoing a profound transformation, endeavoring to align themselves with the dynamic learning needs of students in the modern era. Peace engineering, as an integral part of this transformation, recognizes the significance of nurturing a mindset that goes beyond technical expertise, placing emphasis on ethical considerations, cultural sensitivity, and a commitment to social responsibility [6]. The Peace Engineering Consortium promotes knowledge, tools, and talent, offering a valuable framework for institutions developing their own Peace Engineering program. Educational institutions should expand curricula to educate engineers proficient in technical skills and well-versed in disciplines such as STEM fields, social sciences, arts, diplomacy, health, law, and business [7].

The Peace Engineering Consortium brings together numerous academic institutions worldwide, as well as industry partners and US National Laboratories. PEC seeks collaborative partners to join its mission of making a global impact.

1.3 Case Study: Peace Engineering Minor at the University of New Mexico

UNM-SoE introduced a Peace Engineering Minor, accessible to all disciplines [10]. Understanding peace within this interdisciplinary framework significantly enhances the effectiveness of addressing this intricate and transformative issue within engineering education. The Peace Engineering Minor, developed by New Mexico University, delves deeply into both the intended and unintended consequences of engineering solutions, relying on verifiable and trusted data to generate Actionable Knowledge. This approach forms the foundation of a new mindset for engineering students, empowering them to design innovative solutions while combatting disinformation. PENG fosters transdisciplinary collaboration, breaks down academic barriers, and covers diverse topics. Peace Engineering integrates STEAM principles, sustainability, cultural awareness, and innovation to advance peace, gaining recognition worldwide [9].

Purpose and Goals. The Minor aims to educate adaptable professionals across disciplines, fostering "street smart" global citizens. Students learn digital skills, resilience, ethics, diplomacy, and basic management. Key attributes include systems thinking, social communication, and reliance on verifiable data for informed decision-making [5, 10]. The Peace Engineering Minor comprises 15 credit hours, with two required courses: ENG 220—Engineering, Business, Sustainability, Ethics, and JEDAI, and ENG 320—Design Thinking, Metrics, Analytics, AI, ML, Trusted Data, and JEDAI, each worth 3 credit hours. The remaining 9 h consist of pre-approved electives from any discipline, following ABET guidelines. Course offerings may expand in the future. The PENG Minor collaborates with PEC for content, presentations, and discussions. Team-teaching emphasizes the intersection of engineering, business, and society, covering innovation, entrepreneurship, global standards, and ethics. Business and technical writing are integrated, and professionals offer talks and lead discussions. Goals include providing a broad understanding of discipline practices, fostering workplace skills, addressing ethical issues, and enhancing technical communication. In the Fall 2022 semester, ENG 220 course enrollment was limited to 50 students due to increased interest. Students collaborated in teams focusing on select UN SDGs. The course culminated in an online Peace Engineering Dashboard for New Mexico, aiming to provide verifiable trusted data on Water, Air Quality, Health, and Finance in the Middle Rio Grande. The dashboard's purpose is to empower citizens with accessible information to make informed decisions, fostering transparency and accountability at all levels of government [11]. The PENG Minor promotes interdisciplinary collaboration, reaching diverse future global citizens. Consortium members enhance learning through talks, workshops, and varied perspectives. Curriculum updates are guided by the Consortium's peer review process. The program's growing enrollment over its initial semesters reflects its success.

2 Strategies for Teaching Peace Engineering

2.1 The Inner Foundations of Peace Engineering for Engineering Education Skills

In this undertaking, we expound upon the synergy of three paradigms, forming a robust three-pronged strategy poised to reshape the education landscape for engineers and professionals across diverse domains.

To forge an engineer of peace, the imperative lies in cultivating skills across a threefold spectrum—encompassing social, technological, and emotional competencies. In the realm of social skills, the ability to negotiate, mediate, and reconcile takes center stage, serving as a preventive and resolution-focused approach to conflicts. An ethical understanding of social and environmental diversity is deemed essential, fostering a culture of peace grounded in principles of social justice. Pillars such as tolerance, respect for diversity, gender equality, non-violence, equity, and citizen participation should underpin every creative project envisioned by students.

Regarding technological skills, PENG aims to foster resilient communities and enhance intercultural understanding through technology. Addressing modern challenges such as artificial intelligence, technological advancements, biological threats, species loss, climate change, and global health emergencies, Peace Engineering emerges as a strong solution. It seeks to embed human values into engineering and other professional practices, empowering experts to provide not only advanced technical answers but also solutions that prioritize global welfare. The third facet of this framework is emotional, with Peace Engineering embracing a positive mindset applicable to teaching (see Sect. 3), the creative process, and project execution. This positive mental pattern, achievable through various techniques, aims for harmony between the coherence of the heart and a mental state defined as Peace.

A positive and optimistic mindset opens students to new ideas, fostering an environment conducive to inspiration, creativity and innovation. Mindful breathing practices are often employed to cultivate calm and mental clarity, enabling better focus on the present moment and creating fertile ground for creative thinking and inspiration. In this relaxed and open mental environment, inspiration flourishes, expanding the possibilities for significant contributions to the promotion of peace and global well-being.

In the realm of education, the foundation of Peace Engineering takes the form of a dynamic triad, encapsulating three indispensable dimensions that go beyond traditional paradigms, offering a comprehensive and transformative approach. A punchline for Peace Engineering is "One planet, One Environment, One Chance", Verifiable Trusted Data, and Actionable Knowledge challenges the traditional and established route through which individuals enter and progress within engineering careers [5].

It is imperative that we recalibrate our approach to meet the demands of a contemporary landscape characterized by digital transformation, diversity, globalization, and swift change.

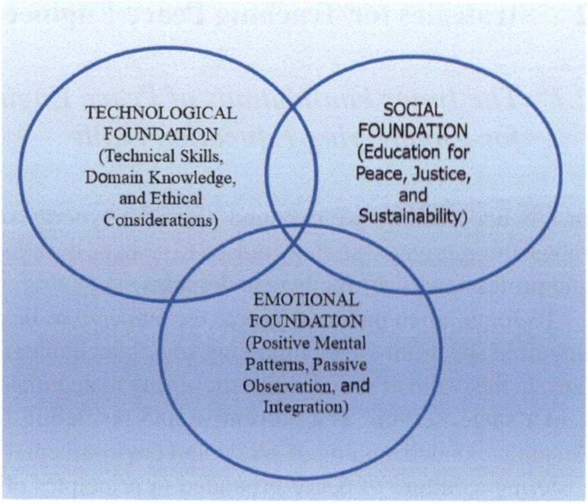

Fig. 1 Three fundamental dimensions of peace engineering education: social, technological, and emotional

The educational approach to Peace Engineering comprises three foundational dimensions (see Fig. 1):

1. **Social Foundation (Education for Peace, Justice, and Sustainability)**: Paramount among these, it is intricately woven with principles of social justice and ethics, fostering collaborative, inclusive, and sustainable approaches to harmonize social dynamics and proactively deter conflicts.
2. **Technological Foundation (Technical Skills, Domain Knowledge, and Ethical Considerations)**: Especially pertinent in the contemporary technological landscape, engineering education plays a pivotal role in instilling an awareness of the ethical implications of technology, countering potential dehumanization associated with total automation, artificial intelligence, blockchain and quantum computing. It is extremely important and vital that all professions and practitioners challenge disinformation by the use and protection of Verifiable Trusted Data. This statement calls for a novel approach to problem-solving: one that emphasizes collaboration and the integration of transdisciplinary expertise and education programs. It advocates for the concurrent application of technological solutions alongside the reinforcement of ethics, policy, and consideration for living systems. Peace Engineering harnesses and acts upon reliable, verified data to guide its actions.
3. **Emotional Foundation (Positive Mental Patterns, Passive Observation, and Integration)**: This foundation becomes a vital consideration, encouraging engineering students to develop a profound respect for the intricate consequences of their actions through passive observation and non-intervention. This mindful approach serves as a catalyst to mitigate economic, social, and environmental impacts, emphasizing the responsibility of future engineers in crafting solutions that reflect a deeper understanding of the interconnectedness between humanity and the natural world.

The evolution of engineering education through Peace Engineering encapsulates a harmonious fusion of technology, social awareness, and emotional intelligence. This approach not only equips students for the challenges of a rapidly changing world but also cultivates a new generation of engineers who are not only proficient in their technical domains but are also socially conscious and capable of contributing meaningfully to the betterment of society. This is also valid for any existing disciplines and new ones to be created to address the global challenges.

3 Integrating Mindfulness, Mind Engineering and EAD in the Framework

3.1 Mindfulness

Accordingly, OECD [13], future schools and curriculums should develop life skills and cultivate competencies like self-awareness, self-regulation, resilience, perspective taking, empathy, gratitude, mindfulness, and leadership. Among the myriad strategies explored, the incorporation of meditation and mindfulness practices has emerged as a beacon of hope for many seeking a holistic approach to mental well-being. The analysis of theory and literature indicates that mindfulness holds promise in fostering various advantages across diverse populations. These benefits encompass the development of both intrapersonal and interpersonal competencies, such as enhancing self-regulation, resilience, well-being, and empathy. The research landscape in the post-pandemic era has witnessed a surge in projects dedicated to understanding and addressing the psychological impact of this unprecedented time. The abrupt shift to remote work and the implementation of virtual learning modalities have significantly altered the dynamics of how people engage with their work and studies. These changes have brought about a host of challenges, including heightened stress levels, blurred boundaries between personal and professional life, and the need for innovative strategies to enhance mental well-being.

However, despite the extensive research demonstrating these advantages, there remains a notable gap in exploring the potential intersections between mindfulness and the field of engineering [14]. Years of research show mindfulness benefits from regular practice. Major corporations like Google, Cisco, P&G, and Facebook embrace mindfulness training to boost creativity, emotional intelligence, and well-being among employees [15]. Adapting these practices to engineering teams, companies should explore how mindfulness enhances divergent thinking crucial in technical design. A mindful openness, not just attention, can ignite innovative mindsets. In this context, the recognition of the effectiveness of brain breaks has taken on a new level of significance. Brain breaks, short pauses or activities designed to provide mental relief and enhance focus, have become essential tools in navigating the challenges posed by the current global landscape [1].

Mindfulness practices offer a powerful means to enhance collaboration and communication among engineering students [14]. Engaging in these practices equips individuals with the ability to appreciate diverse viewpoints and facilitate effective knowledge exchange. Additionally, mindfulness cultivates enhanced focus and concentration, enabling students to delve deeper into their studies and contribute more accurately and insightfully.

Mindfulness risks losing its essence through excessive commercialization, seen as apolitical and overlooking broader social impacts. Assuming ethical behavior emerges naturally simplifies its potential, hindering moral intelligence. True mindfulness entails critical reflection and active engagement with ethics, not passive acceptance.

This notion resonates with Peace Engineering, which integrates technical and broader social, cultural, and ethical dimensions. PENG embodies a positive mindset fostering transformation in teaching and project execution. Embracing this mindset involves harmonizing heart coherence with Peace-defined mental patterns.

3.2 Ethically Aligned Design—EAD

In PENG, ethical considerations hold significant weight. The discipline emphasizes the responsible and sustainable application of engineering knowledge and skills. Students are exposed to topics such as professional ethics, responsible engineering practices, and cultural sensitivity, which are pivotal for fostering an understanding of the ethical dimensions inherent in Peace Engineering work and for nurturing a holistic perspective that values human values.

Ethically Aligned Design (EAD), developed by the Institute of Electrical and Electronics Engineers (IEEE), is a framework guiding the development of emerging technologies with a focus on ethical considerations. It offers principles and guidelines to ensure that technology aligns with human values, respects individual rights, and promotes social well-being. This framework is crucial for engineering students as they are the future innovators and creators of technology.

EAD is significant for engineering students due to several reasons. Firstly, it emphasizes ethical responsibility, urging students to consider the social, cultural, and environmental impacts of their designs. Secondly, it prioritizes user-centric design, encouraging the development of inclusive and user-friendly solutions that cater to diverse societal needs. Thirdly, it underscores accountability and trust, highlighting the importance of transparent and reliable technology design practices that respect privacy.

Moreover, EAD aids in mitigating bias and discrimination, particularly in emerging technologies like artificial intelligence (AI). It prompts students to be vigilant about bias during the design process and develop strategies to identify and eliminate biases in algorithms and systems. Additionally, EAD promotes a long-term perspective on technology design, emphasizing sustainability and environmental impact, thereby enabling students to develop solutions that are environmentally friendly and promote sustainable practices. Integrating EAD principles equips engineering students with the tools and mindset needed to navigate complex ethical dilemmas and design solutions that benefit society as a whole within the context of Peace Engineering. By integrating EAD principles into their education and practice, engineering students can contribute to the responsible development and deployment of technology, making a positive impact on society while upholding ethical standards [17].

3.3 Mind Engineering

Engineers can effectively integrate ethical considerations into peace-building projects through the Mind Engineering methodology [18]. Initially, the team identifies key concepts and sources pertinent to domain evaluation and project selection for PENG initiatives. Subsequently, the team establishes strategic criteria, focusing on diversity and sustainability. Inventories of content, processes, and personnel are then conducted, guided by principles of knowledge sharing, visualization, and relevance definition, employing a controlled vocabulary to describe developments.

This framework should be applied in future Peace Engineering projects to evaluate its tangible effectiveness. Previous applications of the Mind Engineering framework in decision support projects involving AI have showcased its ability to facilitate enduring integration of proficient teams and ensure the ongoing maintenance and enhancement of knowledge within the system [17].

Highlights of the Mind Engineering Framework for Project Management in Peace Engineering—PENG [18]:

1. **Knowledge Sharing**: Identifies subjective aspects within the team and separates knowledge conditionings to enhance learning. Facilitates team integration and encourages specialists to understand their role in the work environment.
2. **Visualization**: Defines similarities among individual images to construct a unified understanding before project implementation. Enhances precision in identifying objectives and context, boosting imagination and capacity.
3. **Relevance Definition**: This phase is critical, as it involves making emotional decisions that reflect the collective unconscious. Defines the project's social dimension and the necessary actions to achieve its goals.

Instilling ethics within engineering demands a holistic approach extending beyond curriculum enhancements. Thus, we have embraced methodologies like EAD [16] and Mind Engineering [18], utilized in various AI developments, to leverage the collective wisdom of engineers and subject matter experts.

4 Examples

Since the early 1980s, there has been a notable rise in organizations dedicated to leveraging engineering for humanitarian and development purposes [19]. These include Engineering for Change, Engineers Against Poverty, Engineers for Overseas Development, Humanitarian Engineering, Contextual Engineering, as well as national chapters of Engineers Without Borders and RedRs. This surge in organizational activity has significantly expanded humanitarian engineering education programs and initiatives worldwide. Humanitarian engineering has long been an integral part of engineering education. Engineers and educators alike have embraced the concept of solidarity with society, especially in times of crisis. However, as we navigate a world increasingly reshaped by engineering and the onset of the Fourth Industrial Revolution, we must confront both the benefits and challenges these advancements bring to society.

Peace Engineering emphasizes engineers' role in fostering peace, promoting social justice, and addressing root causes of conflicts, including environmental justice. According to Trevelyan [20], research in social sciences and humanities can unveil insights guiding interventions to improve engineering practices.

The mission of Peace Engineering is to cultivate a culture of peace, requiring peace engineers, global citizens, and professionals across disciplines to become well-rounded global thinkers and doers. They must be mindful of their ethical obligations, understand their roles within society, and grasp the consequences of their decisions. Peace engineering is characterized not only by its applications but also by its ethos, committed to social justice.

Various academic programs intersect with Peace Engineering, including Humanitarian Engineering, Engineering for Good, Green Engineering, and Contextual Engineering. Establishing Peace Engineering curricula and degree programs is crucial for cultivating new talent capable of addressing global challenges proactively. Integrating PENG into the curriculum transcends traditional humanitarian engineering initiatives, reshaping engineering students' cognitive processes and interpersonal dynamics within our environment. The United Nations' Sustainable Development Goals, adopted in 2015 [21], highlight the interconnectedness of various global challenges, including poverty, inequality, environmental degradation, and conflict. Engineers recognized the importance of their expertise in achieving these goals and began exploring ways to apply engineering principles to promote peace and sustainability. The pursuit of cultivating a culture of peace has long been on the agenda, as recognized by UNESCO [21]. Contrary to the notion that peace is a mere choice or attitude one can adopt effortlessly, it is, in fact, a skill that requires deliberate development. Much like learning mathematics, peace is an internal proficiency, a unique way of comprehending the world that necessitates intentional cultivation and understanding. "One Planet, One Environment, One Chance."

5 Conclusions

In conclusion, the establishment of a framework integrating emotional, technological, and social dimensions is paramount for the advancement of Peace Engineering. Initiatives aiming to develop engineers holistically, beyond technical skills, are gaining momentum globally. Among these, the training of Peace Engineering Ambassadors, spearheaded by IFEES [22] and the Istec Institute [9], stands out, particularly with its launch in Latin America. In a world grappling with conflict and environmental degradation, Peace Engineering Ambassadors play a pivotal role in driving transformative change. They act as advocates, facilitators, and catalysts for peacebuilding and sustainability efforts, bridging divides between communities, industries, and policymakers. Equipped with specialized training and mentorship, these ambassadors possess expertise in conflict resolution, sustainable engineering practices, and community engagement, enabling them to tackle complex challenges with creativity and compassion.

Looking ahead, there is a pressing need to expand Minor programs in Peace Engineering, following the model developed by UNM, to universities worldwide. By broadening access to such programs, we can cultivate a new generation of engineers equipped to address the multifaceted challenges of our time. Through interdisciplinary education and practical experience, students can develop the skills and mindset necessary to contribute meaningfully to peacebuilding, sustainable development, and social justice initiatives. By embracing the principles of Peace Engineering, universities can play a crucial role in shaping a more harmonious and resilient future for generations to come.

References

1. Global trends – 2040. March 2021. NIC 2021–02339 - ISBN 978–1–929667–33–8. To view digital version: https://www.dni.gov/files/ODNI/documents/assessments/GlobalTrends_2040.pdf
2. National Academy of Engineering (2009) Developing metrics for assessing engineering instruction: what gets measured is what gets improved. The National Academies Press, Washington, DC. https://doi.org/10.17226/12636
3. Tharakan J (2020) Disrupting engineering education: beyond peace engineering to educating engineers for justice. Procedia Comput Sci 172:765–769, ISSN 1877–0509, https://doi.org/10.1016/j.procs.2020.05.109.
4. The Global Risks Report 2023 18th Edition. World Economic Forum https://www.weforum.org/publications/global-risks-report-2023/
5. Jordan R, Hamke E, Yousef M, Koechner D (2024) Peace engineering minor: actionable knowledge. towards a hybrid, flexible and socially engaged higher education. In: Proceedings of the 26th international conference on interactive collaborative learning (ICL2023), vol. 2. Springer, pp 189–199. https://trebuchet.public.springernature.app/get_content/380c6f16-50fd-4847-b8d4-06c3f9702c1b?sapoutboundid=2C3CDBC0D820D95605A62C485C69E1A47E64C0E3&utm_source=standard&utm_medium=email&utm_campaign=000_LAN36_0000019083_Book%20author%20congrats%20NEW&utm_content=EN_34155_20240130&mktkey=42010A0D55461EDDB7E8132DA477872E

6. Jordan R, Agi K, Amadei B, Casuse T, Koechner D, Nair I, Nelson M, Olson M, Quihuis M, Tangorra J (2020) Outcomes of the first global peace engineering conference. Procedia Comput Sci 172:803–810
7. Peace Engineering (PENG) for a Sustainable Planet by 2030: Call to Action 2020–2030. Summary of peace engineering biome kickoff workshop. SAND2021–15614 R. Unclassified unlimited release. https://www.ifees.net/wp-content/uploads/2022/02/PENG-WorkshopSummary10Dec21_c.pdf
8. Vesilind PA (2010) Engineering, peace, and justice: the responsibility of engineers to society. Springer. N Y. https://doi.org/10.1007/978-1-84882-674-8
9. Jordan R, Agi K, Maio E, Nair I, Koechner D, Ballard D (2018) Invitation to shape peace engineering, November 12, 2018, WEEF-GEDC-2018 (Albuquerque, NM, November 12–16, 2018)
10. Engineering Projects in Community Service at UNM (EPICS@UNM). https://peaceengineering.unm.edu/academic-programs/epics.html. Accessed 24 May 2023
11. Peace Engineering Dashboard 2023. https://peacedashboard.world/peace-engineering-dashboard-home/dashboard_2023/. Accessed 14 Dec 2023
12. World Economic Forum Annual Meeting 2016 Mastering the Fourth Industrial Revolution. https://www3.weforum.org/docs/WEF_AM16_Report.pdf
13. OECD (2021) Embedding values and attitudes in curriculum: shaping a better future. OECD Publishing, Paris,. https://doi.org/10.1787/aee2adcd-en
14. Huerta MV, Carberry AR, Pipe T, McKenna AF (2021) Inner engineering: evaluating the utility of mindfulness training to cultivate intrapersonal and interpersonal competencies among first-year engineering students. J Eng Educ 110(3):636–670. https://doi.org/10.1002/jee.20407
15. Rieken B, Shapiro S, Gilmartin S, Sheppard SD (2019) How mindfulness can help engineers solve problems. Harvard Business Rev. https://hbr.org/2019/01/how-mindfulness-can-help-engineers-solve-problems
16. The IEEE global initiative on ethics of autonomous and intelligent systems. Ethically aligned design: a vision for prioritizing human well-being with autonomous and intelligent systems, Version 2. IEEE (2017). http://standards.ieee.org/develop/indconn/ec/autonomous_systems.html.
17. Bueno TCD, Hoeschl HC, Hoeschl MB (2018) Developing an ontology-oriented framework to enable peace-building solutions based on fairplay algorithms. In: 2018 world engineering education forum - Global Engineering Deans Council (WEEF-GEDC), Albuquerque, pp 1–6. https://doi.org/10.1109/WEEF-GEDC.2018.8629678
18. Bueno TCD, Hoeschl H (2023) Harmonizing peace engineering and mind engineering: a lifelong learning approach for sustainable societies. In: Manuscript submitted for publication" 2023. World Engineering Education Forum - Global Engineering Deans Council (WEEF-GEDC). Monterrey, México
19. Smith J, Anderson B, Brown N, Colley A, Stoakley A, Turner J (2017) The rise of humanitarian engineering education in Australasia. In: 28th annual conference of the australasian association for engineering education (AAEE 2017), pp 312–320. https://doi.org/10.3316/informit.390763431262433
20. Trevelyan JP (2022) A social science research agenda for engineering practice. In Christensen SH, Buch A, Conlon E, Didier C, Mitcham C, Murphy M (eds) Engineering, social sciences, and the humanities: have their conversations come of age? Springer Science Business Media B.V. https://doi.org/10.1007/978-3-031-11601-8_13
21. United Nations Educational, Scientific and Cultural Organization (UNESCO). The culture of peace program: from national programs to a project of global scope. http://www.culture-of-peace.info/vita/1994/145EX15.pdf (Accessed Oct 2023)
22. International Federation of Engineering Education Societies (IFEES). Initiatives. https://www.ifees.net/peace-engineering/ (Accessed on Feb 2024)

The manufacturer's authorised representative in the EU is Springer Nature Customer Service Centre GmbH, Europaplatz 3, 69115 Heidelberg, Germany. If you have any concerns regarding our products, please contact ProductSafety@springernature.com

Printed and bound by CPI Group (UK) Ltd, Croydon, CR0 4YY
01/10/2026
02225851-0016